Small Fiber Neuropathy and Related Syndromes: Pain and Neurodegeneration

Sung-Tsang Hsieh • Praveen Anand
Christopher H. Gibbons
Claudia Sommer
Editors

Small Fiber Neuropathy and Related Syndromes: Pain and Neurodegeneration

Springer

Editors
Sung-Tsang Hsieh
Department of Neurology
National Taiwan University Hospital
Taipei
Taiwan

Praveen Anand
Department of Neurology
Imperial College London
Hammersmith Hospital
London
UK

Christopher H. Gibbons
Department of Neurology
Harvard Medical School
Beth Israel Deaconess Medical Center
Boston, MA
USA

Claudia Sommer
Neurologische Klinik
Universitätsklinikum Würzburg
Würzburg
Germany

ISBN 978-981-13-3545-7 ISBN 978-981-13-3546-4 (eBook)
https://doi.org/10.1007/978-981-13-3546-4

Library of Congress Control Number: 2019930030

This Springer imprint is published by the registered company Springer Nature Singapore Pte Ltd.
The registered company address is: 152 Beach Road, #21-01/04 Gateway East, Singapore 189721, Singapore

Sung-Tsang Hsieh dedicates this book to
beloved family (Whei-Min, Paul-Chen, and Christine Yi-Chen) and
mentor (late Professor Jack Griffin)

Foreword

This is a very good book focusing on small fiber neuropathy (SFN), an important and interesting disease. When I was a young neurologist in the 1980s, there were few clinicians or researchers whose main theme was SFN; since it was a very complicated disease entity, there were few practical tests for its diagnosis; and it was very difficult or impossible to care for patients with SFN. In fact, as written in Chap. 1 (Overview) by Prof. Hsieh, chief editor of this book, "SFN is a commonly encountered clinical entity that significantly compromises patients' overall quality of life. Patients having SFN are heterogeneous in clinical presentation, underlying causes, and pathophysiology. The presence of symptoms and/or signs of small fiber damage warrant a thorough evaluation for SFN. Furthermore, a diagnosis of SFN should also be considered in patients with chronic pain and autonomic dysfunctions with unclear causes."

However, due to the tireless efforts by some clinicians and researchers including the authors of this book, causes of SFN have been gradually clarified, and treatments for some diseases have been established, as shown clearly in this book. In addition, some clinical testing methods are now used to make a diagnosis of SNF.

I strongly recommend this book to not only experts of SFN but also junior clinicians to study SFN, since this book clearly covers both basic and up-to-date knowledge on the disease.

Ryusuke Kakigi
International Federation
of Clinical Neurophysiology (IFCN)
Vancouver, BC, Canada

Department of Integrative Physiology
National Institute for Physiological Sciences
Okazaki, Japan

The Graduate University for Advanced Studies
of Life Science (SOKENDAI)
Kanagawa, Japan

Foreword

> If your feet hurt, everything hurts. (widely attributed quote)
>
> Springtime in nerveland (one of my patients with HIV neuropathy)

Two illuminating quotes from patients with neuropathic pain of small fiber neuropathy illustrate the tremendous constant impact of this disease entity or syndrome on quality of life. The incidence and prevalence of neuropathic pain continues to rise, and a recent systematic review estimated the prevalence of painful diabetic peripheral neuropathy ranging from 15.3 to 72.3/100,000 PY. Diabetes mellitus is certainly the most common trigger for painful neuropathy, and rates will likely to continue to rise with the epidemic of obesity and overconsumption of refined carbohydrates. Considering the wide range of systemic and primary neurological conditions associated with neuropathic pain related to small fiber neuropathy, this is clearly an affliction that is often misdiagnosed or inadequately treated. In recent years, considerable research has been dedicated to understanding its mechanisms, and new techniques have been developed for diagnosis including the much wider use of genetic testing, at substantially lower costs, and punch skin biopsy that has now entered the field as a simple and reliable tool to assess intra-epidermal nerve fiber densities and sweat gland innervation. Despite these advancements, there have been few advances in definitive treatments, with the exception of ASO therapies silencing therapies (siRNA and antisense oligonucleotide) for amyloid neuropathy, approved in 2018. In contradistinction, we now have a much better understanding of effective use of combination pain-modifying agents and increasingly of neuromodulatory strategies using devices such as the Scrambler™ and Rebuilder™. There are several new elements, however, that have changed the landscape for the management of neuropathic pain in small fiber neuropathy. First, these symptomatic therapies have developed in the setting of the opioid crisis that kills on average 116 people daily in the USA. Since the late 1990s, pharmaceutical companies reassured the medical community that patients would not become addicted to opioid pain relievers and healthcare providers began to prescribe them at greater rates. This led to widespread misuse of both prescription and nonprescription opioids before it became clear that these medications could indeed be highly addictive. According to the 2015 National Survey on Drug Use and Health (NSDUH), approximately 91.8 million adults aged 18 or older were users of prescription pain relievers in 2015, representing more than one-third (37.8%) of the adult US population. About 11.5 million adults misused prescription pain relievers at least once in the past year. The most common reason for their last misuse of pain relievers was to relieve physical pain (63.4%). In 2017, HHS declared a public health emergency and announced a 5-point strategy to combat the crisis. A second element is the spreading legalization of marijuana. Medical marijuana is now legal in 30 states in the USA, and public support reached new levels in 2018 with 64% of Americans favoring legalization. It is highly likely that legal forms of marijuana will increasingly be used to treat small fiber neuropathy-related neuropathic pain. Finally, I believe that the future of developing even more effective therapies for small fiber neuropathy will predictably involve a "precision medicine" approach incorporating genetic testing (e.g., COMT and the variant allele V158M), metabolic parameters (diabetic control, hypertriglyceridemia, and inflammatory markers), and comorbid conditions.

The contributors to this book are internationally renowned leaders in the field of small fiber neuropathy. They discuss clinical approaches to diagnosis and treatment of small fiber neuropathy neuropathic pain and its underlying mechanisms. This book will serve as a useful guide for diagnostic approaches and treatment of small fiber neuropathy for the student, resident, practicing physician or advanced practice provider, researcher, and neuromuscular specialist.

Justin C. McArthur
Department of Neurology
Johns Hopkins University Hospital
Baltimore, MD, USA

Preface by Sung-Tsang Hsieh

Small fiber neuropathy or syndrome of small fiber pathology has become a recognized disease entity due to improvement in diagnostic tools during the past decades. The development of skin biopsy is a key step to revolutionize assessments of small fiber neuropathy by providing objective and quantitative evidence of nociceptive nerve degeneration at the level of pathology. During the past 20 years, additional examinations on psychophysics and physiology such as quantitative sensory testing, pain-evoked potentials including laser-evoked potentials, and contact heat-evoked potentials offer complementary evaluations for functional deficits of small fiber neuropathy. Initially small fiber neuropathy was recognized as the major manifestations of neuropathies mainly affecting nociceptive and autonomic nerves, for example, diabetic neuropathy and familial amyloid polyneuropathy. With the applications of these advanced and integrated examinations including pathology of skin innervation, psychophysics of quantitative sensory testing, and physiology of pain evoked potentials, the spectrum of small fiber pathology actually extended to neuropathies in which large fiber deficits were considered as the main presentations, e.g., Guillain-Barré syndrome and vasculitic neuropathy. In addition to documenting nociceptive nerve degeneration in small fiber neuropathy of sensory type, skin biopsy also provides assessments for small fiber neuropathy of autonomic type including sudomotor, pilomotor, and vasomotor innervation. Furthermore, the applications of these tests expanded our understanding of neurodegenerative disorders and complex pain syndrome: Parkinson's disease, fibromyalgia, etc. This is an intriguing topic which certainly provides foundations for future studies to test whether small fiber pathology in the periphery could serve as a window to neurodegenerative disorders in the central nervous system.

This contributed volume intends to provide updated and concise information in the field of small fiber neuropathy and syndrome. Such a book project would never become a reality without the tremendous expertise and efforts of all coeditors, Professors Anand, Sommer, and Gibbons. I am indebted to the excellent editorial assistance from Springer Nature, particularly Xuewen, who enthusiastically initiated this project. Forewords by Professors Kakigi and McArthur are greatly appreciated which point the significance of this field. This work is in memorial for late Professor Jack Griffin, a great mentor in my research career, and also dedicated to my wife,

Whei-Min, who took care of checking the format of the writing with my son, Paul-Chen, and daughter, Christine Yi-Chen, and offer endless support with my works.

Neurology is in rapid progress, and the information needs update continuously. We look forward to comments from colleagues and readers.

Taipei, Taiwan
October 15, 2018

Sung-Tsang Hsieh

Acknowledgments

Dr. Hsieh's laboratory of nerve degeneration and neuropathic pain has received funding and support from the Ministry of Science and Technology, National Taiwan University College of Medicine, National Taiwan University Hospital, and National Health Research Institute, Taiwan.

Dr. Sommer received grant support for subjects related to the content of the book from Deutsche Forschungsgemeinschaft (SO 328/10-1) and from International Parkinson Fonds; she has received funding (2014–2017) to study neuropathic pain from the European Commission FP7-Health-2013-Innovation, Grant No. 602133.

At the organization stage of this book, colleagues provided valuable suggestions and are highly appreciated: Professor Michael Polydefkis, Michael Shy, and Roy L. Freeman.

Contents

Part III Roles of Small Fiber Pathology: Pain Syndromes and Neurodegeneration

Part IV Neuropathic Pain in Small Fiber Neuropathy: Mechanisms and Therapy

Contributors

Ahmad R. Abuzinadah, MD Neurology Division, Internal Medicine Department, King Abdulaziz University, Jeddah, Saudi Arabia

Praveen Anand, MD Peripheral Neuropathy Unit, Centre for Clinical Translation, Division of Brain Sciences, Imperial College London, Hammersmith Hospital, London, UK

Chi-Chao Chao, MD, PhD Department of Neurology, National Taiwan University Hospital, Taipei, Taiwan

Ya-Yin Cheng, MS Department of Anatomy and Cell Biology, National Taiwan University College of Medicine, Taipei, Taiwan

Ming-Chang Chiang, MD, PhD Department of Biomedical Engineering, National Yang-Ming University, Taipei, Taiwan

Kathrin Doppler, MD Universitätsklinikum Würzburg, Würzburg, Germany

Christopher H. Gibbons, MD, MMSc Neurology Department, Harvard Medical School, Beth Israel Deaconess Medical Center, Boston, MA, USA

Paul-Chen Hsieh, MD Department of Dermatology, National Taiwan University Hospital, Taipei, Taiwan

Sung-Tsang Hsieh, MD, PhD Department of Anatomy and Cell Biology, National Taiwan University College of Medicine, Taipei, Taiwan

Center of Precision Medicine, National Taiwan University College of Medicine, Taipei, Taiwan

Graduate Institute of Brain and Mind Sciences, National Taiwan University College of Medicine, Taipei, Taiwan

Graduate Institute of Clinical Medicine, National Taiwan University College of Medicine, Taipei, Taiwan

Department of Neurology, National Taiwan University Hospital, Taipei, Taiwan

Hung-Wei Kan, PhD Department of Anatomy and Cell Biology, National Taiwan University College of Medicine, Taipei, Taiwan

Chun-Liang Pan, MD, PhD Graduate Institute of Molecular Medicine, National Taiwan University College of Medicine, Taipei, Taiwan

Center of Precision Medicine, National Taiwan University College of Medicine, Taipei, Taiwan

Rosario Privitera, MD Peripheral Neuropathy Unit, Centre for Clinical Translation, Division of Brain Sciences, Imperial College London, Hammersmith Hospital, London, UK

Claudia Sommer, MD Universitätsklinikum Würzburg, Würzburg, Germany

Ming-Tsung Tseng, MD, PhD Graduate Institute of Brain and Mind Sciences, National Taiwan University College of Medicine, Taipei, Taiwan

Nurcan Üçeyler, MD Universitätsklinikum Würzburg, Würzburg, Germany

Ti-Yen Yeh, PhD Department of Anatomy and Cell Biology, National Taiwan University College of Medicine, Taipei, Taiwan

Abbreviations

AAN	American Academy of Neurology
ABCA1	Adenosine triphosphate-binding cassette transporter 1
ACC	Anterior cingulate cortex
ACCORD	Action to Control Cardiovascular Risk in Type 2 Diabetes
ALS	Amyotrophic lateral sclerosis
anti-dsDNA	Anti-double-stranded DNA
ApoA1	Apolipoprotein A1
ASIC	Acid-sensing ion channel
BDNF	Brain-derived neurotrophic factor
BMS	Burning mouth syndrome
BOLD	Blood-oxygen-level-dependent
BoNT/A	Botulinum toxin
BPS/IC	Bladder pain syndrome/interstitial cystitis
CADASIL	Cerebral autosomal dominant arteriopathy with subcortical infarcts and leukoencephalopathy
CAN	Cardiac autonomic neuropathy, cardiovascular autonomic neuropathy
CASPR2	Contactin-associated protein-like 2
CB1	Cannabinoid receptor 1
CBD	Corticobasal degeneration
CCI	Chronic constriction injury
CCM	Corneal confocal microscopy
CGRP	Calcitonin gene-related protein
CHEP	Contact heat-evoked potential
CIDP	Chronic inflammatory demyelinating polyneuropathy
CIPN	Chemotherapy-induced peripheral neuropathy
CMAP	Compound muscle action potential
CMT	Charcot-Marie-Tooth neuropathy (disease)
CMT2	Charcot-Marie-Tooth neuropathy type 2
CNS	Central nervous system
CRPS	Complex regional pain syndrome
CSF	Cerebrospinal fluid
CTCAE	Common Terminology Criteria for Adverse Events
DAB	3,3′-Diaminobenzidine
DBH	Dopamine beta-hydroxylase
DCCT	Diabetes Control and Complications trial
DLPFC	Dorsolateral prefrontal cortex

DLRPN	Diabetic lumbosacral radiculoplexus neuropathy
DRG	Dorsal root ganglion/ganglia
DSPN	Distal symmetric polyneuropathy
EFNS	European Federation of Neurological Societies
EMG	Electromyography
EPO	Erythropoietin
EPSC	Excitatory postsynaptic current
EPSP	Excitatory postsynaptic potential
ESCS	Electrical spinal cord stimulation
FAC	Familial amyloid cardiomyopathy
FACT/GOG-Ntx	FACT/GOG-Neurotoxicity subscale
FAP	Familial amyloid polyneuropathy
FDA	Federal Drug Administration
fMRI	Functional magnetic resonance imaging
GABA	γ-aminobutyric acid
GAP-43	Growth-associated protein 43
Gb3	Globotriaosylceramide
GBS	Guillain-Barré syndrome
GDNF	Glial cell-derived neurotrophic factor
GERD/GORD	Gastroesophageal (esophageal) reflux disease
GL3	Globotriaosylceramide
HbA_{1C}	Hemoglobin A_{1C}
HDL	High-density lipoprotein
HIV	Human immunodeficiency virus
HSN/HSAN	Hereditary sensory (and autonomic) neuropathies
HSV	Herpes simplex virus
IASP	International Association for the Study of Pain
IB4	Isolectin B4
IBD	Inflammatory bowel disease
IBS	Irritable bowel syndrome
IDO	Idiopathic detrusor overactivity
IENF(s)	Intraepidermal nerve fiber(s)
IENFD	Intraepidermal nerve fiber density
IgM	Immunoglobulin M
IL-1β	Interleukin-1β
IL-2	Interleukin-2
iRBD	Idiopathic rapid eye movement (REM) sleep behavior disorder
IVIg	Intravenous immunoglobulin
LDI_{FLARE}	Laser Doppler imager-FLARE
LEP	Laser-evoked potential
LGI-1	Leucine-rich glioma-inactivated 1
LTD	Long-term depression
LTP	Long-term potentiation
M-CSF	Macrophage colony-stimulating factor
mGluR5	Metabotropic glutamate receptor subtype 5
MPEP	2-methyl-6-(phenylethynyl)-pyridine
MSA	Multiple system atrophy

Na_v	Voltage-gated sodium channels
NCS	Nerve conduction studies
NeuPSIG	Neuropathic Pain Special Interest Group of International Association for the study of Pain
NFκB	Nuclear transcription factor κB
NGF	Nerve growth factor
NK1	Neurokinin-1
NMDA	*N*-methyl-D-aspartate or *N*-methyl-D-aspartic acid
NNH	Number needed to cause harm
NNT	Number needed to treat
NPY	Neuropeptide Y
NSAID	Nonsteroidal anti-inflammatory drug
NT-3	Neurotrophin-3
PAG	Periaqueductal gray
PD	Parkinson's disease
PENS	Percutaneous electrical nerve stimulation
PEPD	Paroxysmal extreme pain disorder
PGP9.5	Protein gene product 9.5
PHN	Postherpetic neuralgia
PKA	Protein kinase A
PKC	Protein kinase C
PMP22	Peripheral myelin protein 22
PNQ	Participant Neurotoxicity Questionnaire
PNS	Peripheral nervous system
POEMS	Polyneuropathy associated with organomegaly, endocrinopathy, monoclonal gammopathy, and skin hyperpigmentation
POTS	Postural tachycardia syndrome
PPI	Psychophysiological interaction
PSP	Progressive supranuclear palsy
QSART	Quantitative sudomotor axon reflex test
QST	Quantitative sensory testing
RA	Rheumatoid arthritis
RAGE	Receptors for advanced glycation end products
RBD	Rapid eye movement (REM) sleep behavior disorder, REM sleep behavior disorder
RBP	Retinol-binding protein
REM	Rapid eye movement
RTX	Resiniferatoxin
RVM	Rostroventromedial medulla
SFN	Small fiber neuropathy
SGNFD	Sudomotor nerve fiber density/sweat gland nerve fiber density
SLE	Systemic lupus erythematosus
SNRI	Serotonin noradrenalin reuptake inhibitor
SP	Substance P
SSEP	Somatosensory-evoked potential
SSR	Sympathetic skin response

SSRI	Serotonin specific reuptake inhibitor
T4	Thyroxine
TCA	Tricyclic antidepressant
tDCS	Transcranial direct current stimulation
TGF	Transforming growth factor
TH	Tyrosine hydroxylase
TIND	Treatment-induced neuropathy of diabetes
TMS	Transcranial magnetic stimulation
TNFα	Tumor necrosis factor α
TNSc	Total Neuropathy Score clinical version
TRPV1	Transient receptor potential subfamily vanilloid 1
TST	Thermoregulatory sweat testing
tTMS	Repetitive TMS
TTR	Transthyretin
UC	Ulcerative colitis
VAS	Visual analog scale
VEGF	Vascular endothelial growth factor
VGKC	Voltage-gated potassium channels
VIP	Vasoactive intestinal peptide
VZV	Varicella zoster virus
WHO	World Health Organization
WHOQoL	WHO quality of life
α-Gal A	α-galactosidase A

Part I

Overview and Assessments of Small Fiber Neuropathy

Overview of Small Fiber Neuropathy

1

Ming-Tsung Tseng, Chun-Liang Pan, and Sung-Tsang Hsieh

Abstract

Small fiber neuropathy (SFN) results from impairment of small-diameter myelinated Aδ- and unmyelinated C-fibers. This debilitating condition usually leads to alterations in nociceptive processing, thermal sensations, and autonomic functions. The most common clinical feature of SFN is neuropathic pain, which is often described as burning, shooting, tingling, and even pruritic. Usually, symptoms have a length-dependent distribution, but they may also present in a non-length-dependent manner. The natural course of SFN is highly variable, and, in some cases, large fiber neuropathy may develop. Despite being regarded as a distinct nosologic entity, SFN is either idiopathic or associated with a heterogeneous group of diseases. The pathogenesis of the development and maintenance of SFN is not completely understood. Recently, gain-of-function mutations of the sodium channels have been found to enhance the excitability of dorsal root ganglion neurons, which may explain the presence of neuropathic pain symptoms in patients with SFN. However, the underlying mechanisms leading to the axonal degeneration of small-diameter sensory nerves remain unclear. On neurological examination, impaired small fiber sensations can be detected, including thermal or pinprick hypoesthesia or hyperalgesia, and allodynia to mechanical stimulations. At present, the diagnosis of SFN relies upon clinical signs of small fiber damage, abnormality in small fiber neurophysiological testing, and reduced intraepidermal nerve fiber density. Non-length-dependent SFN is underdiagnosed due to the absence of a typical topographic pattern. In conclusion, patients having chronic pain and autonomic dysfunctions with unclear

M.-T. Tseng (✉)
Graduate Institute of Brain and Mind Sciences, National Taiwan University College of Medicine, Taipei, Taiwan
e-mail: mingtsungtseng@ntu.edu.tw

C.-L. Pan
Graduate Institute of Molecular Medicine, National Taiwan University College of Medicine, Taipei, Taiwan

S.-T. Hsieh
Graduate Institute of Brain and Mind Sciences, National Taiwan University College of Medicine, Taipei, Taiwan

Department of Anatomy and Cell Biology, National Taiwan University College of Medicine, Taipei, Taiwan

Graduate Institute of Clinical Medicine, National Taiwan University College of Medicine, Taipei, Taiwan

Center of Precision Medicine, National Taiwan University College of Medicine, Taipei, Taiwan

Department of Neurology, National Taiwan University Hospital, Taipei, Taiwan

S.-T. Hsieh et al. (eds.), *Small Fiber Neuropathy and Related Syndromes: Pain and Neurodegeneration*, https://doi.org/10.1007/978-981-13-3546-4_1

causes warrant a diagnostic consideration of SFN. After the diagnosis is confirmed, underlying etiologies that are potentially treatable should be investigated.

Keywords
Painful neuropathy · Autonomic neuropathy Neuropathic pain · Pruritus · Skin biopsy Intraepidermal nerve fiber · Clinical presentations · Pain evoked potential (pain-related evoked potential) · Dysautonomia

1.1 Introduction

Small fiber neuropathy (SFN) is characterized by the impairment of small-diameter myelinated Aδ- and unmyelinated C-fibers. The term SFN commonly involves "painful neuropathy" and "autonomic neuropathy," and frequently neuropathic pain or autonomic symptoms dominate the clinical picture, respectively [1, 2]. From clinical and research' points of view, SFN is defined according to a set of criteria at three levels of diagnostic certainty after exclusion of large fiber involvement: (1) clinical symptoms/signs, (2) abnormal psychophysical and/or neurophysiological examinations, and (3) reduced skin innervation as pathological evidence (see Sect. 1.4 and Sect. 2.8 for details in Chap. 2) [1–4]. Although Paul Langerhans described the presence of small-diameter sensory nerves in the epidermis in 1868 [5], evaluating the integrity of small-diameter sensory nerves, especially the intraepidermal nerve fibers (IENFs) systematically and quantitatively at the light microscopic level, was not feasible until the 1990s, when immunohistochemical procedures using pan-neuronal markers (particularly protein gene product 9.5) were established [6, 7].

SFN accounts for approximately 50% of sensory neuropathy [1]. The precise prevalence of SFN remains unclear, because routine nerve conduction and evoked potential studies fail to detect the integrity of small nerve fibers. Moreover, different diagnostic criteria have been used in different studies, and the confirmation of reduced skin innervation requires taking consideration of age- and gender-adjusted normative values for interpreting IENF density (IENFD). SFN has been reported to affect at least 52.95 cases per 100,000 populations per year [8].

1.2 Clinical Manifestation

Typical symptoms of SFN stem from impairment in nociceptive processing, thermal sensation, and autonomic function (Table 1.1) [2, 9]. The most prominent clinical feature of SFN is neuropathic pain, which is noted in more than 80% of patients [1, 2, 8]. Pain quality in SFN is commonly described as burning, shooting, tingling, and even pruritic. Among different characters of pain, burning pain is a frequent symptom, affecting about two-thirds of patients, which is followed by sharp pain. In addition to spontaneous pain, patients with SFN may also complain of allodynia (i.e., pain elicited by a stimulus that does not normally provoke pain) evoked by static light touch, dynamic mechanical stimulation, or innocuous heat. Most patients complain of both spontaneous and evoked pain, although some may have either of them alone.

Pruritus has been reported to occur frequently in patients with SFN, involving up to 68.3% of

Table 1.1 Presentations of small fiber neuropathy

Sensory system	Spontaneous pain (burning, sharp) Evoked pain (allodynia, hyperalgesia) Pruritus (alloknesis, hyperknesis) Paraesthesias Dysesthesias Hypoesthesia (thermal and pinprick) Restless legs syndrome
Autonomic system	Sudomotor (hypohidrosis, anhidrosis) Cardiovascular (orthostatic hypotension) Gastrointestinal (diarrhea, constipation, dysphagia) Genitourinary (retention, incontinence, sexual dysfunction) Visual (blurred vision, light hypersensitivity) Mucocutaneous (dry eye, dry mouth, skin discoloration)

patients [10]. It is usually accompanied by neuropathic pain and rarely occurs alone, given that pruriceptors and nociceptors are usually coexpressed in the same peripheral nerve fibers [11]. It is more severe in the evening and often present in a distal-to-proximal gradient. Like allodynia and hyperalgesia in neuropathic pain, alloknesis (itch sensation provoked by non-itching stimuli such as light touch) and punctate hyperknesis (itch sensation provoked by punctate mechanical stimuli) may occur in SFN patients with neuropathic pruritus [12]. It remains unknown why some SFN patients manifest pain but others had pruritus. An interesting observation is that herpes zoster is associated with a high incidence of neuropathic pruritus [13].

During the disease course, about half of the patients exhibit manifestations related to the autonomic nervous system. The most common autonomic presentation is hypohidrosis or anhidrosis, occurring in about 25% of patients [1]. Other autonomic symptoms include dysfunctions in the cardiovascular (orthostatic hypotension), gastrointestinal (diarrhea, constipation, dysphagia, gastroparesis, increased gastric motility, and early satiety), and genitourinary systems (retention, incontinence, and sexual dysfunction) [9]. Some patients may also complain of blurred vision, light hypersensitivity, skin discoloration, dry eyes, dry mouth, and dizziness. Although small-diameter sensory fibers mediate both somatosensory and autonomic functions, no evidence suggests a clear relationship between manifestations attributed to these two classes of small fiber nerve functions [14, 15], suggesting that the involvement of autonomic and somatic functions could be independent.

Other SFN symptoms include dysesthesia, numbness, and coldness sensations. Restless legs syndrome is present in about 40% of patients with painful neuropathy [16]. Negative symptoms can include thermal and pinprick hypoesthesia and insensitivity to pain. Symptoms usually present in a length-dependent distribution, i.e., "stocking-glove" pattern, starting from the toes and slowly progressing to the distal legs, at which point the distal parts of the upper extremities may also become affected [17]. However, symptoms may also develop in a non-length-dependent distribution, involving face, trunk, or proximal limbs in the early phase of disease. Patchy involvement of small fibers has been proposed to be the underlying etiology for some focal burning pain syndromes, such as burning mouth syndrome [18] and notalgia paresthetica [19]. Compared to length-dependent SFN, patients with non-length-dependent SFN appear to have younger age of onset, report more pruritic symptoms and allodynia, and have more immune-mediated but less dysglycemic etiologies [20, 21].

SFN is now considered as a distinct nosologic entity after exclusion of large fiber dysfunctions. Symptoms and signs of large fiber dysfunctions, such as imbalance, abnormal joint position sense, weakness, and muscle wasting, may develop together with above described clinical presentation of SFN. Furthermore, small fiber pathology may coexist with a disease status such as fibromyalgia [22, 23] or neurodegenerative disease of Parkinson's disease [24]. A general nomenclature of small fiber pathology or syndrome [23] is designated for small fiber nerve degeneration and corresponding functional impairments on a background of larger fiber neuropathy (described in Chap. 9) or pain syndromes (discussed in Chap. 11) and neurodegenerative disease (described in Chap. 13). In many systemic diseases, such as diabetes (please see Chap. 6), amyloidosis (detailed in Chap. 8), paraneoplastic syndromes, etc., pure SFN may be rare or variable in frequency depending on study populations. Damage of small fibers is frequently encountered in some forms of neuropathies at the early stage, such as amyloid neuropathy and large fiber involvement developed at the late stage. About 10% of patients with a first diagnosis of SFN will develop symptoms and signs indicating the involvement of large fibers in the following 2 years [1].

The natural course of SFN is highly variable. Clinical symptoms may remain stationary, remit spontaneously, or deteriorate during the disease course [1]. In patients with idiopathic SFN and SFN associated with diabetes or impaired glucose intolerance, recent evidence suggests that the small fiber degeneration progresses in a non-length-dependent manner at a 2–3-year follow-up [25].

1.3 Etiology

Although many conditions are associated with SFN (Table 1.2), the largest proportion of patients is categorized as idiopathic [2]. In a large case series, up to 41.8% of SFN patients have no identifiable cause, and the underlying causes could only be confirmed in 25% of SFN patients during a follow-up period of 2 years [1]. For SFNs secondary to other causes, diabetes and impaired glucose tolerance are the most common causes, accounting for about 36% of SFN patients in total [1].

Table 1.2 Etiology of small fiber neuropathy

Metabolic	Glucose metabolism-related (DM, IGT) Chronic kidney disease Hypothyroidism Hyperlipidemia Vitamin B_{12} deficiency
Immune-mediated	Autoimmune diseases (SLE, SS) Inflammatory neuropathies (GBS, CIDP) Monoclonal gammopathy (AL, MGUS) Inflammatory bowel disease Vasculitis Paraneoplastic syndrome Sarcoidosis
Infectious	Human immunodeficiency virus Hepatitis C virus Leprosy Lyme disease
Drugs and toxic	Ethanol Bortezomib Metronidazole Nitrofurantoin
Hereditary	Hereditary sensory and autonomic neuropathies Channelopathy (SCM) Familial amyloidosis Fabry disease Tangier disease
Neurodegenerative	Parkinson's disease Pure autonomic failure Motor neuron disease
Idiopathic	Idiopathic Fibromyalgia Complex regional pain syndrome Focal burning syndrome (BMS, NP)

AL light chain amyloidosis, *BMS* burning mouth syndrome, *CIDP* chronic inflammatory demyelinating polyneuropathy, *DM* diabetes mellitus, *GBS* Guillain–Barré syndrome, *IGT* impaired glucose tolerance, *MGUS* monoclonal gammopathy of undetermined significance, *NP* notalgia paresthetica, *SCM* sodium channel mutations, *SLE* systemic lupus erythematosus, *SS* Sjögren's syndrome

The pathogenesis of the development and maintenance of SFN is not completely understood. For SFN associated with hyperglycemia, evidence indicates that dysregulated axonal transport and reduced vascular growth play important roles in the impairment of axonal regeneration [26, 27]. With regard to immune-related SFN, evidence suggests that both autoantibodies and proinflammatory cytokines are elevated in patients with SFN [28, 29], and vasculitis contributes to the degeneration of small-diameter nerve fibers [30, 31]. In alcohol-related SFN, the direct neurotoxic effects of ethanol or its metabolites had been implicated [32]. Recently, altered nerve excitability due to gain-of-function mutations in $Na_v1.7$, $Na_v1.8$, and $Na_v1.9$, three voltage-gated sodium channels, had been reported in painful neuropathy [33–35]. Mutations of these sodium channels have been found in up to 30% of carefully selected cases diagnosed as idiopathic SFN. Gain-of-function mutations of $Na_v1.7$, $Na_v1.8$, and $Na_v1.9$ enhance the excitability of dorsal root ganglion neurons, which likely contributes to pain in SFN. Nevertheless, whether and how these mutations lead to axonal degeneration of small-diameter sensory nerves remains unclear.

1.4 Diagnosis

On neurological examination, the major findings are sensory deficits related to small fiber dysfunction. About half of the patients demonstrate thermal and/or pinprick hypoesthesia, 10–20% with hyperalgesia, and about half of them with allodynia to mechanical stimulation [1, 8]. Large fiber motor (muscle strength and deep tendon reflexes) and sensory functions (light touch, vibration, and proprioception) are relatively preserved. With regard to dysautonomia, common manifestations include sluggish or absent light reflexes, orthostatic hypotension, skin color changes, and warmth or coldness of the skin. Patients with suspected dysautonomia should receive a complete assessment of the autonomic nervous system, including the cardiovascular adrenergic (blood pressure response to postural changes), gastrointestinal (constipation, gastroparesis), genitourinary (libido, erectile function), sudomotor (sweat output), and pupillary functions (light reflex) [36].

Irrespective of the underlying causes, a diagnosis of SFN can be made when at least two of the following criteria are met [1–4]:

1. Clinical symptoms and signs of small fiber damage (pinprick and thermal sensory loss and/or allodynia and/or hyperalgesia), with a length-dependent or non-length-dependent distribution
2. Abnormal warm and/or cooling threshold in the foot based on quantitative sensory testing, or abnormalities in small fiber neurophysiological testing, such as nociceptive evoked potential (pain evoked potential or pain-related evoked potential, which will be discussed in Chap. 3) by laser, electrical, or contact heat stimulators [37, 38]
3. Reduced IENFD in the distal leg

Clinically, these criteria are used to define three levels of diagnostic certainty:

1. Possible SFN: the presence of length-dependent symptoms and/or clinical signs of small fiber damage
2. Probable SFN: the presence of length-dependent symptoms, clinical signs of small fiber damage, and normal sural nerve conduction study
3. Definite SFN: the presence of length-dependent symptoms, clinical signs of small fiber damage, normal sural nerve conduction study, and reduced IENFD at the ankle and abnormal quantitative sensory testing thermal thresholds in the foot and/or abnormal small fiber neurophysiological findings [17, 39]

These graded diagnostic criteria were developed for length-dependent SFN and also applicable for non-length-dependent SFN. Note that, due to the absence of a typical topographic pattern of symptoms in non-length-dependent SFN, it is conceivable that non-length-dependent SFN is an underdiagnosed condition. Serious comorbidities may complicate the diagnosis of isolated SFN. Moreover, malfunctions of the small fibers, especially overactivity, can occur even without evidence of loss of function or reduction of IENFs on skin biopsies.

1.5 Treatment

Treatment of SFN should be aimed at the underlying etiology if identifiable [40]. Nevertheless, whether and how disease-modifying therapy halts small fiber degeneration remains unclear. In diabetes-related SFN, persistent glycemic control appears to reduce SFN symptoms [39], but rapid improvement in glycemic control paradoxically induces painful neuropathy [41]. In Fabry disease, although preliminary evidence suggests that enzyme replacement therapy improves clinical manifestations of SFN [42, 43], its effects on small-diameter sensory nerves remain to be clarified. For idiopathic SFN, treatment is mainly directed to the relief of symptoms, especially the control of neuropathic pain, which will be detailed in Chap. 15 of this monograph. Antidepressants and antiepileptic drugs that have shown their efficacy for neuropathic pain in general (pregabalin, gabapentin, tricyclics, duloxetine, etc.) are considered as first-line treatment. Other options include opioids, lidocaine, and capsaicin patches [44, 45]. $Na_v1.7$ and $Na_v1.8$ blockers are potential therapeutic options in the future. With regard to the autonomic neuropathy, most symptoms are mild to moderate and do not require medical treatment [1, 14]. There might be regional variations in the degree of dysautonomia. For severe postural hypotension affecting quality of life, inotropic agents (midodrine and pyridostigmine) and mineral corticosteroids (fludrocortisone) are effective, although patients usually need additional non-pharmacological management to raise their standing blood pressure (e.g., postural adjustment and compression garments) [46–48]. For gastrointestinal mobility disorders, traditional prokinetic drugs can enhance cholinergic function, and new medications, such as new-generation serotonin 5-HT_4 receptor agonists and secretagogues, are under development [49, 50].

Currently, available pharmacological treatment for SFN is often partially effective in managing pain and dysautonomia. Future studies investigating the pathophysiology of neuropathic pain and autonomic dysfunctions in SFN will facilitate the development of new therapeutic strategies.

1.6 Conclusion

SFN is a commonly encountered clinical entity that significantly compromises patients' overall quality of life [51]. Patients with SFN are heterogeneous in clinical presentation, underlying causes, and pathophysiology. The presence of symptoms and/or signs of small fiber damage warrants a thorough evaluation for SFN. Furthermore, a diagnosis of SFN should also be considered in patients with chronic pain and autonomic dysfunction of unclear cause. Since currently the diagnosis of SFN mainly relies upon clinical presentation, quantitative sensory testing, specialized neurophysiology, and the evaluation of IENFD, future researches should evaluate the diagnostic value of these and of newer or neglected techniques, such as assessment of corneal nerve fiber density by corneal confocal microscopy [52] or axon reflex flare [53]. For idiopathic SFN, treatment mainly focuses on symptomatic relief, especially of neuropathic pain. Up to now, the treatment of neuropathic pain in SFN is generally inadequate and awaits new technology and molecular targets for specific etiology, pathophysiology, and mechanisms tailored individually from the perspective of precision medicine.

References

1. Devigili G, Tugnoli V, Penza P, Camozzi F, Lombardi R, Melli G, et al. The diagnostic criteria for small fiber neuropathy: from symptoms to neuropathology. Brain. 2008;131:1912–25.
2. Terkelsen AJ, Karlsson P, Lauria G, Freeman R, Finnerup NB, Jensen TS. The diagnostic challenge of small fiber neuropathy: clinical presentations, evaluations, and causes. Lancet Neurol. 2017;16:934–44.
3. Lacomis D. Small-fiber neuropathy. Muscle Nerve. 2002;26:173–88.
4. Stewart JD, Low PA, Fealey RD. Distal small fiber neuropathy: results of tests of sweating and autonomic cardiovascular reflexes. Muscle Nerve. 1992;15:661–5.
5. Langerhans P. Über die Nerven der menschlichen Haut. Virchows Arch Pathol Anat. 1868;44:325–37.
6. Dalsgaard CJ, Rydh M, Haegerstrand A. Cutaneous innervation in man visualized with protein gene product 9.5 (PGP 9.5) antibodies. Histochemistry. 1989;92:385–90.
7. Kennedy WR, Wendelschafer-Crabb G. The innervation of human epidermis. J Neurol Sci. 1993;115:184–90.
8. Üçeyler N, Vollert J, Broll B, Riediger N, Langjahr M, Saffer N, et al. Sensory profiles and skin innervation of patients with painful and painless neuropathies. Pain. 2018;159:1867–76.
9. Chan AC, Wilder-Smith EP. Small fiber neuropathy: getting bigger! Muscle Nerve. 2016;53:671–82.
10. Brenaut E, Marcorelles P, Genestet S, Menard D, Misery L. Pruritus: an underrecognized symptom of small-fiber neuropathies. J Am Acad Dermatol. 2015;72:328–32.
11. Ikoma A, Steinhoff M, Stander S, Yosipovitch G, Schmelz M. The neurobiology of itch. Nat Rev Neurosci. 2006;7:535–47.
12. Misery L, Brenaut E, Le Garrec R, Abasq C, Genestet S, Marcorelles P, et al. Neuropathic pruritus. Nat Rev Neurol. 2014;10:408–16.
13. Oaklander AL. Neuropathic itch. Semin Cutan Med Surg. 2011;30:87–92.
14. Novak V, Freimer ML, Kissel JT, Sahenk Z, Periquet IM, Nash SM, et al. Autonomic impairment in painful neuropathy. Neurology. 2001;56:861–8.
15. Thaisetthawatkul P, Fernandes JA, Herrmann DN. Autonomic evaluation is independent of somatic evaluation for small fiber neuropathy. J Neurol Sci. 2014;344:51–4.
16. Gemignani F, Brindani F, Vitetta F, Marbini A. Restless legs syndrome and painful neuropathy-retrospective study. A role for nociceptive deafferentation? Pain Med. 2009;10:1481–6.
17. Lauria G, Merkies IS, Faber CG. Small fiber neuropathy. Curr Opin Neurol. 2012;25:542–9.
18. Lauria G, Majorana A, Borgna M, Lombardi R, Penza P, Padovani A, et al. Trigeminal small-fiber sensory

neuropathy causes burning mouth syndrome. Pain. 2005;115:332–7.
19. Lauria G, Devigili G. Skin biopsy as a diagnostic tool in peripheral neuropathy. Nat Clin Pract Neurol. 2007;3:546–57.
20. Gemignani F, Giovanelli M, Vitetta F, Santilli D, Bellanova MF, Brindani F, et al. Non-length dependent small fiber neuropathy. A prospective case series. J Peripher Nerv Syst. 2010;15:57–62.
21. Khan S, Zhou L. Characterization of non-length-dependent small-fiber sensory neuropathy. Muscle Nerve. 2012;45:86–91.
22. Oaklander AL, Herzog ZD, Downs HM, Klein MM. Objective evidence that small-fiber polyneuropathy underlies some illnesses currently labeled as fibromyalgia. Pain. 2013;154:2310–6.
23. Üçeyler N, Zeller D, Kahn AK, Kewenig S, Kittel-Schneider S, Schmid A, et al. Small fiber pathology in patients with fibromyalgia syndrome. Brain. 2013;136:1857–67.
24. Doppler K, Ebert S, Üçeyler N, Trenkwalder C, Ebentheuer J, Volkmann J, et al. Cutaneous neuropathy in Parkinson's disease: a window into brain pathology. Acta Neuropathol. 2014;128:99–109.
25. Khoshnoodi MA, Truelove S, Burakgazi A, Hoke A, Mammen AL, Polydefkis M. Longitudinal assessment of small fiber neuropathy: evidence of a non-length-dependent distal axonopathy. JAMA Neurol. 2016;73:684–90.
26. Ebenezer GJ, O'Donnell R, Hauer P, Cimino NP, McArthur JC, Polydefkis M. Impaired neurovascular repair in subjects with diabetes following experimental intracutaneous axotomy. Brain. 2011;134:1853–63.
27. Polydefkis M, Hauer P, Sheth S, Sirdofsky M, Griffin JW, McArthur JC. The time course of epidermal nerve fiber regeneration: studies in normal controls and in people with diabetes, with and without neuropathy. Brain. 2004;127:1606–15.
28. Dabby R, Weimer LH, Hays AP, Olarte M, Latov N. Antisulfatide antibodies in neuropathy: clinical and electrophysiologic correlates. Neurology. 2000;54:1448–52.
29. Üçeyler N, Kafke W, Riediger N, He L, Necula G, Toyka KV, et al. Elevated proinflammatory cytokine expression in affected skin in small fiber neuropathy. Neurology. 2010;74:1806–13.
30. Chao CC, Hsieh ST, Shun CT, Hsieh SC. Skin denervation and cutaneous vasculitis in eosinophilia-associated neuropathy. Arch Neurol. 2007;64:959–65.
31. Tseng MT, Hsieh SC, Shun CT, Lee KL, Pan CL, Lin WM, et al. Skin denervation and cutaneous vasculitis in systemic lupus erythematosus. Brain. 2006;129:977–85.
32. Koike H, Iijima M, Sugiura M, Mori K, Hattori N, Ito H, et al. Alcoholic neuropathy is clinicopathologically distinct from thiamine-deficiency neuropathy. Ann Neurol. 2003;54:19–29.
33. Faber CG, Hoeijmakers JGJ, Ahn HS, Cheng XY, Han CY, Choi JS, et al. Gain of function $Na_v1.7$ mutations in idiopathic small fiber neuropathy. Ann Neurol. 2012;71:26–39.
34. Faber CG, Lauria G, Merkies IS, Cheng X, Han C, Ahn HS, et al. Gain-of-function $Na_v1.8$ mutations in painful neuropathy. Proc Natl Acad Sci U S A. 2012;109:19444–9.
35. Huang J, Han C, Estacion M, Vasylyev D, Hoeijmakers JG, Gerrits MM, et al. Gain-of-function mutations in sodium channel $Na_v1.9$ in painful neuropathy. Brain. 2014;137:1627–42.
36. Vinik AI, Maser RE, Mitchell BD, Freeman R. Diabetic autonomic neuropathy. Diabetes Care. 2003;26:1553–79.
37. Casanova-Molla J, Grau-Junyent JM, Morales M, Valls-Sole J. On the relationship between nociceptive evoked potentials and intraepidermal nerve fiber density in painful sensory polyneuropathies. Pain. 2011;152:410–8.
38. Chao CC, Hsieh SC, Tseng MT, Chang YC, Hsieh ST. Patterns of contact heat evoked potentials (CHEP) in neuropathy with skin denervation: correlation of CHEP amplitude with intraepidermal nerve fiber density. Clin Neurophysiol. 2008;119:653–61.
39. Tesfaye S, Boulton AJ, Dyck PJ, Freeman R, Horowitz M, Kempler P, et al. Diabetic neuropathies: update on definitions, diagnostic criteria, estimation of severity, and treatments. Diabetes Care. 2010;33:2285–93.
40. Chiang MC, Tseng MT, Pan CL, Chao CC, Hsieh ST. Progress in the treatment of small fiber peripheral neuropathy. Expert Rev Neurother. 2015;15:305–13.
41. Tesfaye S, Malik R, Harris N, Jakubowski JJ, Mody C, Rennie IG, et al. Arterio-venous shunting and proliferating new vessels in acute painful neuropathy of rapid glycaemic control (insulin neuritis). Diabetologia. 1996;39:329–35.
42. Hilz MJ, Brys M, Marthol H, Stemper B, Dutsch M. Enzyme replacement therapy improves function of C-, Adelta-, and Abeta-nerve fibers in Fabry neuropathy. Neurology. 2004;62:1066–72.
43. Schiffmann R, Floeter MK, Dambrosia JM, Gupta S, Moore DF, Sharabi Y, et al. Enzyme replacement therapy improves peripheral nerve and sweat function in Fabry disease. Muscle Nerve. 2003;28:703–10.
44. Attal N, Cruccu G, Baron R, Haanpaa M, Hansson P, Jensen TS, et al. EFNS guidelines on the pharmacological treatment of neuropathic pain: 2010 revision. Eur J Neurol. 2010;17:1113–e88.
45. Finnerup NB, Attal N, Haroutounian S, McNicol E, Baron R, Dworkin RH, et al. Pharmacotherapy for neuropathic pain in adults: a systematic review and meta-analysis. Lancet Neurol. 2015;14:162–73.
46. Gupta V, Lipsitz LA. Orthostatic hypotension in the elderly: diagnosis and treatment. Am J Med. 2007;120:841–7.
47. Low PA, Singer W. Management of neurogenic orthostatic hypotension: an update. Lancet Neurol. 2008;7:451–8.
48. Singer W, Sandroni P, Opfer-Gehrking TL, Suarez GA, Klein CM, Hines S, et al. Pyridostigmine

treatment trial in neurogenic orthostatic hypotension. Arch Neurol. 2006;63:513–8.
49. Camilleri M. Pharmacological agents currently in clinical trials for disorders in neurogastroenterology. J Clin Invest. 2013;123:4111–20.
50. Janssen P, Harris MS, Jones M, Masaoka T, Farre R, Tornblom H, et al. The relation between symptom improvement and gastric emptying in the treatment of diabetic and idiopathic gastroparesis. Am J Gastroenterol. 2013;108:1382–91.
51. Bakkers M, Faber CG, Hoeijmakers JG, Lauria G, Merkies IS. Small fibers, large impact: quality of life in small-fiber neuropathy. Muscle Nerve. 2014;49:329–36.
52. Tavakoli M, Marshall A, Pitceathly R, Fadavi H, Gow D, Roberts ME, et al. Corneal confocal microscopy: a novel means to detect nerve fiber damage in idiopathic small fiber neuropathy. Exp Neurol. 2010;223:245–50.
53. Namer B, Pfeffer S, Handwerker HO, Schmelz M, Bickel A. Axon reflex flare and quantitative sudomotor axon reflex contribute in the diagnosis of small fiber neuropathy. Muscle Nerve. 2013;47:357–63.

2 Pathology of Small Fiber Neuropathy: Skin Biopsy for the Analysis of Nociceptive Nerve Fibers

Claudia Sommer

Abstract

Reduced skin innervation as shown by immunostaining in skin biopsies is a sensitive and specific indicator of small fiber neuropathy (SFN). Standard methods for staining and quantification have been established, and normative values are available. However, not every condition with reduced skin innervation is a SFN, and not all types of disorders with pathological small fiber function manifest with reduced skin innervation. Identification of nociceptor subpopulations in humans in the skin is only beginning to yield data. Detection of inflammatory cells and pathologic deposits are additional diagnostic benefits of skin biopsy.

Keywords

Skin biopsy · Small fiber neuropathy · Protein gene product 9.5 · Intraepidermal nerve fiber density · Subepidermal nerve fibers · Ranvier nodes · Amyloidosis · Fabry disease

2.1 Introduction

Reduced skin innervation is the hallmark of small fiber neuropathy (SFN). In recent years, methods have been refined, normative cohorts have been published, and the techniques have found widespread clinical use. Immunostaining with antibodies against the pan-neuronal ubiquitin hydrolase protein gene product 9.5 (PGP 9.5) has become the standard technique. Usually, the number of epidermal nerve fibers per length of epidermis, i.e., the intraepidermal nerve fiber density (IENFD), is assessed. Some laboratories offer additional analyses from the skin biopsies, e.g., amyloid stains, immunohistochemistry for inflammatory cells, or the analysis of subepidermal nerve fibers. Some new questions have arisen, for example, whether SFN can be diagnosed with normal skin innervation as assessed by PGP 9.5 immunostaining. Also, the question of the importance of morphological changes in the intraepidermal nerve fibers (IENF), for example, axonal swellings or branching, has been discussed. This chapter will give an overview of currently used methods for the analysis of small nerve fibers in the skin and will summarize current findings on SFN and skin innervation.

C. Sommer (✉)
Universitätsklinikum Würzburg, Würzburg, Germany
e-mail: sommer@uni-wuerzburg.de

S.-T. Hsieh et al. (eds.), *Small Fiber Neuropathy and Related Syndromes: Pain and Neurodegeneration*,
https://doi.org/10.1007/978-981-13-3546-4_2

2.2 Normal Innervation of the Skin

Nerve fibers deriving from neurons in the dorsal root ganglia as well as the sympathetic and parasympathetic ganglia terminate in the skin, innervating the epidermis as "free nerve endings," the mechanoreceptors such as Meissner's corpuscles and Merkel cells, and the sweat glands, hair follicles, and blood vessels. The majority of the axons in the skin are unmyelinated (C-fibers); others are thinly myelinated (Aδ-fibers) and may lose their myelin at their distal endings [1]. C-fibers transmit the sensation of warmth, heat, and slow pain, and Aδ-fibers transmit sharp pain and the sensation of cold. Several C-fibers are wrapped into single Schwann cells, forming the Remak bundles, that can be visualized by electron microscopy [2].

In the epidermis, the IENF meander between keratinocytes, with which they are closely associated and probably connected by chemical synapses. These fibers are derived from the subepidermal nerve plexus that can be observed as horizontally oriented nerve fiber bundles located just below the epidermis. From here, the IENF traverse the epidermal basement membrane and ascend vertically between the keratinocytes. The IENF can be easily seen after immunoreaction with antibodies to PGP 9.5 (Fig. 2.1). This ubiquitin hydrolase is considered a pan-axonal marker, i.e., all axons are supposed to stain positively. However, there remains the possibility that some axons do not express PGP9.5 in a sufficient amount to be detectable by immunostaining, which could only be solved by parallel assessment of the same section with immunostaining and electron microscopy. Below the epidermis, thick nerve bundles including unmyelinated and myelinated nerves run horizontally to the surface. These are particularly numerous in glabrous skin, where the mechanoreceptors are innervated [3]. Blood vessels, arrector pilorum muscles (in hairy skin), and sweat glands are also densely innervated.

2.3 Standard Methods

2.3.1 Biopsy, Choice of Site, and Method

Biopsies are usually performed by disposable circular punch devices with diameters between 2 mm for the fingertip and up to 6 mm for the back. The most commonly used 3-mm punches taken at the lower and upper leg yield enough tissue for reliable IENFD quantification [4–6]. Biopsies are performed under sterile conditions and with local anesthesia. Local infections, severe wound healing deficits, and bleeding disorders including anticoagulation are contraindications.

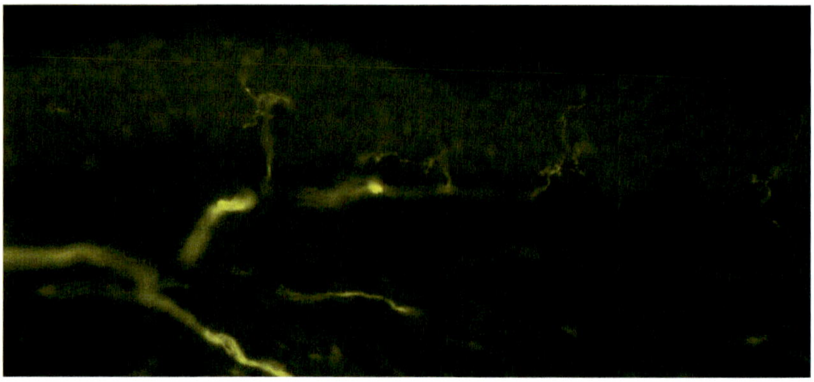

Fig. 2.1 Skin innervation in small fiber neuropathy (SFN). Photomicrograph showing immunofluorescence for PGP 9.5 in a 50-μm section of the skin from the upper leg of a patient with SFN. Note the subepidermal nerve plexus and some intraepidermal nerve fibers traversing the basement membrane, most with numerous branches, which may be a sign of pathological sprouting

The resulting wound may be closed by a suture or a tight adhesive, and patients are advised to avoid strain of the area or bathing until the wound has closed, which usually occurs within 7–10 days. In the diagnostic workup of SFN, biopsies are usually taken from the lateral distal leg (10 cm above the external malleolus) and from the lateral upper thigh (20 cm below the anterior iliac spine). This choice of biopsy sites allows not only to confirm a suspected SFN, if IENFD in the distal biopsy is below the lower limits of normal, but also to classify the neuropathy as length-dependent or non-length-dependent [7]. Furthermore, in severe SFN, the distal site may be completely denervated, and any pathology (e.g., axon swellings or prolonged Ranvier nodes) may only be seen in the proximal sample. Other biopsy areas may be chosen depending on the indication. However, data can only be judged if normative values are available, preferably within the laboratory that is evaluating the sample.

2.3.2 Fixation and Staining

Fixation is necessary to preserve PGP 9.5 immunoreactivity. In unfixed samples, as often used in dermatology to assess inflammatory cells, PGP 9.5 immunoreactivity is lost, and no nerve fibers can be detected. On the other hand, over-fixation may also lead to loss of immunoreactivity. Several fixation protocols have been established. A standard protocol involves fixation overnight in fresh 4% paraformaldehyde or 2% paraformaldehyde-lysine periodate (2% PLP), or Zamboni's fixative (2% paraformaldehyde and picric acid), followed by cryoprotection in 10% sucrose (or in glycerol) overnight, washing and keeping in buffer at 4 °C until use [8]. These protocols work very well for PGP 9.5 immunohistochemistry or immunofluorescence, and normative values have been established using them, but they may need to be adapted when other antigens are looked for. In particular, some antigens can only be visualized in samples fixed for very short time, e.g., 30 min.

For IENFD quantification, the most frequently used and best established method is cutting 50-μm cryosections and immunoreacting the sections with antibodies to PGP 9.5, either on glass slides or with the free-floating method. The primary antibody can be visualized with a fluorescence-tagged secondary antibody (e.g., with Cy2, Cy3), this can be done on the glass slide and with the free-floating method, and these sections can be used for fluorescence or confocal microscopy. Visualization of a secondary antibody with 3,3′-diaminobenzidine (DAB) or a similar non-fluorescent chromophore usually requires the free-floating technique but has the advantage that it can be seen under a bright-field microscope and that the stain is stable over time. However, although some fluorescent stains rapidly fade, we have kept PGP 9.5-immunoreacted skin samples with Cy3-coupled secondary antibodies for years and counted very similar IENFD numbers after several years of storage.

Other antibodies against specific cytoskeletal and axonal membrane epitopes [9] have been used and were shown to yield very similar IENFD values as PGP 9.5-antibodies. However, they have not gained widespread use in clinical routine.

Different somatic and autonomic fiber subtypes can be distinguished by immunoreactions; see also Chap. 5. Innervation of sweat glands, blood vessels, and arrector pilorum muscles may be visualized by antibodies to adrenergic sympathetic fibers (tyrosine hydroxylase, TH, and dopamine beta-hydroxylase, DBH), noradrenergic sympathetic fibers (neuropeptide Y, NPY), cholinergic sympathetic fibers (vasointestinal peptide, VIP), and vasodilatory peptidergic fibers (calcitonin gene-related protein (CGRP) and Substance P (SP)) [10]. Sweat glands are innervated by fibers expressing VIP, CGRP, SP, and DBH [11–13]. The majority of fibers innervating sweat glands are VIP immunoreactive; few express the neuropeptides CGRP and SP [12, 14]. Sympathetic adrenergic fibers innervate arrector pilorum muscles and vascular structures, including arteriovenous anastomoses and arterioles, and can be shown by TH and DBH antibodies [12]. Blood vessels are innervated by sensory and autonomic, sympathetic, as well as parasympathetic fibers.

2.3.3 Microscopy, Quantification, and Documentation of IENFD

A minimum of three sections should be analyzed. Sections are viewed using a light or fluorescence microscope, depending on the staining method. It is important to measure the epidermal length. This can be done by digitalization of the images with a program allowing simple morphometry. A microscope intraocular lens ruler can also be used and has produced similar results [15]. It has been suggested that IENFD can also be calculated by dividing the number of counted fibers by 3, if the biopsy was done with a 3-mm punch. This was shown to correlate well with IENFD measured using computer software [16]. However, this would not account for variations in sectioning and would not be possible if, for example, 5-mm punches are used, and sections are divided to be processed in different ways for several analyses, for example, for histology and for gene expression.

The number of immunoreactive axons is counted under the microscope (not on the digitized image) using a 40x objective, allowing scanning through the planes of the 50-μm-thick section. By convention, only axons crossing the basement membrane are counted [17, 18]. If an axon branches within the epidermis, this is counted as one. The IENFD is then calculated and expressed as number of fibers per millimeter. Others have suggested counting also isolated nerve fragments in the epidermis that do not cross the basement membrane, with the argument that this leads to lower error in cases where only a few fibers are present [19]. Most work has been done using the abovementioned counting rules [17, 18]. If they are strictly followed by well-trained observers, a high degree of inter- and intrarater reliability can be reached. In biopsy samples from 106 healthy volunteers, using the free-floating method and bright-field microscopy, mean differences in IENFD by intraobserver analysis of 0.2 ± 1.2 ENF/mm and by interobserver analysis of 0.4 ± 1.5 fibers/mm were found [20]. This cohort was also used to establish normative values (see below). Another group, using biopsies from 188 healthy volunteers and bright-field microscopy, also found a high intra- and interrater reliability [21]. That this may require a lot of training and experience is shown by the observation that intra- and interobserver variability can indeed be high if the basement membrane cannot be clearly seen [22]. Since assessment of IENFD is time-intensive, semiautomatic methods have been proposed. One semiautomatic method was found to have high experimenter-independent reproducibility and high correlations with conventional counts, when immunofluorescence was used [23].

2.4 Intraepidermal Nerve Fiber Density: Normative Values

Reference values are available for immunohistochemistry using bright-field microscopy and for immunofluorescence. Values for the lower leg, forearm, trunk, and finger have been established. The largest collection of data is available for the lower leg with several large studies providing values. In a study with 98 healthy volunteers, using bright-field immunohistochemistry, McArthur et al. [7] reported a mean IENFD of 13.8 ± 6.7 fibers/mm in the distal leg. This study also showed higher values in the younger age group, gave date for the upper leg, and defined the ratio between thigh and calf IENFD (see below). In samples from 106 healthy volunteers, Gøransson et al. [20] found an IENFD of 12.4 ± 4.6, Umapathi et al. [24] found 11.7 ± 4.1 fibers in 84 volunteers, and Pan et al. [25] found 13.0 ± 0.3 fibers in 55 (all using bright-field microscopy).

Nolano et al. [26] compared bright-field immunohistochemistry and immunofluorescence (non-confocal) using free-floating sections from lower-leg samples from 55 healthy persons and 63 patients with possible or probable SFN. Mean control values seen in bright-field immunohistochemistry were 6.9 ± 2.6 and those in immunofluorescence 13.2 ± 4.2. In SFN, values were 4.6 ± 2.8 and 9.4 ± 4.8, respectively. Numbers from both methods correlated well, and the diagnostic efficacy was very much comparable. The authors do not discuss why values for bright-field immunohistochemistry in normal people in this study are lower than in previous cohorts (see above and [27]).

The influence of age, gender, BMI, and height on IENFD has been debated. McArthur et al. [7] did not find an overall effect of age; only the youngest subgroup (16–20 years) had higher values than the others. Others [20, 24, 28] report a mild decrease in IENFD with age. All report a higher IENFD in women, but height and weight seemed to be confounders in the earlier studies [24]. However, the higher IENFD in women is confirmed by three large studies [21, 27, 28] with 188, 550, and 300 participants, respectively. In these studies, height and weight were not of significant impact. The recent study with 300 healthy subjects using bright-field microscopy also showed a higher number of small swellings and an irregular distribution along the dermal-epidermal junction with age. This study included an ultrastructural analysis of a subset of 100 samples and found a diameter of dermal axons of 0.2–0.6 μm, which did not change with age [28]. This is in accordance with a small control cohort collected in our laboratory [2].

Which cutoff to use to diagnose a pathological biopsy has been debated. Commonly, values below the fifth percentile are considered pathological [7, 21]. We used ROC analysis to determine the cutoff with optimal sensitivity and specificity and found a cutoff value of 8.8 that had a specificity of 80% and sensitivity of 77% [29]. Nebuchennykh et al. [30] used lower-leg biopsies from 45 patients with SFN and 134 healthy persons to compare three methods of determining a pathological sample. They calculated Z-scores from multiple regression analysis and established cutoff values for each patient adjusted for age and gender. They used the fifth percentile, which gave them a 6.7 nerve fibers/mm cutoff value, and they used ROC analysis, which yielded a cutoff value of 10.3. Z-scores and the fifth percentile method showed high specificity (98% and 95%, respectively) but lower sensitivity (31% and 35%), while ROC analysis gave a specificity of 64% and sensitivity of 78%. Nolano et al. [26] suggest to rate a biopsy as pathological if the value is 1 IENF below the 5% cutoff, because 1 IENF represents the inter-rater variability as calculated on the same section [20]. The authors confirm the view [31] that IENFD values very close to the cutoff should be considered with caution. While this sounds reasonable, it may also lead to an increase in numbers of false negatives. In our experience, IENFD in healthy people varies widely, in particular in the elderly. This raises the question, what needs to be done to define a healthy person? This has recently been discussed [32].

2.5 Intraepidermal Nerve Fiber Density: Findings in SFN

Many investigators have shown the utility of IENFD measurements in the evaluation of SFN [5–7, 16, 33–46]. It has been generally accepted that a decrease in IENFD is a very good test, if not the gold standard for the diagnosis of SFN [47]. Many studies over the years have shown a good sensitivity and specificity of IENFD for the diagnosis of SFN, if compared to integrated clinical judgment [48]. This has recently been confirmed for diabetic SFN [49]. However, it has to be considered that reduction in IENFD does not automatically make a diagnosis of SFN and that gain-of-function neuropathies may not be detected by IENFD quantification [50] (see below).

IENFD is a good measure for the follow-up of SFN. We prospectively observed a cohort of patients with the x-linked lysosomal storage disease Fabry disease, who are known to develop IENFD reduction and who have a particular pain phenotype [51]. This phenotype is characterized by attacks of burning pain in the extremities that are triggered by heat or exercise and less frequently chronic pain. IENFD was greatly reduced in men with Fabry disease, more so in those with impaired renal function [52]. During the observation period of 4 years, IENFD deteriorated, and the causally oriented treatment, enzyme replacement therapy, only rescued proximal (samples were taken from the back), but not distal, skin innervation.

There are only scarce data on the relation of IENFD in SFN and pain. While in mixed neuropathy, those patients with lower IENFD have been shown to have more pain in several small cohorts [29, 53, 54], no such data are available for SFN. In a group of 20 patients with vitamin B12 deficiency and no large fiber involvement, those with pain had reduced IENFD; those without pain did not [55]. However, in most studies of SFN, the number of PGP 9.5 immunoreactive IENF is not related to measures of pain. We recently found large overlaps in IENFD values in painful and painless neuropathy and in SFN [56].

An important issue is the distinction between length-dependent and non-length-dependent SFN. In a healthy person, the ratio between IENFD at the thigh and at the lower leg is between 1.5 and 2.3, with an increase of the ratio with age [7]. In length-dependent SFN, this ratio is increased, because IENFD at the lower leg goes down, while that at the thigh may remain normal. If IENFD is reduced in parallel at both sites, the ratio will stay normal or even decrease. This is an indication of a pathophysiological process at the DRG neurons, i.e., a ganglionopathy. The differential diagnosis of ganglionopathy is much narrower than that of length-dependent SFN and includes potentially treatable causes like Sjögren's syndrome and celiac disease. This has recently been confirmed [57].

It has been discussed whether reduced IENFD by itself justifies the diagnosis of SFN. If only reduced IENFD is considered, about 75% of patients with Parkinson's disease or ALS would fall under this diagnosis [58]. However, most of these patients do not have the typical SFN complaints. Most authors agree that the term SFN should only be used if the typical symptoms are present. Another point of ongoing discussion is how to classify syndromes that entail typical SFN-like distal burning pain and small fiber abnormalities in functional tests like microneurography but in which IENFD is normal. This can be the case in genetically caused small fiber overactivity, like in mutations in genes coding for voltage-gated sodium channels (Na_v). We identified a family with SFN-like pain, normal IENFD (although values were at the lower limit of normal), and the $Na_v1.7$ functional variant R1150W [50]. The Dutch/US collaboration that was the first to identify Na_v mutations in patients with otherwise typical SFN included only patients with reduced IENFD. Thus, this approach does not tell us how many patients with otherwise unexplained SFN pain and normal IENFD harbor such mutations [59].

2.6 Subtyping Nerve Fibers

While immunolabeling with the antibody to PGP 9.5 has been extremely reliable, immunostaining for nerve fiber subclasses in human skin has been challenging. Data from animal tissue, including cats and primates, indicate that there are peptidergic and non-peptidergic free nerve endings in the epidermis [60]. Peptidergic axons have mostly been labeled with antibodies to CGRP. Non-peptidergic axons have either been identified by the absence of CGRP or SP (while immunoreactive for PGP 9.5) or by the presence of prurinergic receptors like $P2X_3$ and binding to isolectin B4 (IB4). It has been suggested that non-peptidergic axons reach the very external layers of the keratinocytes, while peptidergic axons terminate between the proximal third to half of the epidermis [61]. Evidence for this statement mainly comes from data derived from rat skin [62]. In human skin, using the commercially available antibodies, CGRP-immunoreactive fibers are very scarce [10, 63]. In our experience, using either immunofluorescence or bright-field microscopy, scarce CGRP-immunoreactive fibers can be seen meandering between the keratinocytes using thinner sections than for IENFD counts (i.e., 10 or 20 μm), and then they can be traced up to the middle of the epidermis (Fig. 2.2a), while in mouse foot pad, CGRP-immunoreactive fibers are frequent and reach the upper layer of the epidermis (Fig. 2.2b). Also in the rat, purinergic fibers in the epidermis are found much more abundant than peptidergic fibers, and it has been suggested that these are the main pain-sensing neurons [62]. The predominant innervation of the

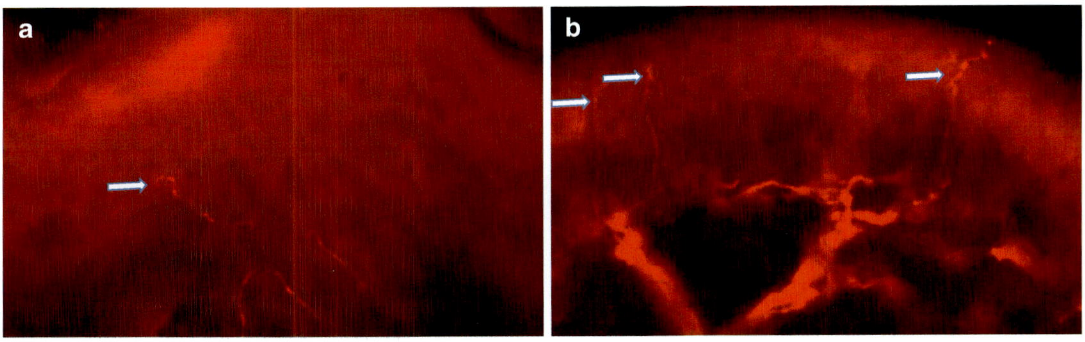

Fig. 2.2 Peptidergic innervation of calcitonin gene-related peptide (CGRP) in the skin. Photomicrograph showing immunofluorescence for CGRP in intraepidermal nerve fibers in human (**a**) and rat (**b**) skin. Note that human CGRP-immunoreactive fibers are scarce and can only be traced in the lower layers of the epidermis (*arrow*), while in the mouse foot pad, CGRP-immunoreactive fibers are more frequent and reach the upper epidermal layers (*arrows*)

epidermis by non-peptidergic fibers has led to the assumption that these are responsible for the particular characteristics of neuropathic pain [64]. This was supported by the finding that the density of purinoceptor $P2X_3$ but not of CGRP-immunoreactive intraepidermal axons was correlated to measures of pain behavior in a rat model of nerve injury [64]. In contrast, peptidergic fibers were increased in rat models of skin inflammation leading to itch, indicating a role of these fibers in this condition [65] and supporting the notion of a role of CGRPergic fibers in itch [66]. In mice, the depletion of neurons immunoreactive for the transient receptor potential vanilloid 1 (TRPV1) channel, which colocalized with SP, and their peripheral axons led to thermal hypoalgesia [67].

While morphologists differentiate between peptidergic and non-peptidergic C-fibers in the skin, neurophysiologists describe mechanosensitive and mechanoinsensitive, polymodal, and heat- or mechano-specific fibers [68]. Studies combining morphological and neurophysiological characterization are rare. Using a combination of retrograde tracing, immunohistochemistry, and electrophysiology in mice, it was shown that about 50% of cutaneous neurons respond to TRPV1 [69]. This study showed a predominance of peptidergic fibers; however, whether they terminate in the epidermis or in deeper layers was not shown, such that the data may not contradict the finding of fewer peptidergic fibers in the epidermis. In many species including primates, mechanoinsensitive C-fibers are the most common nociceptor type encountered in electrophysiological recordings [70]. In primates they were found to belong to the non-peptidergic subtypes [70].

Given the pivotal role of TRPV1 receptors in pain sensation, many efforts have been made to identify the TRPV1-positive nerve fibers in human skin. While commercially available TRPV1 antibodies work reliably on rodent DRG neurons, reliable staining of skin TRPV1 has been difficult. One study showed that all intraepidermal nerve fibers are TRPV1 positive in humans [71]. Since the TRPV1 antibody used was from a private source, this study could not be replicated by others. Another study, using a commercially available antibody, showed scarce TRPV1 immunoreactive fibers colocalizing with CGRP and SP in the dermis, but not in the epidermis [72]. Others could not detect TRPV1 immunoreactive intraepidermal nerve fibers at all but only found TRPV1 on keratinocytes in human skin [73, 74]. Using the Alomone rabbit anti-TRPV1 antibody, we very rarely detected single TRPV1-immunoreactive fibers in human epidermis. In conclusion, it is likely that some IENF express the TRPV1 receptor, but the amount of antigen exposed may be too low to be detected by the presently available antibodies.

2.7 Subepidermal Nerves

In a skin biopsy section processed with anti-PGP 9.5 antibodies, apart from the IENF, extensive nerve structures can be seen in the subepidermis and dermis. The subepidermal nerve plexus consists of horizontally oriented nerve fiber bundles running in parallel to the epidermis. Sometimes large nerve bundles coming from the deep dermis and running almost perpendicularly to the dermis can be seen, giving rise to the subepidermal plexus. In many studies of SFN of different etiologies, those subepidermal nerve fibers were described as normal. In contrast, the subepidermal nerve fiber bundles are pathologically altered in mixed neuropathy, as, for example, with type 2 diabetes [75]. Others also found involvement of the subepidermal nerve plexus in SFN of different etiologies, including chemotherapy toxicity and autoimmune disorders [76, 77]. It would be interesting to know if these patients with subepidermal nerve plexus rarefication later developed a mixed neuropathy. Methods to quantify subepidermal and nerve bundles in the area comprising 200 μm adjacent to the dermal-epidermal junction have been described, including stereology [78–80], and a reduction of these fibers was found in SFN. One reason for the discrepancy in the abovementioned study may be the different methods. It is possible that a reduction in the subepidermal nerve fiber plexus escapes the eye of the observer but is detectable with morphometric methods. Another reason may be the variability between patient groups. Within SFN, there may be subgroups with distal sensory neuropathy manifesting as SFN or early stages of mixes neuropathy. Whether the finding of a rarified subepidermal plexus in a patient with SFN has a prognostic value remains to be determined.

Nerve bundles in the deeper dermis have not been formally studied in SFN. We studied 35 patients with diabetic neuropathy and 17 with diabetes and no clinical neuropathy including normal findings in quantitative sensory testing (QST) and found reduced IENFD also in this asymptomatic group, compared to controls. Furthermore, we found a reduction of myelinated fibers in the dermis (excluding the subepidermal nerve plexus from the analysis) both in patients with diabetic neuropathy and in those with diabetes but no signs of neuropathy or SFN [81]. While these patients may develop diabetic neuropathy over time and did not have the clinical features of SFN, the example shows that morphologically, profound changes can occur in the skin before neuropathy signs even appear. Elongated nodes of Ranvier and dispersion of paranodal proteins could be found in both diabetic groups [81]. In 19 patients with SFN (16 idiopathic), we found a reduced number of dermal myelinated fibers compared to controls, but no difference in dermal nerve fiber bundles in total [82]. A subgroup of SFN patients even had elongated Ranvier nodes, which is considered a hallmark of demyelination. While such methods might be developed into tools to better differentiate SFN subgroups, the findings still have to be confirmed in larger patient series.

By definition, the innervation of touch receptors like Merkel cells and Meissner bodies should not be altered in SFN, but no data on this question are available. Sweat glands, arrector pili muscles, and arterioles are innervated by C-fibers, and reduced innervation of all of these structures has been shown in SFN [83, 84]. We saw reduced sweat gland innervation in women with Fabry disease who suffered from anhidrosis [85].

As in other body regions, nerve fibers run alongside blood vessels [86, 87] (Fig. 2.3a). Small subepidermal blood vessels are innervated by peptidergic C-fibers [88]. Deeper dermal blood vessels have a rich network of both CGRPergic and sympathetic neuropeptide Y-positive innervation [87]. Interestingly, patients with fibromyalgia syndrome were found to have increased peptidergic innervation of the dermal arteriovenous shunts [89]. This leads to an imbalance with higher numbers of vasodilatatory compared to vasoconstrictive fibers. The authors concluded that blood flow dysregulation as a consequence of this hyperinnervation might contribute to pain and fatigue in the fibromyalgia syndrome. Whether there are changes in subepidermal and dermal innervation in the different types of SFN remains to be determined.

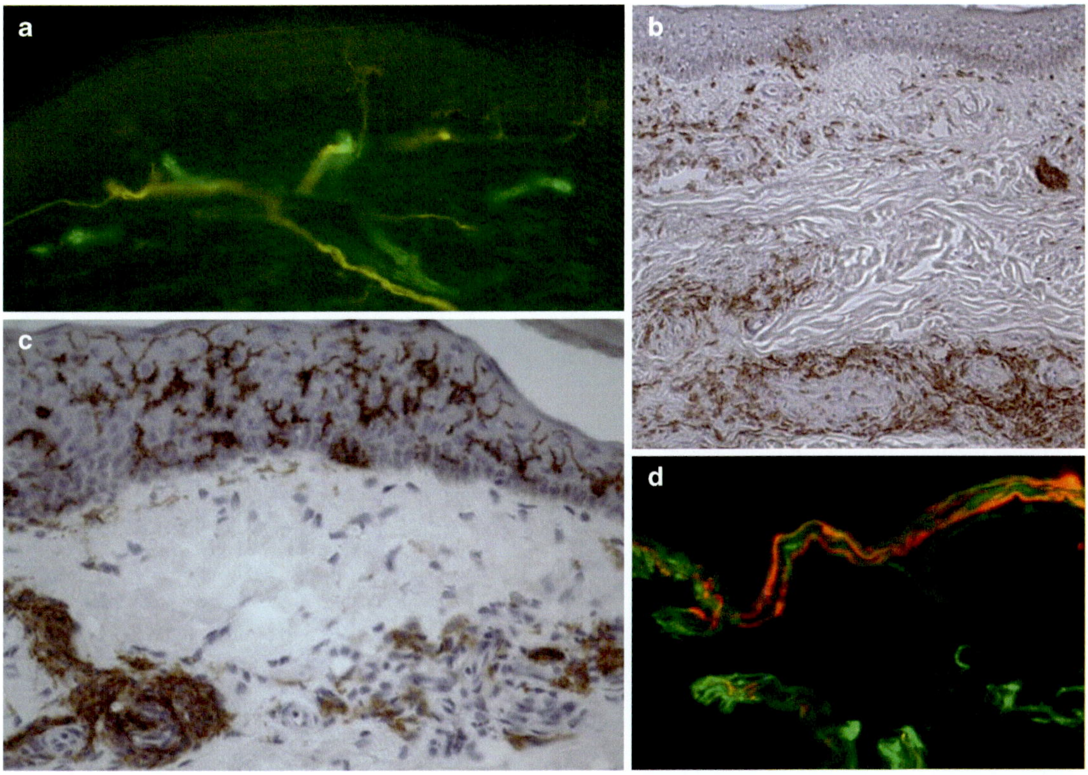

Fig. 2.3 Pathology of the skin in small fiber neuropathy (SFN) and rapid eye movement sleep behavior disorder. Nerve fiber running alongside a dermal blood vessel, double immunofluorescence with antibodies to PGP 9.5 (*yellow*) and to CD31 (*green*) (**a**). Dense infiltrates of T-lymphocytes, as immunoreacted with antibodies to CD3, around subepidermal blood vessels in a patient with SFN (**b**). Langerhans cells, immunoreacted with CD1a, are numerous in the epidermis of controls (not shown) and of patients with SFN (**c**). Deposit of phosphorylated alpha-synuclein (*red*) in nerve fibers (*green*) of a patient with idiopathic rapid eye movement sleep behavior disorder (**d**)

2.8 Further Yield of Skin Biopsy

2.8.1 Inflammation

While the assessment of inflammation is standard in dermatological histopathology, it has been neglected by most neurological skin biopsy laboratories. Thus, data on inflammatory cells in skin biopsies from SFN patients are scarce. However, immunostaining for macrophages and T-lymphocytes can be useful to detect vasculitis in skin biopsies and thus gives a hint that a vasculitic neuropathy might be the underlying cause of a patient's complaints [90]. Further than that, we found increased inflammatory cells in skin of most types of neuropathy; however, only few cases of SFN were included in this study [91]. On an individual basis, increased numbers of macrophages and T-cells can be seen in SFN patients (Fig. 2.3b), but the diagnostic and therapeutic consequences of such findings are unknown. Epidermal Langerhans cells, the major epidermal immune cells, are easily visible in skin biopsies, using, for example, immunohistochemistry for CD1a. In SFN, one can find a wide variety of Langerhans cell density and distribution, without obvious correlation to clinical parameters (own observations, Fig. 2.3c). One group reported increased numbers of Langerhans cells in a small cohort of patients with painful diabetic SFN [92]. These data have not yet been reproduced, nor have other SFN cohorts with data on Langerhans cells been published.

2.8.2 Amyloid and Other Deposits in Skin

Familial amyloid neuropathy associated with mutations in the transthyretin gene (TTR-FAP) usually starts with a SFN, before large fibers are involved. Early diagnosis of TTR-FAP has come into the focus of interest with the availability of new therapies [93–95]. Cutaneous amyloid can be detected in skin biopsies, and the amyloid burden correlates with IENFD [19]. The authors reported a sensitivity of 70% for the detection of amyloid in skin of TTR-FAP patients, such that the test cannot be used to rule out the disease.

Skin biopsy can be diagnostic in other types of SFN leading to intracellular deposits. Fabry disease is one example, where the sphingolipid globotriaosylceramide (Gb3) can be detected in the skin of affected men [96]. However, this can be less invasively done in blood samples [97], such that skin biopsy might not be the method of choice.

In cerebral autosomal dominant arteriopathy with subcortical infarcts and leukoencephalopathy (CADASIL), a disorder caused by mutations of the NOTCH3 gene that manifests, among other symptoms, with migraine and strokes, skin biopsy is diagnostic by showing granular osmiophilic material around dermal vessels. These patients also have non-length-dependent reduced IENFD and a severe loss of autonomic innervation [98].

Patients with idiopathic rapid eye movement (REM) sleep behavior disorder (iRBD) have an increased risk of developing Parkinson's disease (PD). It has therefore been asked whether, like in PD, phosphorylated alpha-synuclein deposits can be detected in the skin of iRBD patients. This is indeed the case in more than 50% of RBD patients [99] (Fig. 2.3d). These patients also have reduced IENFD without suffering from clinical signs of SFN [100].

2.9 Conclusion

Skin biopsy is a minimally invasive, repeatable diagnostic procedure that, since its introduction into clinical practice, has yielded great advances in our understanding of SFN and related disorders. Some major open questions still prevail. IENFD is reduced in many conditions that do not fulfill the current criteria of SFN. On the other hand, skin biopsy may not detect disorders with small fiber hyperactivity, like the genetic sodium channel-related SFN, which is important to consider in SFN classifications [101]. There is a major gap between neurophysiological concepts nociceptor dysfunction and their morphological correlates, which needs to be closed by further research.

References

1. McGlone F, Reilly D. The cutaneous sensory system. Neurosci Biobehav Rev. 2010;34:148–59.
2. Doppler K, Rittner HL, Deckart M, Sommer C. Reduced dermal nerve fiber diameter in skin biopsies of patients with fibromyalgia. Pain. 2015;156:2319–25.
3. Nolano M, Provitera V, Crisci C, Stancanelli A, Wendelschafer-Crabb G, Kennedy WR, et al. Quantification of myelinated endings and mechanoreceptors in human digital skin. Ann Neurol. 2003;54:197–205.
4. Wang L, Hilliges M, Jernberg T, Wiegleb-Edstrom D, Johansson O. Protein gene product 9.5-immunoreactive nerve fibres and cells in human skin. Cell Tissue Res. 1990;261:25–33.
5. Kennedy WR, Wendelschafer-Crabb G. The innervation of human epidermis. J Neurol Sci. 1993;115:184–90.
6. McCarthy BG, Hsieh ST, Stocks A, Hauer P, Macko C, Cornblath DR, et al. Cutaneous innervation in sensory neuropathies: evaluation by skin biopsy. Neurology. 1995;45:1848–55.
7. McArthur JC, Stocks EA, Hauer P, Cornblath DR, Griffin JW. Epidermal nerve fiber density: normative reference range and diagnostic efficiency. Arch Neurol. 1998;55:1513–20.
8. Lauria G, Cornblath DR, Johansson O, McArthur JC, Mellgren SI, Nolano M, et al. EFNS guidelines on the use of skin biopsy in the diagnosis of peripheral neuropathy. Eur J Neurol. 2005;12:747–58.
9. Lauria G, Borgna M, Morbin M, Lombardi R, Mazzoleni G, Sghirlanzoni A, et al. Tubule and neurofilament immunoreactivity in human hairy skin: markers for intraepidermal nerve fibers. Muscle Nerve. 2004;30:310–6.
10. Albrecht PJ, Hines S, Eisenberg E, Pud D, Finlay DR, Connolly MK, et al. Pathologic alterations of cutaneous innervation and vasculature in affected limbs from patients with complex regional pain syndrome. Pain. 2006;120:244–66.

11. Uno H, Parker F. Autonomic innervation of the skin in primary erythermalgia. Arch Dermatol. 1983;119:65–71.
12. Donadio V, Nolano M, Provitera V, Stancanelli A, Lullo F, Liguori R, et al. Skin sympathetic adrenergic innervation: an immunofluorescence confocal study. Ann Neurol. 2006;59:376–81.
13. Nolano M, Provitera V, Perretti A, Stancanelli A, Saltalamacchia AM, Donadio V, et al. Ross syndrome: a rare or a misknown disorder of thermoregulation? A skin innervation study on 12 subjects. Brain. 2006;129:2119–31.
14. Kennedy WR, Wendelschafer-Crabb G, Brelje TC. Innervation and vasculature of human sweat glands: an immunohistochemistry-laser scanning confocal fluorescence microscopy study. J Neurosci. 1994;14:6825–33.
15. Wilder-Smith EP, Chow A. Comparison of a simple method for quantitation of intraepidermal nerve fibres with a standard image analysis method using hypothenar skin. J Neurol. 2006;253:1011–5.
16. Chien HF, Tseng TJ, Lin WM, Yang CC, Chang YC, Chen RC, et al. Quantitative pathology of cutaneous nerve terminal degeneration in the human skin. Acta Neuropathol (Berl). 2001;102:455–61.
17. Kennedy WR, Wendelschafer-Crabb G, Polydefkis M, McArthur J. Pathology and quantitation of cutaneous nerves. In: Dyck PJ, Thomas PK, editors. Peripheral neuropathy. 4th ed. Philadelphia: Saunders; 2005. p. 869–96.
18. Lauria G, Hsieh ST, Johansson O, Kennedy WR, Leger JM, Mellgren SI, et al. European Federation of Neurological Societies/Peripheral Nerve Society Guideline on the use of skin biopsy in the diagnosis of small fiber neuropathy. Report of a joint task force of the European Federation of Neurological Societies and the Peripheral Nerve Society. Eur J Neurol. 2010;17:903–12, e44–9.
19. Ebenezer GJ, Hauer P, Gibbons C, McArthur JC, Polydefkis M. Assessment of epidermal nerve fibers: a new diagnostic and predictive tool for peripheral neuropathies. J Neuropathol Exp Neurol. 2007;66:1059–73.
20. Goransson LG, Mellgren SI, Lindal S, Omdal R. The effect of age and gender on epidermal nerve fiber density. Neurology. 2004;62:774–7.
21. Bakkers M, Merkies IS, Lauria G, Devigili G, Penza P, Lombardi R, et al. Intraepidermal nerve fiber density and its application in sarcoidosis. Neurology. 2009;73:1142–8.
22. Burns TM, Smith AG. "Measure twice, cut once": improving diagnostic accuracy of skin biopsy. Neurology. 2012;79:2164–5.
23. Seger S, Stritt M, Doppler K, Frank S, Panaite A, Kuntzer T, et al. A semi-automated method to assess intraepidermal nerve fibre density in human skin biopsies. Histopathology. 2016;68:657–65.
24. Umapathi T, Tan WL, Tan NC, Chan YH. Determinants of epidermal nerve fiber density in normal individuals. Muscle Nerve. 2006;33:742–6.
25. Pan CL, Lin YH, Lin WM, Tai TY, Hsieh ST. Degeneration of nociceptive nerve terminals in human peripheral neuropathy. Neuroreport. 2001;12:787–92.
26. Nolano M, Biasiotta A, Lombardi R, Provitera V, Stancanelli A, Caporaso G, et al. Epidermal innervation morphometry by immunofluorescence and bright-field microscopy. J Peripher Nerv Syst. 2015;20:387–91.
27. Lauria G, Bakkers M, Schmitz C, Lombardi R, Penza P, Devigili G, et al. Intraepidermal nerve fiber density at the distal leg: a worldwide normative reference study. J Peripher Nerv Syst. 2010;15:202–7.
28. Collongues N, Samama B, Schmidt-Mutter C, Chamard-Witkowski L, Debouverie M, Chanson JB, et al. Quantitative and qualitative normative dataset for intraepidermal nerve fibers using skin biopsy. PLoS One. 2018;13:e0191614.
29. Vlckova-Moravcova E, Bednarik J, Dusek L, Toyka KV, Sommer C. Diagnostic validity of epidermal nerve fiber densities in painful sensory neuropathies. Muscle Nerve. 2008;37:50–60.
30. Nebuchennykh M, Loseth S, Lindal S, Mellgren SI. The value of skin biopsy with recording of intraepidermal nerve fiber density and quantitative sensory testing in the assessment of small fiber involvement in patients with different causes of polyneuropathy. J Neurol. 2009;256:1067–75.
31. Engelstad JK, Taylor SW, Witt LV, Hoebing BJ, Herrmann DN, Dyck PJ, et al. Epidermal nerve fibers: confidence intervals and continuous measures with nerve conduction. Neurology. 2012;79:2187–93.
32. Donadio V, Incensi A, Giannoccaro MP, Cortelli P, Di Stasi V, Pizza F, et al. Peripheral autonomic neuropathy: diagnostic contribution of skin biopsy. J Neuropathol Exp Neurol. 2012;71:1000–8.
33. Kennedy WR, Wendelschafer-Crabb G. Utility of skin biopsy in diabetic neuropathy. Semin Neurol. 1996;16:163–71.
34. Holland NR, Crawford TO, Hauer P, Cornblath DR, Griffin JW, McArthur JC. Small-fiber sensory neuropathies: clinical course and neuropathology of idiopathic cases. Ann Neurol. 1998;44:47–59.
35. Holland NR, Stocks A, Hauer P, Cornblath DR, Griffin JW, McArthur JC. Intraepidermal nerve fiber density in patients with painful sensory neuropathy. Neurology. 1997;48:708–11.
36. Lauria G, Holland N, Hauer P, Cornblath DR, Griffin JW, McArthur JC. Epidermal innervation: changes with aging, topographic location, and in sensory neuropathy. J Neurol Sci. 1999;164:172–8.
37. Lauria G, Sghirlanzoni A, Lombardi R, Pareyson D. Epidermal nerve fiber density in sensory ganglionopathies: clinical and neurophysiologic correlations. Muscle Nerve. 2001;24:1034–9.
38. Herrmann DN, Griffin JW, Hauer P, Cornblath DR, McArthur JC. Epidermal nerve fiber density and sural nerve morphometry in peripheral neuropathies. Neurology. 1999;53:1634–40.

39. Periquet MI, Novak V, Collins MP, Nagaraja HN, Erdem S, Nash SM, et al. Painful sensory neuropathy: prospective evaluation using skin biopsy. Neurology. 1999;53:1641–7.
40. Scott LJ, Griffin JW, Luciano C, Barton NW, Banerjee T, Crawford T, et al. Quantitative analysis of epidermal innervation in Fabry disease. Neurology. 1999;52:1249–54.
41. Wakamoto H, Hirai A, Manabe K, Hayashi M. Idiopathic small-fiber sensory neuropathy in childhood: a diagnosis based on objective findings on punch skin biopsy specimens. J Pediatr. 1999;135:257–60.
42. Hsieh ST, Chiang HY, Lin WM. Pathology of nerve terminal degeneration in the skin. J Neuropathol Exp Neurol. 2000;59:297–307.
43. Polydefkis M, Allen RP, Hauer P, Earley CJ, Griffin JW, McArthur JC. Subclinical sensory neuropathy in late-onset restless legs syndrome. Neurology. 2000;55:1115–21.
44. Polydefkis M, Yiannoutsos CT, Cohen BA, Hollander H, Schifitto G, Clifford DB, et al. Reduced intraepidermal nerve fiber density in HIV-associated sensory neuropathy. Neurology. 2002;58:115–9.
45. Verze L, Viglietti-Panzica C, Plumari L, Calcagni M, Stella M, Schrama LH, et al. Cutaneous innervation in hereditary sensory and autonomic neuropathy type IV. Neurology. 2000;55:126–8.
46. Hoitsma E, Marziniak M, Faber CG, Reulen JP, Sommer C, De Baets M, et al. Small fibre neuropathy in sarcoidosis. Lancet. 2002;359:2085–6.
47. Boruchow SA, Gibbons CH. Utility of skin biopsy in management of small fiber neuropathy. Muscle Nerve. 2013;48:877–82.
48. Devigili G, Tugnoli V, Penza P, Camozzi F, Lombardi R, Melli G, et al. The diagnostic criteria for small fibre neuropathy: from symptoms to neuropathology. Brain. 2008;131:1912–25.
49. Abuzinadah AR, Kluding P, Wright D, D'Silva L, Ryals J, Hendry B, et al. Less is more in diabetic neuropathy diagnosis: comparison of quantitative sudomotor axon reflex and skin biopsy. J Clin Neuromuscul Dis. 2017;19:5–11.
50. Harrer JU, Uceyler N, Doppler K, Fischer TZ, Dib-Hajj SD, Waxman SG, et al. Neuropathic pain in two-generation twins carrying the sodium channel Nav1.7 functional variant R1150W. Pain. 2014;155:2199–203.
51. Üçeyler N, Ganendiran S, Kramer D, Sommer C. Characterization of pain in Fabry disease. Clin J Pain. 2014;30:915–20.
52. Üçeyler N, He L, Schönfeld D, Kahn AK, Reiners K, Hilz MJ, et al. Small fibers in Fabry disease: baseline and follow-up data under enzyme replacement therapy. J Peripher Nerv Syst. 2011;16:304–14.
53. Vlckova-Moravcova E, Bednarik J, Belobradkova J, Sommer C. Small-fibre involvement in diabetic patients with neuropathic foot pain. Diabet Med. 2008;25:692–9.
54. Sorensen L, Molyneaux L, Yue DK. The relationship among pain, sensory loss, and small nerve fibers in diabetes. Diabetes Care. 2006;29:883–7.
55. Günes HN, Bekircan-Kurt CE, Tan E, Erdem-Ozdamar S. The histopathological evaluation of small fiber neuropathy in patients with vitamin B12 deficiency. Acta Neurol Belg. 2018;118:405–10.
56. Üçeyler N, Vollert J, Broll B, Riediger N, Langjahr M, Saffer N, et al. Sensory profiles and skin innervation of patients with painful and painless neuropathies. Pain. 2018;159:1867–76.
57. Provitera V, Gibbons CH, Wendelschafer-Crabb G, Donadio V, Vitale DF, Loavenbruck A, et al. The role of skin biopsy in differentiating small-fiber neuropathy from ganglionopathy. Eur J Neurol. 2018;25:848–53.
58. Cazzato D, Lauria G. Small fibre neuropathy. Curr Opin Neurol. 2017;30:490–9.
59. Waxman SG, Merkies IS, Gerrits MM, Dib-Hajj SD, Lauria G, Cox JJ, et al. Sodium channel genes in pain-related disorders: phenotype-genotype associations and recommendations for clinical use. Lancet Neurol. 2014;13:1152–60.
60. Herrmann DN, O'Connor AB, Schwid SR, Da Y, Goodman AD, Rafferty J, et al. Broadening the spectrum of controls for skin biopsy in painful neuropathies. Muscle Nerve. 2010;42:436–8.
61. Ringkamp M, Raja SN, Campbell A, Meyer RA. Peripheral mechanisms of cutaneous nociception. In: McMahon SB, Koltzenburg M, Tracey I, Turk DC, editors. Wall and Melzack's textbook of pain. Philadelphia: Elsevier; 2013. p. 1–30.
62. Taylor AM, Peleshok JC, Ribeiro-da-Silva A. Distribution of P2X(3)-immunoreactive fibers in hairy and glabrous skin of the rat. J Comp Neurol. 2009;514:555–66.
63. Nolano M, Provitera V, Caporaso G, Stancanelli A, Leandri M, Biasiotta A, et al. Cutaneous innervation of the human face as assessed by skin biopsy. J Anat. 2013;222:161–9.
64. Bechakra M, Schuttenhelm BN, Pederzani T, van Doorn PA, de Zeeuw CI, Jongen JLM. The reduction of intraepidermal P2X3 nerve fiber density correlates with behavioral hyperalgesia in a rat model of nerve injury-induced pain. J Comp Neurol. 2017;525:3757–68.
65. Schuttenhelm BN, Duraku LS, Dijkstra JF, Walbeehm ET, Holstege JC. Differential changes in the peptidergic and the non-peptidergic skin innervation in rat models for inflammation, dry skin itch, and dermatitis. J Invest Dermatol. 2015;135:2049–57.
66. McCoy ES, Taylor-Blake B, Street SE, Pribisko AL, Zheng J, Zylka MJ. Peptidergic CGRPalpha primary sensory neurons encode heat and itch and tonically suppress sensitivity to cold. Neuron. 2013;78:138–51.
67. Hsieh YL, Lin CL, Chiang H, Fu YS, Lue JH, Hsieh ST. Role of peptidergic nerve terminals in the skin: reversal of thermal sensation by calcitonin gene-related peptide in TRPV1-depleted neuropathy. PLoS One. 2012;7:e50805.

68. Dubin AE, Patapoutian A. Nociceptors: the sensors of the pain pathway. J Clin Invest. 2010;120:3760–72.
69. da Silva Serra I, Husson Z, Bartlett JD, Smith ES. Characterization of cutaneous and articular sensory neurons. Mol Pain. 2016;12. https://doi.org/10.1177/1744806916636387.
70. Wooten M, Weng HJ, Hartke TV, Borzan J, Klein AH, Turnquist B, et al. Three functionally distinct classes of C-fibre nociceptors in primates. Nat Commun. 2014;5:4122.
71. Lauria G, Morbin M, Lombardi R, Capobianco R, Camozzi F, Pareyson D, et al. Expression of capsaicin receptor immunoreactivity in human peripheral nervous system and in painful neuropathies. J Peripher Nerv Syst. 2006;11:262–71.
72. Axelsson HE, Minde JK, Sonesson A, Toolanen G, Hogestatt ED, Zygmunt PM. Transient receptor potential vanilloid 1, vanilloid 2 and melastatin 8 immunoreactive nerve fibers in human skin from individuals with and without Norrbottnian congenital insensitivity to pain. Neuroscience. 2009;162:1322–32.
73. Wilder-Smith EP, Ong WY, Guo Y, Chow AW. Epidermal transient receptor potential vanilloid 1 in idiopathic small nerve fibre disease, diabetic neuropathy and healthy human subjects. Histopathology. 2007;51:674–80.
74. Han SB, Kim H, Cho SH, Lee JD, Chung JH, Kim HS. Transient Receptor Potential Vanilloid-1 in Epidermal Keratinocytes May Contribute to Acute Pain in Herpes Zoster. Acta Derm Venereol. 2016;96:319–22.
75. Chao CC, Tseng MT, Lin YJ, Yang WS, Hsieh SC, Lin YH, et al. Pathophysiology of neuropathic pain in type 2 diabetes: skin denervation and contact heat-evoked potentials. Diabetes Care. 2010;33:2654–9.
76. Tseng MT, Hsieh SC, Shun CT, Lee KL, Pan CL, Lin WM, et al. Skin denervation and cutaneous vasculitis in systemic lupus erythematosus. Brain. 2006;129:977–85.
77. Casanova-Molla J, Morales M, Garrabou G, Sola-Valls N, Soriano A, Calvo M, et al. Mitochondrial loss indicates early axonal damage in small fiber neuropathies. J Peripher Nerv Syst. 2012;17:147–57.
78. Sommer C, Lindenlaub T, Zillikens D, Toyka KV, Naumann M. Selective loss of cholinergic sudomotor fibers causes anhidrosis in Ross syndrome. Ann Neurol. 2002;52:247–50.
79. Lauria G, Cazzato D, Porretta-Serapiglia C, Casanova-Molla J, Taiana M, Penza P, et al. Morphometry of dermal nerve fibers in human skin. Neurology. 2011;77:242–9.
80. Karlsson P, Porretta-Serapiglia C, Lombardi R, Jensen TS, Lauria G. Dermal innervation in healthy subjects and small fiber neuropathy patients: a stereological reappraisal. J Peripher Nerv Syst. 2013;18:48–53.
81. Doppler K, Frank F, Koschker AC, Reiners K, Sommer C. Nodes of Ranvier in skin biopsies of patients with diabetes mellitus. J Peripher Nerv Syst. 2017;22:182–90.
82. Doppler K, Werner C, Henneges C, Sommer C. Analysis of myelinated fibers in human skin biopsies of patients with neuropathies. J Neurol. 2012;259:1879–87.
83. Gibbons CH, Illigens BM, Wang N, Freeman R. Quantification of sweat gland innervation: a clinical-pathologic correlation. Neurology. 2009;72:1479–86.
84. Dabby R, Vaknine H, Gilad R, Djaldetti R, Sadeh M. Evaluation of cutaneous autonomic innervation in idiopathic sensory small-fiber neuropathy. J Peripher Nerv Syst. 2007;12:98–101.
85. Kokotis P, Uceyler N, Werner C, Tsivgoulis G, Papanikola N, Katsanos AH, et al. Quantification of sweat gland innervation in patients with Fabry disease: a case-control study. J Neurol Sci. 2018;390:135–8.
86. Pare M, Smith AM, Rice FL. Distribution and terminal arborizations of cutaneous mechanoreceptors in the glabrous finger pads of the monkey. J Comp Neurol. 2002;445:347–59.
87. Rice FL, Albrecht PJ. Cutaneous mechanisms of tactile perception: morphological and chemical organization of the innervation to the skin. In: Basbaum A, Kaneko A, Shepherd GM, Westheimer G, editors. The senses. San Diego: Academic Press; 2008. p. 1–32.
88. Rice FL, Rasmusson DD. Innervation of the digit on the forepaw of the raccoon. J Comp Neurol. 2000;417:467–90.
89. Albrecht PJ, Hou Q, Argoff CE, Storey JR, Wymer JP, Rice FL. Excessive peptidergic sensory innervation of cutaneous arteriole-venule shunts (AVS) in the palmar glabrous skin of fibromyalgia patients: implications for widespread deep tissue pain and fatigue. Pain Med. 2013;14:895–915.
90. Üçeyler N, Devigili G, Toyka KV, Sommer C. Skin biopsy as an additional diagnostic tool in non-systemic vasculitic neuropathy. Acta Neuropathol. 2010;120:109–16.
91. Üçeyler N, Braunsdorf S, Kunze E, Riediger N, Scheytt S, Divisova S, et al. Cellular infiltrates in skin and sural nerve of patients with polyneuropathies. Muscle Nerve. 2017;55:884–93.
92. Casanova-Molla J, Morales M, Planas-Rigol E, Bosch A, Calvo M, Grau-Junyent JM, et al. Epidermal Langerhans cells in small fiber neuropathies. Pain. 2012;153:982–9.
93. Plante-Bordeneuve V. Transthyretin familial amyloid polyneuropathy: an update. J Neurol. 2018;265:976–83.
94. Adams D, Gonzalez-Duarte A, O'Riordan WD, Yang CC, Ueda M, Kristen AV, et al. Patisiran, an RNAi therapeutic, for hereditary transthyretin amyloidosis. N Engl J Med. 2018;379:11–21.
95. Benson MD, Waddington-Cruz M, Berk JL, Polydefkis M, Dyck PJ, Wang AK, et al. Inotersen treatment for patients with hereditary transthyretin amyloidosis. N Engl J Med. 2018;379:22–31.

96. Üçeyler N, Schröter N, Kafke W, Kramer D, Wanner C, Weidemann F, et al. Skin globotriaosylceramide 3 load is increased in men with advanced fabry disease. PLoS One. 2016;11:e0166484.
97. Üçeyler N, Böttger J, Henkel L, Langjahr M, Mayer C, Nordbeck P, et al. Detection of blood Gb3 deposits as a new tool for diagnosis and therapy monitoring in patients with classic Fabry disease. J Intern Med. 2018;284:427–38.
98. Nolano M, Provitera V, Donadio V, Caporaso G, Stancanelli A, Califano F, et al. Cutaneous sensory and autonomic denervation in CADASIL. Neurology. 2016;86:1039–44.
99. Doppler K, Jentschke HM, Schulmeyer L, Vadasz D, Janzen A, Luster M, et al. Dermal phospho-alpha-synuclein deposits confirm REM sleep behaviour disorder as prodromal Parkinson's disease. Acta Neuropathol. 2017;133:535–45.
100. Schrempf W, Katona I, Dogan I, Felbert VV, Wienecke M, Heller J, et al. Reduced intraepidermal nerve fiber density in patients with REM sleep behavior disorder. Parkinsonism Relat Disord. 2016;29:10–6.
101. Levine TD. Small fiber neuropathy: disease classification beyond pain and burning. J Cent Nerv Syst Dis. 2018;10:1179573518771703.

Neurophysiological Assessments in Small Fiber Neuropathy: Evoked Potentials

3

Rosario Privitera and Praveen Anand

Abstract

The syndromes of small fiber neuropathies (SFN) affect thinly myelinated Aδ-fibers and unmyelinated C-fibers, which are assessed by a range of techniques including cerebral potentials evoked by noxious electrical, laser, or contact heat stimuli. These specialized neurophysiological techniques contribute to the diagnosis, monitoring, and evaluation of treatment effects in SFN and understanding their pathophysiology. The standard clinical neurophysiological tests (nerve conduction studies, NCS, and somatosensory-evoked potentials, SSEPs) assess large myelinated Aβ-nerve fibers and their associated pathways in the central nervous system (CNS), hence may be normal in SFN. Nociceptor dysfunction assessment with electrical stimuli has been used in organs such as tooth pulp, which are exclusively innervated by small fibers, mainly to investigate the efficacy of analgesic agents in acute experimental pain. Laser-evoked potentials (LEPs) are obtained after rapid heating of the skin with laser pulses, can help to identify lesions of small sensory nerve fibers and/or their CNS pathways, and represent a sensitive, objective, and noninvasive method for the assessment of SFN. Contact heat-evoked potentials (CHEPs) are similar to LEPs with some practical advantages for clinical use, particularly when the region affected precludes skin biopsy, as it has been correlated with other techniques for assessing SFN including intraepidermal nerve fiber density (IENFD) in skin biopsies. They help in assessing patients unable to perform quantitative sensory testing (QST). However, LEPs/CHEPs may be reduced or absent in patients with CNS pain conditions and do not localize the abnormality to peripheral nerves. They can be helpful in distinguishing SFN from other conditions such as nonorganic pain or non-neuropathic hypersensitivity disorders, in which LEPs/CHEPs are preserved or even enhanced. Pain-evoked potentials are a useful tool for studying endogenous processing of emotional-motivational responses related to pain. They have been recommended in guidelines for the assessment of neuropathic pain.

Keywords

Pain-evoked potential · Laser-evoked potential (LEP) · Contact heat-evoked potential (CHEP) · Tooth pulp-evoked potential (TPEP)

R. Privitera · P. Anand (✉)
Peripheral Neuropathy Unit, Centre for Clinical Translation, Division of Brain Sciences, Imperial College London, Hammersmith Hospital, London, UK
e-mail: p.anand@imperial.ac.uk

S.-T. Hsieh et al. (eds.), *Small Fiber Neuropathy and Related Syndromes: Pain and Neurodegeneration*,
https://doi.org/10.1007/978-981-13-3546-4_3

3.1 Introduction

Small fiber or nociceptive-evoked potentials have been recommended by various authors for the diagnosis and monitoring of neuropathic pain conditions, including small fiber neuropathy (SFN) [1–4]. There is a substantial literature with evidence supporting the utility of pain-evoked potentials in a range of clinical conditions affecting small sensory nerve fibers, particularly in patients presenting with neuropathic pain [5–13]. These techniques have also been used in comparative studies with other techniques for assessing SFN, the effects of treatments, and for the experimental study of pain in healthy volunteers [14–16].

Both conventionally evoked somatosensory potentials and pain-evoked potentials are time-locked to a trigger stimulus and reflect the integrity of the nervous system and its capacity to respond to external stimuli. The waveform obtained has predictable and reproducible peaks for latency and amplitude.

Standard clinical neurophysiological tests (nerve conduction studies, NCS, and somatosensory-evoked potentials, SSEPs) are commonly used; they assess the integrity of large myelinated nerve fibers and their associated central nervous system pathways. Such studies may be normal in SFN, as they preferentially activate large myelinated Aβ-nerve fibers [17]. Somatosensory-evoked potentials correspond to the electrical activity recorded from the sensorimotor cortex and other sites along the pathway [18–24] and generated after stimulation using tactile [25, 26], vibration [27], electrical [28], or thermal [14, 29] stimuli.

Nociceptor dysfunction may be assessed using electrical stimuli [28, 30], including their application in organs such as tooth pulp which are exclusively innervated by small fibers. Other methods to assess small nerve fiber function use radiant or contact heat delivered, respectively, through laser beams [31, 32] or thermodes [33]. Evoked potentials obtained after rapid heating of the skin have been shown to be also a reliable neurophysiological tool for the clinical assessment of pain syndromes; the neural pathways mediating the transmission of pain and thermal information are in common, via activation of thinly myelinated Aδ-fibers and unmyelinated C-fibers, and differ in many important aspects from those related to large fibers [1, 29, 34]. Contact heat-evoked potentials (CHEPs) show a very similar topography of evoked potentials to that evoked by radiant heat (laser) stimulation [35]. However, laser-evoked potentials (LEPs) require precautions, e.g., to avoid eye damage and skin burns. CHEPs represent a safe method of assessing neurological deficits in the clinic and particularly when the region affected precludes skin biopsy as undesirable to determine intraepidermal nerve fiber density (IENFD), e.g., face and palms. The decrease of CHEPs amplitude correlates with the reduction of IENFD in neuropathic conditions [10, 14, 36].

3.2 Pain-Evoked Potentials

3.2.1 Electrical Stimuli-Evoked Potentials

Conventional neurophysiological studies have limited utility in supporting the diagnosis of SFN or neuropathic pain as they only provide a precise topographical localization of the peripheral lesions when large and small fibers are both affected [37]. Pain-evoked potentials are generated when electrical stimuli activate Aδ- and/or C-fibers, using current applied peripherally at a defined minimum, and increased in steps until the subject reports the perception of the electrical stimulus and then pain [1, 38, 39]. Such neurophysiological pain studies activate not only the nociceptive Aδ- and C-fibers but also A-beta fibers, because the latter have a lower electrical threshold. Amplitudes of pain-evoked potentials have been useful in assessment of SFN [40–43].

Electrical stimulation of the tooth pulp has been used to overcome the lack of nociceptive selectivity in neurophysiological studies, when cortical potentials are evoked by transcutaneous electrical stimuli [44]. For this purpose, electrical stimuli have been applied to the tooth pulp or delivered intracutaneously through special electrodes [45]. Tooth pulp-evoked potentials (TPEPs) have been associated with the perception of noxious stimuli and are described as objective, quantifiable events [44] composed of earlier and later TPEP components. These reflect, respectively, the sensory transmission processes and further brain activities generated by neural structures anatomically [46] not specifically related to the pain processes [47]. In healthy volunteers, the amplitude but not the latency of TPEPs was found to increase with the increase of pain reported [48]. TPEP amplitudes corresponded to the reduction of subjective pain ratings with meperidine, while only the TPEP amplitudes, but not the pain scores, were reduced with nalbuphine. This was attributed to a depressive effect on the brain activity of nalbuphine, which was not linked to pain-mediating pathways [49]. Tooth pain-evoked potentials have been largely utilized to investigate the efficacy of different analgesic agents and obtain a quantitative comparison of their analgesic potency in the acute experimental pain model setting. TPEP amplitude was reduced when analgesia was obtained using acetylsalicylic acid (aspirin) [50], local anesthetics (lidocaine) [51], metamizole (NSAID) and tramadol [52], codeine [53], hydromorphone [54], and alfentanil [55, 56]. Electrical stimuli for localized pain sensations have been obtained using a cathode inserted into a small bore of the epidermis, placed in the superficial layer of the skin [45]. Based on this technique, Inui and colleagues developed a methodology based on a pushpin-like electrode that did not require any special preparation of the skin as previously described by Bromm and Meier [45]. High current density at low current intensities using a planar concentric stimulating electrode applied through the epidermis have been utilized to achieve depolarization limited to the superficial layer of the dermis [57, 58]. Intraepidermal electrical stimulation has been shown to produce well-defined pricking pain without definite tactile sensation, along with a conduction velocity of evoked potentials in the range of the Aδ-fibers, and correlated with that obtained by the CO_2 laser stimulation. During intraepidermal electrical stimulation obtained in healthy volunteers, lidocaine increased pain thresholds and the abolished the evoked potentials [59]. Stimuli applied using intraepidermal electrical stimulation can be increased in steps, and threshold easily assessed in neuropathic pain conditions such as amyloid neuropathy [60] and diabetic neuropathy [9, 61, 62], which could be extended to other small fiber neuropathies.

3.2.2 Laser-Evoked Potentials

Brief (millisecond) laser stimuli selectively activate Aδ- and C-nociceptors in the epidermal layer of the skin [31] and evoke time-locked brain responses, i.e., laser-evoked brain potentials (LEPs) [63, 64]. CO_2 and argon lasers have been reported to be useful methods, which can be applied for clinical purposes [65].

Several prospective studies in patients with peripheral and central nervous system disease, including SFN, have shown the usefulness of LEPs, such as in familial amyloid neuropathy [66], Charcot-Marie-Tooth 1A [67], Fabry's disease [68], and diabetic neuropathy [69, 70]. The severity of latency and amplitude changes were found to correlate with the loss of thinly myelinated nerve fibers in sural nerve biopsy and with the degree of hypoalgesia [71]. In early stages of diabetic neuropathy, changes in LEPs had higher diagnostic sensitivity compared to clinical examination in detecting small fiber dysfunction.

Patients with peripheral nerve, plexus, or root lesions affecting small nerve fibers show loss of temperature and pain sensitivity. In such conditions, the Aδ-mediated LEPs may be absent, attenuated in amplitude or delayed in latency. In nerve root lesions, the sensory deficit was accompanied by the absence or significant attenuation of LEP amplitude at the affected site [72]. In patients with carpal tunnel syndrome, LEP amplitude was reduced from the region of nerve compression [73]. However, LEPs are also reduced or absent in patients with spinal cord and/or other areas in the CNS and so do not localize the lesions to the periphery, e.g., LEPs provide a sensitive and specific clinical neurophysiological correlate for dissociated sensory loss in patients with syringomyelia.

Several studies have been performed in healthy volunteers and patients to evaluate the magnitude of brain responses to painful laser stimulation in relation to subjective pain perception. Overall, in neuropathic pain patients with hyperalgesia or allodynia showed significant attenuation of Aδ wave amplitude evoked by stimulation over the painful territory, relative to stimulation of the homologous normal territory [5, 74, 75]. In contrast, LEPs were never attenuated in patients with nonorganic pain, in whom LEPs could even be enhanced. In fibromyalgia, an increase in LEP amplitude has been documented [76, 77]. Patients suffering from migraine headache exhibit abnormally high LEP amplitudes, even when tested between attacks [78].

Laser-evoked potentials have also been shown to be useful, in an experimental human volunteer model, to monitor regenerating cutaneous small nerve fibers noninvasively. Repeated topical application of capsaicin has been shown to produce a selective and reversible chemical axotomy of nerve fibers superficially located in the skin [79, 80], to elevate warm detection and heat pain thresholds [81–83] and to cause reduction of LEPs [84], followed by their restoration when topical capsaicin is discontinued. In a topical capsaicin study, nerve degeneration was followed by regeneration of small nerve fibers marked by growth-associated protein 43 (GAP-43) in sequential skin biopsies; the LEP amplitudes closely followed the latter over time and not the structural marker protein gene product 9.5 (PGP9.5) [85].

3.2.3 Contact Heat-Evoked Potentials

CHEP has been used to selectively excite Aδ- and C-fibers in the glabrous and hairy skin of human volunteers [86]. Noxious stimulation-activated afferents show a conduction velocity consistent with that of Aδ-fibers (approximately 10 m/s) and with the psychophysical attributes of first pain [87]. The precise moment of afferent-fiber activation during the thermode heating ramp is hard to establish, as it may change from one individual to another and from one trial to the following; this may explain the variability reported in the CHEP latencies from different laboratories [33, 87].

The decrease of CHEP amplitude correlates with the reduction of IENFD in neuropathic conditions [10, 14, 36]. Normative data are now available from an international study [88]. The clinical applications of CHEPs in a patient when other assessments have failed to show abnormal results, or when other techniques are not reliable, are illustrated in Figs. 3.1 and 3.2, respectively. CHEP has been studied in volunteer models, where Aδ-amplitudes particularly in the vertex correlated with pain intensity [33, 89]. Functional magnetic resonance imaging during painful and non-painful CHEPS stimulation showed distinct cerebral responses [90]. CHEP and LEP have

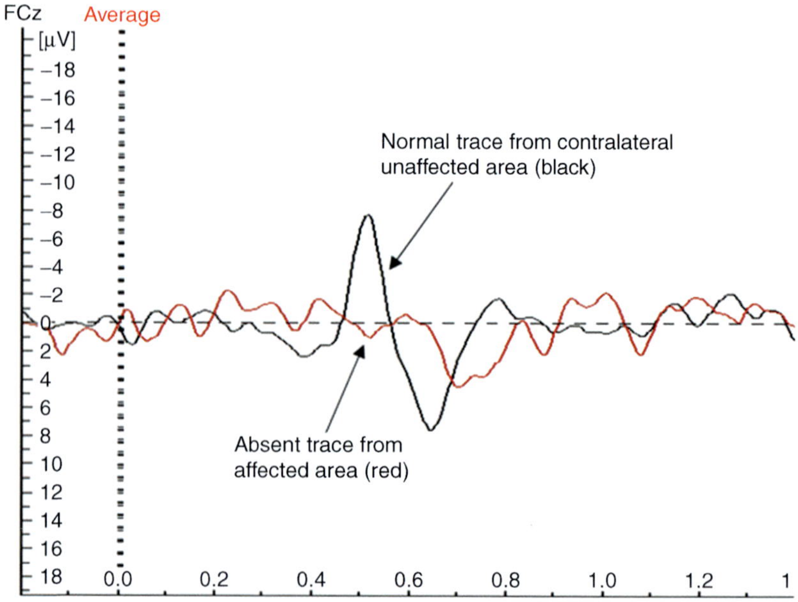

Fig. 3.1 Contact heat-evoked potentials (CHEPs) in a 38-year-old male following an interscalene block. Patient complained of continuing pain, paresthesia, and hypersensitivity in the forearm. Imaging and nerve conduction studies, monofilament, vibration, and thermal thresholds were all normal. CHEPs Aδ-amplitude from the affected region (*red trace*) was absent, while evoked potentials on stimulation of the same site on the contralateral unaffected arm (*black trace*) were preserved

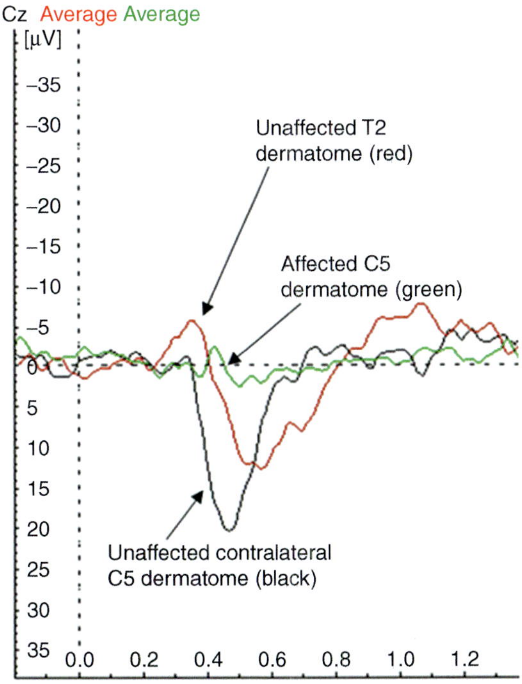

Fig. 3.2 Contact heat-evoked potentials (CHEPs) in a 12-year-old girl with obstetric brachial plexus injury. CHEPs trace was absent in the affected C5 dermatome (*green trace*), while a normal trace was recorded in the unaffected T2 dermatome (*red trace*) and contralateral C5 dermatome (*black trace*)

been shown to give similar results [35]. CHEP is easier to use in the clinic and enables repetitive stimulation for "windup."

3.3 Conclusion

Neurophysiological biomarkers to assess dysfunction or density of Aδ- and C-fibers in SFN include pain-evoked potentials using electrical, laser, and contact heat stimuli. Together with skin biopsies, they provide practical objective assessments for SFN and to distinguish SFN from other chronic pain conditions.

References

1. Cruccu G, Aminoff MJ, Curio G, Guerit JM, Kakigi R, Mauguiere F, et al. Recommendations for the clinical use of somatosensory-evoked potentials. Clin Neurophysiol. 2008;119:1705–19.
2. Haanpää M, Attal N, Backonja M, Baron R, Bennett M, Bouhassira D, et al. NeuPSIG guidelines on neuropathic pain assessment. Pain. 2011;152:14–27.
3. Lagerburg V, Bakkers M, Bouwhuis A, Hoeijmakers JGJ, Smit AM, Van Den Berg SJM, et al. Contact heat evoked potentials: normal values and use in small-fiber neuropathy. Muscle Nerve. 2015;51:743–9.
4. Madsen CS, Finnerup NB, Baumgärtner U. Assessment of small fibers using evoked potentials. Scand J Pain. 2014;5:111–8.
5. Truini A, Haanpaa M, Zucchi R, Galeotti F, Iannetti GD, Romaniello A, et al. Laser-evoked potentials in post-herpetic neuralgia. Clin Neurophysiol. 2003;114:702–9.
6. Cruccu G, Leandri M, Iannetti GD, Mascia A, Romaniello A, Truini A, et al. Small-fiber dysfunction in trigeminal neuralgia: carbamazepine effect on laser-evoked potentials. Neurology. 2001;56:1722–6.
7. Truini A, Galeotti F, Pennisi E, Casa F, Biasiotta A, Cruccu G. Trigeminal small-fibre function assessed with contact heat evoked potentials in humans. Pain. 2007;132:102–7.
8. Chao CC, Tseng MT, Lin YJ, Yang WS, Hsieh SC, Lin YH, et al. Pathophysiology of neuropathic pain in type 2 diabetes: skin denervation and contact heat-evoked potentials. Diabetes Care. 2010;33:2654–9.
9. Mueller D, Obermann M, Koeppen S, Kavuk I, Yoon MS, Sack F, et al. Electrically evoked nociceptive potentials for early detection of diabetic small-fiber neuropathy. Eur J Neurol. 2010;17:834–41.
10. Chao CC, Hsieh SC, Tseng MT, Chang YC, Hsieh ST. Patterns of contact heat evoked potentials (CHEP) in neuropathy with skin denervation: correlation of CHEP amplitude with intraepidermal nerve fiber density. Clin Neurophysiol. 2008;119:653–61.
11. Pazzaglia C, Valeriani M. Brain-evoked potentials as a tool for diagnosing neuropathic pain. Expert Rev Neurother. 2009;9:759–71.
12. Wong MC, Chung JWY. Feasibility of contact heat evoked potentials for detection of diabetic neuropathy. Muscle Nerve. 2011;44:902–6.
13. Pluijms WA, Slangen R, Joosten EA, Kessels AG, Merkies IS, Schaper NC, et al. Electrical spinal cord stimulation in painful diabetic polyneuropathy, a systematic review on treatment efficacy and safety. Eur J Pain. 2011;15:783–8.
14. Atherton DD, Facer P, Roberts KM, Misra VP, Chizh BA, Bountra C, et al. Use of the novel Contact Heat Evoked Potential Stimulator (CHEPS) for the assessment of small fibre neuropathy: correlations with skin flare responses and intra-epidermal nerve fibre counts. BMC Neurol. 2007;7:21.
15. Kodaira M, Inui K, Kakigi R. Evaluation of nociceptive Aδ- and C-fiber dysfunction with lidocaine using intraepidermal electrical stimulation. Clin Neurophysiol. 2014;125:1870–7.
16. Treede RD, Apkarian AV, Bromm B, Greenspan JD, Lenz FA. Cortical representation of pain: functional characterization of nociceptive areas near the lateral sulcus. Pain. 2000;87:113–9.

17. Brennum J, Jensen TS. Relationship between vertex potentials and magnitude of pre-pain and pain sensations evoked by electrical skin stimuli. Electroencephalogr Clin Neurophysiol. 1992;82:387–90.
18. Boor R, Li L, Goebel B, Reitter B. Subcortical somatosensory evoked potentials after posterior tibial nerve stimulation in children. Brain Dev. 2008;30:493–8.
19. Prestor B, Gnidovec B, Golob P. Long sensory tracts (cuneate fascicle) in cervical somatosensory evoked potential after median nerve stimulation. Electroencephalogr Clin Neurophysiol. 1997;104:470–9.
20. Restuccia D, Valeriani M, Di Lazzaro V, Tonali P, Mauguiere F. Somatosensory evoked potentials after multisegmental upper limb stimulation in diagnosis of cervical spondylotic myelopathy. J Neurol Neurosurg Psychiatry. 1994;57:301–8.
21. Restuccia D, Insola A, Valeriani M, Santilli V, Bedini L, Le Pera D, et al. Somatosensory evoked potentials after multisegmental lower limb stimulation in focal lesions of the lumbosacral spinal cord. J Neurol Neurosurg Psychiatry. 2000;69:91–5.
22. Yamada T, Machida M, Kimura J. Far-field somatosensory evoked potentials after stimulation of the tibial nerve. Neurology. 1982;32:1151–8.
23. Naguszewski WK, Naguszewski RK, Gose EE. Dermatomal somatosensory evoked potential demonstration of nerve root decompression after VAX-D therapy. Neurol Res. 2001;23:706–14.
24. Nakamura R, Noritake M, Hosoda Y, Kamakura K, Nagata N, Shibasaki H. Somatosensory conduction delay in central and peripheral nervous system of diabetic patients. Diabetes Care. 1992;15:532–5.
25. Soininen K, Järvilehto T. Somatosensory evoked potentials associated with tactile stimulation at detection threshold in man. Electroencephalogr Clin Neurophysiol. 1983;56:494–500.
26. Kakigi R, Shibasaki H, Neshige R, Ikeda A, Mamiya K, Kuroda Y. Pain-related somatosensory evoked potentials in cortical reflex myoclonus. J Neurol Neurosurg Psychiatry. 1990;53:44–8.
27. Yamauchi N, Fujitani Y, Oikawa T. Somatosensory evoked potentials elicited by mechanical and electrical stimulation of each single pain or tactile spot of the skin. Tohoku J Exp Med. 1981;133: 81–92.
28. Inui K, Tran TD, Hoshiyama M, Kakigi R. Preferential stimulation of Adelta fibers by intra-epidermal needle electrode in humans. Pain. 2002;96:247–52.
29. Treede RD, Lorenz J, Baumgärtner U. Clinical usefulness of laser-evoked potentials. Neurophysiol Clin. 2003;33:303–14.
30. Katsarava Z, Ayzenberg I, Sack F, Limmroth V, Diener HC, Kaube H. A novel method of eliciting pain-related potentials by transcutaneous electrical stimulation. Headache. 2006;46:1511–7.
31. Bromm B, Jahnke MT, Treede RD. Responses of human cutaneous afferents to CO_2 laser stimuli causing pain. Exp Brain Res. 1984;55:158–66.
32. Bromm B, Lorenz J. Neurophysiological evaluation of pain. Electroencephalogr Clin Neurophysiol. 1998;107:227–53.
33. Chen AC, Niddam DM, Arendt-Nielsen L. Contact heat evoked potentials as a valid means to study nociceptive pathways in human subjects. Neurosci Lett. 2001;316:79–82.
34. Treede RD. Neurophysiological studies of pain pathways in peripheral and central nervous system disorders. J Neurol. 2003;250:1152–61.
35. Valeriani M, Le Pera D, Niddam D, Chen AC, Arendt-Nielsen L. Dipolar modelling of the scalp evoked potentials to painful contact heat stimulation of the human skin. Neurosci Lett. 2002;318:44–8.
36. Casanova-Molla J, Grau-Junyent JM, Morales M, Valls-Solé J. On the relationship between nociceptive evoked potentials and intraepidermal nerve fiber density in painful sensory polyneuropathies. Pain. 2011;152:410–8.
37. Garcia-Larrea L. Objective pain diagnostics: clinical neurophysiology. Neurophysiol Clin. 2012;42:187–97.
38. Passmore SR, Murphy B, Lee TD. The origin and application of somatosensory evoked potentials as a neurophysiological technique to investigate neuroplasticity. J Can Chiropr Assoc. 2014;58:170–83.
39. Parhizgar SE, Ekhtiari H. A review on experimental assessments of pain threshold in healthy human subjects. Basic Clin Neurosci. 2010;1:62–7.
40. Hansen N, Kahn AK, Zeller D, Katsarava Z, Sommer C, Üçeyler N. Amplitudes of pain-related evoked potentials are useful to detect small fiber involvement in painful mixed fiber neuropathies in addition to quantitative sensory testing – an electrophysiological study. Front Neurol. 2015;6:244.
41. Üçeyler N, Kahn AK, Kramer D, Zeller D, Casanova-Molla J, Wanner C, et al. Impaired small fiber conduction in patients with Fabry disease: a neurophysiological case-control study. BMC Neurol. 2013;13:47.
42. Hansen N, Obermann M, Üçeyler N, Zeller D, Mueller D, Yoon MS, et al. [Clinical application of pain-related evoked potentials]. Schmerz. 2012;26:8–15.
43. Obermann M, Katsarava Z, Esser S, Sommer C, He L, Selter L, et al. Correlation of epidermal nerve fiber density with pain-related evoked potentials in HIV neuropathy. Pain. 2008;138:79–86.
44. Chatrian GE, Canfield RC, Knauss TA, Eegt EL. Cerebral responses to electrical tooth pulp stimulation in man. An objective correlate of acute experimental pain. Neurology. 1975;25:745–57.
45. Bromm B, Meier W. The intracutaneous stimulus: a new pain model for algesimetric studies. Methods Find Exp Clin Pharmacol. 1984;6:405–10.
46. Chen AC, Chapman CR, Harkins SW. Brain evoked potentials are functional correlates of induced pain in man. Pain. 1979;6:365–74.
47. Fernandes de Lima VM, Chatrian GE, Lettich E, Canfield RC, Miller RC, Soso MJ. Electrical stimulation of tooth pulp in humans. I. Relationships among

physical stimulus intensities, psychological magnitude estimates and cerebral evoked potentials. Pain. 1982;14:207–32.
48. Harkins SW, Chapman CR. Cerebral evoked potentials to noxious dental stimulation: relationship to subjective pain report. Psychophysiology. 1978;15:248–52.
49. Klement W, Medert HA, Arndt JO. Nalbuphine does not act analgetically in electrical painful tooth pulp stimulation in man. Pain. 1992;48:269–74.
50. Chen AC, Chapman CR. Aspirin analgesia evaluated by event-related potentials in man: possible central action in brain. Exp Brain Res. 1980;39:359–64.
51. Gehrig JD, Colpitts YH, Chapman CR. Effects of local anesthetic infiltration on brain potentials evoked by painful dental stimulation. Anesth Analg. 1981;60:779–82.
52. Rohdewald P, Granitzki HW, Neddermann E. Comparison of the analgesic efficacy of metamizole and tramadol in experimental pain. Pharmacology. 1988;37:209–17.
53. Suri A, Kaltenbach ML, Grundy BL, Derendorf H. Pharmacodynamic evaluation of codeine using tooth pulp evoked potentials. J Clin Pharmacol. 1996;36:1126–31.
54. Coda B, Tanaka A, Jacobson RC, Donaldson G, Chapman CR. Hydromorphone analgesia after intravenous bolus administration. Pain. 1997;71:41–8.
55. Hill H, Walter MH, Saeger L, Sargur M, Sizemore W, Chapman CR. Dose effects of alfentanil in human analgesia. Clin Pharmacol Ther. 1986;40:178–86.
56. Chapman CR, Hill HF, Saeger L, Gavrin J. Profiles of opioid analgesia in humans after intravenous bolus administration: alfentanil, fentanyl and morphine compared on experimental pain. Pain. 1990;43:47–55.
57. Katsarava Z, Yaldizli O, Voulkoudis C, Diener HC, Kaube H, Maschke M. Pain related potentials by electrical stimulation of skin for detection of small-fiber neuropathy in HIV. J Neurol. 2006;253:1581–4.
58. Inui K, Kakigi R. Pain perception in humans: use of intraepidermal electrical stimulation. J Neurol Neurosurg Psychiatry. 2012;83:551–6.
59. Otsuru N, Inui K, Yamashiro K, Miyazaki T, Takeshima Y, Kakigi R. Assessing Aδ fiber function with lidocaine using intraepidermal electrical stimulation. J Pain. 2010;11:621–7.
60. Obayashi K, Yamashita T, Tasaki M, Ueda M, Shono M, Jono H, et al. Amyloid neuropathy in a younger domino liver transplanted recipient. Muscle Nerve. 2011;43:449–50.
61. Kukidome D, Nishikawa T, Sato M, Igata M, Kawashima J, Shimoda S, et al. Measurement of small fibre pain threshold values for the early detection of diabetic polyneuropathy. Diabet Med. 2016;33:62–9.
62. Suzuki C, Kon T, Funamizu Y, Ueno T, Haga R, Nishijima H, et al. Elevated pain threshold in patients with asymptomatic diabetic neuropathy: an intraepidermal electrical stimulation study. Muscle Nerve. 2016;54:146–9.
63. Carmon A, Mor J, Goldberg J. Evoked cerebral responses to noxious thermal stimuli in humans. Exp Brain Res. 1976;25:103–7.
64. Plaghki L, Delisle D, Godfraind JM. Heterotopic nociceptive conditioning stimuli and mental task modulate differently the perception and physiological correlates of short CO_2 laser stimuli. Pain. 1994;57:181–92.
65. Arendt-Nielsen L, Bjerring P. Sensory and pain threshold characteristics to laser stimuli. J Neurol Neurosurg Psychiatry. 1988;51:35–42.
66. Ng Wing Tin S, Plante-Bordeneuve V, Salhi H, Goujon C, Damy T, Lefaucheur JP. Characterization of pain in familial amyloid polyneuropathy. J Pain. 2015;16:1106–14.
67. Pazzaglia C, Vollono C, Ferraro D, Virdis D, Lupi V, Le Pera D, et al. Mechanisms of neuropathic pain in patients with Charcot-Marie-Tooth 1 A: a laser-evoked potential study. Pain. 2010;149:379–85.
68. Valeriani M, Mariotti P, Le Pera D, Restuccia D, De Armas L, Maiese T, et al. Functional assessment of A delta and C fibers in patients with Fabry's disease. Muscle Nerve. 2004;30:708–13.
69. Agostino R, Cruccu G, Romaniello A, Innocenti P, Inghilleri M. Dysfunction of small myelinated afferents in diabetic polyneuropathy, as assessed by laser evoked potentials. Clin Neurophysiol. 2000;111:270–6.
70. Agostino R, Cruccu G, Iannetti GD, Innocenti P, Romaniello A, Truini A, et al. Trigeminal small-fibre dysfunction in patients with diabetes mellitus: a study with laser evoked potentials and corneal reflex. Clin Neurophysiol. 2000;111:2264–7.
71. Kakigi R, Shibasaki H, Tanaka K, Ikeda T, Oda K, Endo C, et al. CO_2 laser-induced pain-related somatosensory evoked potentials in peripheral neuropathies: correlation between electrophysiological and histopathological findings. Muscle Nerve. 1991;14:441–50.
72. Lorenz J, Hansen HC, Kunze K, Bromm B. Sensory deficits of a nerve root lesion can be objectively documented by somatosensory evoked potentials elicited by painful infrared laser stimulations: a case study. J Neurol Neurosurg Psychiatry. 1996;61:107–10.
73. Arendt-Nielsen L, Gregersen H, Toft E, Bjerring P. Involvement of thin afferents in carpal tunnel syndrome: evaluated quantitatively by argon laser stimulation. Muscle Nerve. 1991;14:508–14.
74. Wu Q, Garcia-Larrea L, Mertens P, Beschet A, Sindou M, Mauguiere F. Hyperalgesia with reduced laser evoked potentials in neuropathic pain. Pain. 1999;80:209–14.
75. Garcia-Larrea L, Convers P, Magnin M, Andre-Obadia N, Peyron R, Laurent B, et al. Laser-evoked potential abnormalities in central pain patients: the influence of spontaneous and provoked pain. Brain. 2002;125:2766–81.
76. Gibson SJ, Littlejohn GO, Gorman MM, Helme RD, Granges G. Altered heat pain thresholds and cerebral event-related potentials following painful CO_2 laser stimulation in subjects with fibromyalgia syndrome. Pain. 1994;58:185–93.
77. Lorenz J, Grasedyck K, Bromm B. Middle and long latency somatosensory evoked potentials after

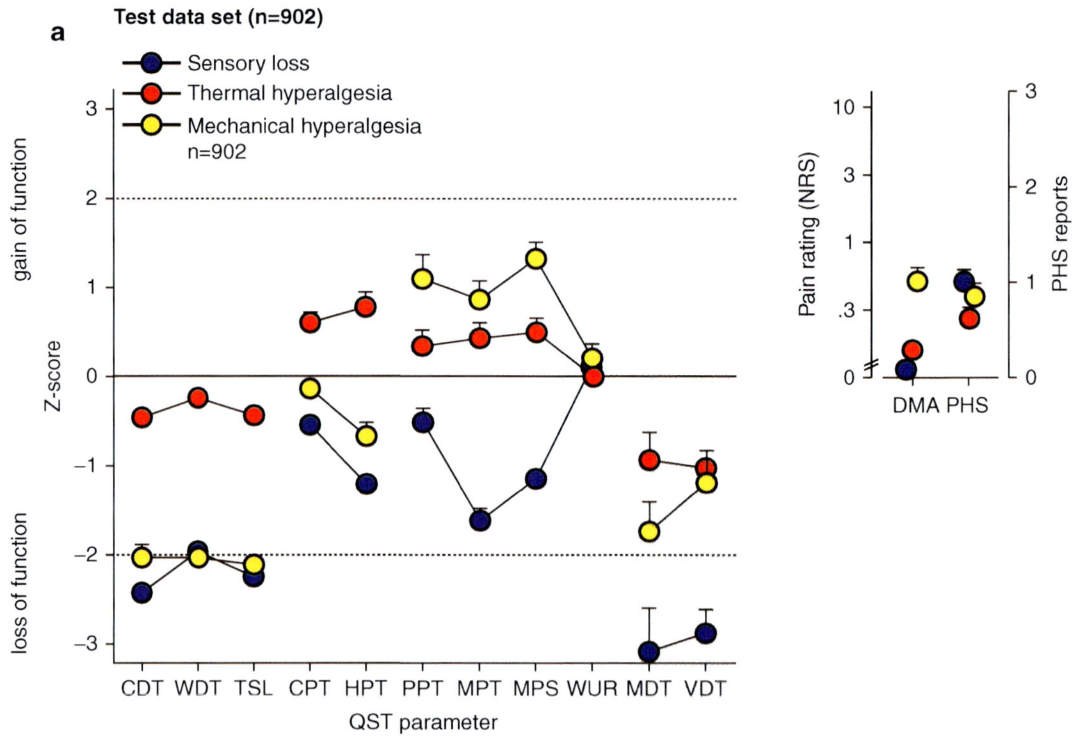
a
Test data set (n=902)
Sensory loss
Thermal hyperalgesia
Mechanical hyperalgesia
n=902
gain of function
Z-score
loss of function
3
2
1
0
−1
−2
−3
CDT WDT TSL CPT HPT PPT MPT MPS WUR MDT VDT
QST parameter
Pain rating (NRS)
PHS reports
10
3
1
.3
0
DMA PHS

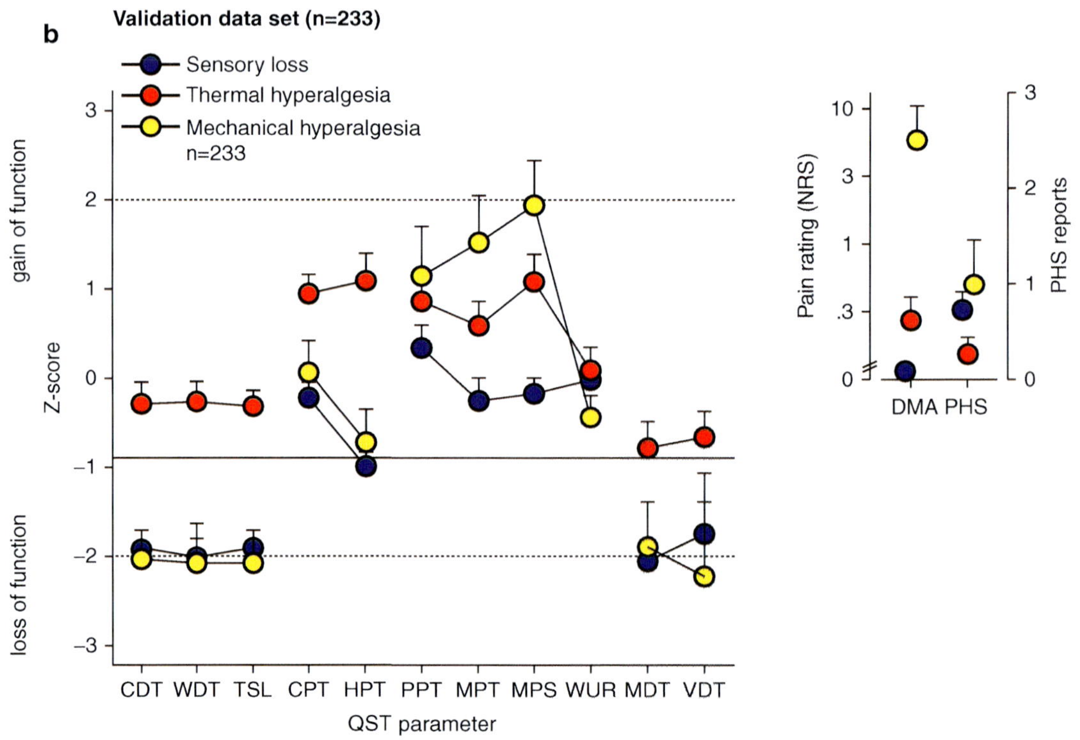
b
Validation data set (n=233)
Sensory loss
Thermal hyperalgesia
Mechanical hyperalgesia
n=233
gain of function
Z-score
loss of function
3
2
1
0
−1
−2
−3
CDT WDT TSL CPT HPT PPT MPT MPS WUR MDT VDT
QST parameter
Pain rating (NRS)
PHS reports
10
3
1
.3
0
DMA PHS

by QST may allow conclusions on the question, which type of nerve fibers or central tracts are affected. In the context of SFN, mostly the small fiber tests are of relevance, but stimuli assessing large fiber functions may be used to exclude an additional large fiber affection.

4.2 The Method

QST has been used since the 1970s in research settings, mainly for diagnosing, assessing, and monitoring sensory neuropathies and pain disorders ([3] and see [4] for review). In QST, being a psychophysical method, similar to tests of hearing acuity, the stimulus is controlled by the investigator; the response, however, depends on active participation of the subject. The investigated person will either press a button to indicate a detection threshold or will comment on whether or not a stimulus is painful or estimate how painful the stimulus is. Therefore, attention, motivation, cognitive impairment, or malingering may influence QST results. A QST result, like that of sensory-evoked potentials, provides information on the entire somatosensory system from the cutaneous receptor to the cortex and thus cannot differentiate whether an abnormality is peripherally or centrally generated. Abnormalities may consist in both loss and gain of function.

Different QST protocols are available. Recently, many researchers and clinicians have adopted the QST protocol described by the German Research Network on Neuropathic Pain (Deutsches Forschungsnetzwerk Neuropathischer Schmerz, DFNS) [5], and this standardization has been of great value for multinational studies. Data are usually given as z-scores, and comprehensive sensory profiles can be depicted for individual patients, for groups of patients, or for different conditions [6, 7] (see Fig. 4.1). In addition to providing information on loss or gain of function for certain qualities, the comprehensive protocol may uncover sensory profiles predictive for specific treatment responses [7, 8].

Using a thermal sensory testing device (e.g., from Somedic, Sweden, or Medoc, Israel), thermodes are applied at standardized points (back of the foot, back of the hand), and the following thresholds are determined: warm and cold detection thresholds and cold and heat thresholds. The temperature of the thermode is increased by 1 °C/sec from an initial temperature of 32 °C. When the respective thresholds are reached, the patient switches off the thermode by pressing a button, and the respective temperature is registered. The tests are repeated and averaged to obtain more stable values. The determination of the warm and cold thresholds is sufficient to determine the integrity of the C- and Aδ-fibers, and the measurement of pain thresholds does not improve the detection of sensory neuropathy [9]. However, the measurement of the pain thresholds helps to phenotype the neuropathic pain syndrome, and it has therefore been recommended to include it [4]. The thermode is also used to measure the detection thresholds for alternating cold and warm stimuli and to count the number of paradoxical heat sensations during this procedure [10].

Fig. 4.1 Quantitative sensory testing (QST) profiles. This graph illustrates three clusters of QST profiles derived from two very large (>5000 participants) independent datasets collected by the German Research Network on Neuropathic Pain (Deutsches Forschungsnetzwerk Neuropathischer Schmerz, DFNS). Positive z-scores indicate positive sensory signs (hyperalgesia), and negative z values indicate negative sensory signs (hypoesthesia and hypoalgesia). Values are significantly different from those of healthy subjects if their 95% confidence interval does not cross the zero line. The *blue symbols* show cluster 1, which includes patient with mainly sensory loss (42% of the population in A and 53% in B). The *red symbols* show cluster 2, patients with prominent thermal hyperalgesia (33% of the cohort in A and B). The *yellow symbols* show cluster 3, patients with mechanical hyperalgesia (24% of cohort A and 14% of cohort B). *CDT* cold detection threshold, *CPT* cold pain threshold, *HPT* heat pain threshold, *MDT* mechanical detection threshold, *MPS* mechanical pain sensitivity, *MPT* mechanical pain threshold, *NRS* Numerical Rating Scale, *PPT* pressure pain threshold, *QST* quantitative sensory testing, *TSL* thermal sensory limen, *VDT* vibration detection threshold, *WDT* warm detection threshold, *WUR* windup ratio. Reproduced with permission from [7]

Pinprick stimulators (e.g., from MRC Systems, USA) are used in the testing of the mechanical pain threshold. These blunt needles can cause a sensation ranging from dull to sharp according to their different weight. The transition from dull to sharp represents the detection threshold for a sharp pain stimulus. The painfulness of the stimulus on a scale of 0–100 is then asked for to record the mechanical pain sensitivity. For testing the phenomenon of temporal summation, the painfulness of a single pinprick stimulus is compared to that at the end of a train of ten pinprick stimuli of the same force repeated at a 1/s rate. A force of 128 mN is used for testing on the face and of 256 mN when testing the hand or foot. The stimuli are applied within an area of 1 cm^2. Single pinprick stimuli are alternated with the train of ten stimuli five times at different skin sites in the same body region. The mean pain rating at the end of the trains is divided by the mean pain rating for the single stimuli; the result is documented as the "windup ratio."

The presence of allodynia is tested for with three light tactile stimulators: a cotton wisp (3 mN), a cotton wool tip (100 mN), and a brush (Somedic, Sweden; 200–400 mN). These three tactile stimuli are applied five times each over the skin with a single stroke of approximately 1–2 cm. They are intermingled with the pinprick stimuli in a balanced order, and subjects are asked to rate pain on the same scale as for the pinprick stimuli.

The detection thresholds of Aß-fibers are tested with a range of thin nylon filaments of ascending strength, the von Frey hairs (e.g., Ugo Basile, Italy). The contact area of the von Frey hairs with the skin should be uniform in size and shape (rounded tip, 0.5 mm diameter) to avoid nociceptor activation by sharp edges. The threshold is calculated as the geometric mean of five series of ascending and descending stimuli [2].

The vibration detection threshold is tested, as in the bedside testing, with a Rydel-Seiffer tuning fork (64 Hz, 8/8 scale) placed over a bony prominence (cheek, processus styloideus ulnae, malleolus medialis), and the vibration detection threshold is determined with three series of descending stimulus intensities.

In the full DFNS protocol, the pressure pain threshold over the musculature is performed with a pressure gauge device (e.g., Wagner Instruments) with a probe area of 1 cm^2 that exerts pressure up to 20 kg/cm^2 [11]. The pressure pain threshold is determined with three series of ascending stimuli applied as a slowly increasing ramps of 50 kPa/s. Usual testing sites are the masseter muscle mid cheek in the face, the thenar at the hand, and the instep at the foot [12].

4.3 Normal Values, Reproducibility, and Sex Differences

Age- and sex-dependent normal values for different body regions have been published [2, 12, 13] for adults and also for children [14]. Quality control procedures have been implemented [15]; a good reproducibility [16] and the homogeneity of data obtained in different European countries have been shown [17]. In children and adolescents, the only difference between girls and boys was that girls were more sensitive to the detection of warm stimuli [18]. A similar trend is also present in adult women at the legs [2]. QST data are age-dependent; this is why age-adapted normative data should be used [12].

4.4 QST in Consensus Papers and Guidelines

The special interest group on neuropathic pain (NeuPSIG) of the International Association for the Study of Pain (IASP) has published recommendations on the assessment of neuropathic pain [19]. Here, the recommendation is that QST can be used along with bedside testing to document the sensory profile. There is no specific statement regarding QST in the diagnosis of SFN. The same is true for the guideline of the European Federation of Neurological Societies (EFNS) [20]. In a consensus paper from 2013, the NeuPSIG recommends the use of QST as a screening test in the assessment of small and large fiber neuropathies and to monitor sensory deficits over time. Additionally, in patients with pain disorders, QST is recommended for documenting and monitoring the magnitude of sensory abnormalities [4]. The authors state that the test site should be in the center of the symptomatic area as determined by bedside investigation and that control sites for comparison should be asymptomatic skin areas, e.g., at the face or the contralateral mirror side, depending on the disorder studied and the distribution of symptoms. They also recommend that QST should not be used in patients with cognitive deficits, language problems, affective disorders, and currently undergoing a litigation process. Also, QST should also not be used as a single test for diagnosing a nerve lesion but should be used in the context of other investigations.

Several problems have been identified concerning the clinical practice of QST and its distribution [4, 21]. The equipment, in particular for thermotesting, is very expensive. Although defined test batteries have been recommended, there are no studies demonstrating that one particular test battery is really necessary in a certain indication or to answer a specific question.

4.5 QST in Small Fiber Neuropathy

Patients with suspected SFN usually see several physicians before a diagnosis is made. SFN is assumed in the constellation of distal symmetric burning pain with the absence of signs and symptoms of large fiber neuropathy and with normal nerve conduction studies and electromyography (EMG). According to currently accepted criteria [22–24], in addition to the typical medical history, the detection of a functional or morphological disturbance of the small nerve fibers is required in two of the following three tests: (a) clinical-neurological examination, (b) QST or special small fiber neurophysiology, and (c) skin punch biopsy. If only one additional criterion is met, one can speak of a possible SFN, with two criteria of probable SFN, and with three additional criteria of definite SFN [23]. Thus, given the typical history and clinical syndrome, QST can help make the diagnosis of a probable SFN, in the absence of skin biopsy.

A comprehensive sensory profile in idiopathic SFN was shown in [25]. As expected, the detection thresholds for warm and cold stimuli were increased (i.e., the sensitivity to these stimuli was diminished). Mechanical detection thresholds were mildly increased, and the pressure pain thresholds were slightly increased compared to controls. The latter is not surprising, since deep tissues are also innervated by C-fibers which may be hyperexcitable in SFN [26, 27]. The change in mechanical detection thresholds, which are classically thought to depend on Aß-fibers, is more difficult to explain. Reasons may be a subclinical involvement of very distal Aß-fibers or their receptors or the involvement of C-tactile afferents in the mechanical detection thresholds [28, 29]. The cold detection threshold was more pathologically affected in those SFN patients who had a diminished IENFD in both the distal and the proximal leg skin (non-length-dependent SFN) than in those with a distal reduction of IENFD only [25]. Whether this is indicative of a more prominent αδ-fiber involvement in the non-length-dependent SFN (ganglionopathy, ganglionitis [30]) needs to be confirmed in a larger sample.

Several groups have investigated the question of sensitivity and specificity of QST for making the diagnosis of SFN. Although this has not been formally tested, the tests assessing small fiber function (warm and cold detection, less pain thresholds, pinprick detection (see [31])) are considered most useful in the diagnosis of SFN. The typical "large fiber tests" like the vibration detection threshold and mechanical detection threshold should remain normal (as shown in [25]), but it is not obligatory to show it for the diagnosis of small fiber neuropathy if it has already been done by clinical examination. Therefore, most investigators used only a restricted battery of "small fiber tests." Furthermore, the studies are hampered by the lack of a gold standard against which to assess the performance of QST. For example, one small study found a high sensitivity but low specificity of QST if skin biopsy was taken as the gold standard [32]. Other small studies also found a relatively high sensitivity [33, 34].

However, the range of normal for IENFD is large [35], such that marginal values may account for false negatives. In another study, QST had a low sensitivity compared to IENFD assessment [36]. In patients with subclinical diabetic neuropathy, QST performed well compared to laser-evoked potentials (LEP), taking IENFD as the gold standard [37]. In a study using a composite gold standard, QST had considerably lower diagnostic efficiency than skin biopsy and even lower than clinical examination [24]. In a mixed cohort with large and small fiber neuropathy, QST also had a low diagnostic performance compared to skin biopsy [31]. However, the positive predictive value of QST for reduced IENFD was high.

Recently, an algorithm was suggested that allocates patients into one of four sensory phenotypes [7]. These phenotypes are characterized by (1) loss of thermal and mechanical detection ("sensory loss"); (2) intact sensory function, often combined with thermal hyperalgesia or allodynia ("thermal hyperalgesia"); (3) loss of thermal detection, but not mechanical detection, accompanied by mechanical hyperalgesia or allodynia ("mechanical hyperalgesia"); and (4) a mostly normal sensory profile resembling that of healthy subjects ("healthy sensory profile") [38]. Using this algorithm, a study including 292 patients with mixed polyneuropathy and 58 patients with pure SFN found that the most common phenotype in patients with SFN was "thermal hyperalgesia" (40%), followed by "mechanical hyperalgesia" (28%). Only 19% was of the "sensory loss" phenotype [39]. Interestingly, patients in the "thermal hyperalgesia" group differed from the other SFN patients by lower IENFD at the thigh.

In summary, the majority of studies show a relatively low sensitivity of QST for the diagnosis of SFN, but it is still useful to support the diagnosis in the setting of the typical clinical picture and to provide a patient sensory phenotype. Given the importance of thermal testing for the detection of SFN, smaller devices allowing just this, and allowing bedside testing, are under development [40] and *BMBF 1316062B*. An attempt to reduce SFN diagnostics to pinprick testing did not reveal sufficient sensitivity [41].

4.6 SFN in Diabetes and Impaired Glucose Tolerance

SFN is prominent in diabetes and its precursors, and this subject is discussed in Chap. 6 of this volume. As to the value of QST in this condition, the method has been used in a number of studies, with or without concomitant skin biopsy, to identify and monitor small fiber involvement in diabetic and prediabetic neuropathy. For example, small fiber involvement was detected by QST in 81% of patients with long-standing type 1 diabetes [42]. Both SFN and small fiber dysfunction were prevalent in a cohort of patients with type 2 diabetes and impaired glucose tolerance [43]. The prevalence and intensity of pain in diabetic neuropathy have been related to the extent of small fiber involvement as assessed by QST [44, 45].

4.7 SFN in Fabry Disease

Patients with Fabry disease suffer from a very specific type of neuropathic pain [46] which is thought to be due to the SFN that is typical for this disorder. Fabry disease is a lysosomal storage disorder with an X-linked inheritance. An enzyme defect leads to accumulation of globotriaosylceramide (Gb3) in various tissues such as the heart, kidneys, or nervous system, which makes it a multisystem disease. Pain is often the first symptom [47] and should alert clinicians to the diagnosis. Early treatment with enzyme replacement therapy (ERT) can delay organ failure [48, 49].

In QST, Fabry patients are characterized by particularly reduced cold sensitivity [50]. We monitored a cohort of Fabry patients with and without ERT closely for the development of SFN using QST and skin biopsy (Fig. 4.2). Of all QST parameters, cold detection thresholds were most affected and increased over time, as did warm detection thresholds [51]. Cold detection thresholds deteriorated most in men with impaired renal function; only women with normal renal function were spared from the deterioration over time. Interestingly, as in the cohort of

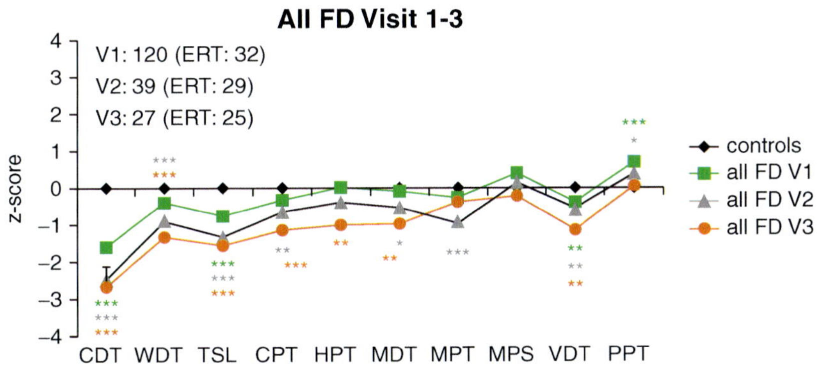

Fig. 4.2 Quantitative sensory testing profiles in Fabry disease. Quantitative sensory testing (QST) profiles over time of Fabry patients (colored lines) compared to controls (*black line*). The QST profile of all Fabry patients at baseline (V1), 1-year follow-up (V2), and 2-year follow-up (V3) is shown. Compared to controls (*black line*), patients have impaired small fiber function and develop further small fiber impairment (CDT, TSL) and impairment of vibration sense (VDT) after 2 years (orange line). $^{*}p < 0.05$, $^{**}p < 0.01$, $^{***}p < 0.001$. Reproduced with permission from [51]

idiopathic SFN [25], pressure pain thresholds were reduced, indicating enhanced sensitivity of deep nociceptors.

4.8 SFN in Sodium Channel Mutations

Mutations in the genes for the voltage-gated sodium channels may lead to different disorders, among them are SFN and erythromelalgia. Erythromelalgia is characterized by attacks of redness, intense heat, and severe burning pain in the hands and feet [52]. Using QST, detection thresholds for cold and warmth were significantly increased in the affected skin of patients with hereditary erythromelalgia [53]. The patients also had heat hyperalgesia, but, surprisingly, this was present more often in otherwise unaffected skin areas. Half of the patients had tactile hypoesthesia in unaffected skin areas.

In selected patients with a clinical picture of SFN and both abnormal QST and IENFD, mutations in genes coding for the voltage-gated sodium channels $Na_v1.7$, $Na_v1.8$, and $Na_v1.9$ were found in a small percentage of the cohort [54–56]. These studies do not answer the question: How many patients with genetically proven sodium channel mutations and clinical SFN have abnormal QST findings? Given the example of a family carrying the $Na_v1.7$ polymorphism R1150W, IENFD and QST may be normal in these patients [57]. Thus, QST is not an adequate screening test for this group of disorders.

4.9 SFN in Sarcoidosis

In a busy sarcoidosis outpatient clinic, it was observed that many patients complained of unexplained pain, mostly of the feet. QST was performed as a screening test for SFN and revealed a disturbed temperature sensitivity in 25 of 31 patients (81%). Thus, QST uncovered a possible SFN as the cause of this pain, which was later confirmed by skin biopsy analysis [58]. The SFN in sarcoidosis has since been extensively studied and has been used as a paradigm for treatment trials against neuropathic pain [59, 60].

4.10 Small Fiber Involvement in Large Fiber Neuropathies or Neuronopathies

QST has uncovered involvement of small nerve fibers (i.e., nociceptors) in a number of disorders where it was unexpected, like in neuropathies that were so far considered of the large fiber type only and in fibromyalgia syndrome [26, 61].

45. Themistocleous AC, Ramirez JD, Shillo PR, Lees JG, Selvarajah D, Orengo C, et al. The Pain in Neuropathy Study (PiNS): a cross-sectional observational study determining the somatosensory phenotype of painful and painless diabetic neuropathy. Pain. 2016;157:1132–45.
46. Üçeyler N, Ganendiran S, Kramer D, Sommer C. Characterization of pain in Fabry disease. Clin J Pain. 2014;30:915–20.
47. Burlina AP, Sims KB, Politei JM, Bennett GJ, Baron R, Sommer C, et al. Early diagnosis of peripheral nervous system involvement in Fabry disease and treatment of neuropathic pain: the report of an expert panel. BMC Neurol. 2011;11:61.
48. Schiffmann R, Pastores GM, Lien YH, Castaneda V, Chang P, Martin R, et al. Agalsidase alfa in pediatric patients with Fabry disease: a 6.5-year open-label follow-up study. Orphanet J Rare Dis. 2014;9:169.
49. Biegstraaten M, Arngrimsson R, Barbey F, Boks L, Cecchi F, Deegan PB, et al. Recommendations for initiation and cessation of enzyme replacement therapy in patients with Fabry disease: the European Fabry Working Group consensus document. Orphanet J Rare Dis. 2015;10:36.
50. Dütsch M, Marthol H, Stemper B, Brys M, Haendl T, Hilz MJ. Small fiber dysfunction predominates in Fabry neuropathy. J Clin Neurophysiol. 2002;19:575–86.
51. Üçeyler N, He L, Schonfeld D, Kahn AK, Reiners K, Hilz MJ, et al. Small fibers in Fabry disease: baseline and follow-up data under enzyme replacement therapy. J Peripher Nerv Syst. 2011;16:304–14.
52. Tang Z, Chen Z, Tang B, Jiang H. Primary erythromelalgia: a review. Orphanet J Rare Dis. 2015;10:127.
53. McDonnell A, Schulman B, Ali Z, Dib-Hajj SD, Brock F, Cobain S, et al. Inherited erythromelalgia due to mutations in SCN9A: natural history, clinical phenotype and somatosensory profile. Brain. 2016;139:1052–65.
54. Faber CG, Hoeijmakers JG, Ahn HS, Cheng X, Han C, Choi JS, et al. Gain of function Nanu1.7 mutations in idiopathic small fiber neuropathy. Ann Neurol. 2012;71:26–39.
55. Faber CG, Lauria G, Merkies IS, Cheng X, Han C, Ahn HS, et al. Gain-of-function Nav1.8 mutations in painful neuropathy. Proc Natl Acad Sci U S A. 2012;109:19444–9.
56. Han C, Yang Y, de Greef BT, Hoeijmakers JG, Gerrits MM, Verhamme C, et al. The domain II S4-S5 linker in Nav1.9: a missense mutation enhances activation, impairs fast inactivation, and produces human painful neuropathy. NeuroMolecular Med. 2015;17:158–69.
57. Harrer JU, Uceyler N, Doppler K, Fischer TZ, Dib-Hajj SD, Waxman SG, et al. Neuropathic pain in two-generation twins carrying the sodium channel Nav1.7 functional variant R1150W. Pain. 2014;155:2199–203.
58. Hoitsma E, Marziniak M, Faber CG, Reulen JP, Sommer C, De Baets M, et al. Small fibre neuropathy in sarcoidosis. Lancet. 2002;359:2085–6.
59. Saito H, Yamaguchi T, Adachi Y, Yamashita T, Wakai Y, Saito K, et al. Neurological symptoms of sarcoidosis-induced small fiber neuropathy effectively relieved with high-dose steroid pulse therapy. Intern Med. 2015;54:1281–6.
60. van Velzen M, Heij L, Niesters M, Cerami A, Dunne A, Dahan A, et al. ARA 290 for treatment of small fiber neuropathy in sarcoidosis. Expert Opin Investig Drugs. 2014;23:541–50.
61. Nolano M, Provitera V, Estraneo A, Selim MM, Caporaso G, Stancanelli A, et al. Sensory deficit in Parkinson's disease: evidence of a cutaneous denervation. Brain. 2008;131:1903–11.
62. Martinez V, Fletcher D, Martin F, Orlikowski D, Sharshar T, Chauvin M, et al. Small fibre impairment predicts neuropathic pain in Guillain-Barre syndrome. Pain. 2010;151:53–60.
63. Reimer M, Rempe T, Diedrichs C, Baron R, Gierthmuhlen J. Sensitization of the nociceptive system in complex regional pain syndrome. PLoS One. 2016;11:e0154553.
64. Gierthmühlen J, Maier C, Baron R, Tolle T, Treede RD, Birbaumer N, et al. Sensory signs in complex regional pain syndrome and peripheral nerve injury. Pain. 2012;153:765–74.
65. Üçeyler N, Eberle T, Rolke R, Birklein F, Sommer C. Differential expression patterns of cytokines in complex regional pain syndrome. Pain. 2007;132:195–205.
66. Oaklander AL, Fields HL. Is reflex sympathetic dystrophy/complex regional pain syndrome type I a small-fiber neuropathy? Ann Neurol. 2009;65:629–38.
67. Kim DH, Zeldenrust SR, Low PA, Dyck PJ. Quantitative sensation and autonomic test abnormalities in transthyretin amyloidosis polyneuropathy. Muscle Nerve. 2009;40:363–70.
68. Heldestad V, Nordh E. Quantified sensory abnormalities in early genetically verified transthyretin amyloid polyneuropathy. Muscle Nerve. 2007;35:189–95.
69. Adams D, Suhr OB, Hund E, Obici L, Tournev I, Campistol JM, et al. First European consensus for diagnosis, management, and treatment of transthyretin familial amyloid polyneuropathy. Curr Opin Neurol. 2016;29(Suppl 1):S14–26.
70. Weis J, Katona I, Muller-Newen G, Sommer C, Necula G, Hendrich C, et al. Small-fiber neuropathy in patients with ALS. Neurology. 2011;76:2024–9.
71. Isak B, Pugdahl K, Karlsson P, Tankisi H, Finnerup NB, Furtula J, et al. Quantitative sensory testing and structural assessment of sensory nerve fibres in amyotrophic lateral sclerosis. J Neurol Sci. 2017;373:329–34.
72. Truini A, Biasiotta A, Onesti E, Di Stefano G, Ceccanti M, La Cesa S, et al. Small-fibre neuropathy related to bulbar and spinal-onset in patients with ALS. J Neurol. 2015;262:1014–8.
73. Gröne E, Üçeyler N, Abahji T, Fleckenstein J, Irnich D, Mussack T, et al. Reduced intraepidermal nerve fiber density in patients with chronic ischemic pain in peripheral arterial disease. Pain. 2014;155:1784–92.

Autonomic Testing and Nerve Fiber Pathology

5

Ahmad R. Abuzinadah
and Christopher H. Gibbons

Abstract

The autonomic nervous system is a diffuse, complicated component of both the peripheral and central nervous system. Due to the extensive innervation of most organs, there are a number of different techniques that have been developed to study the autonomic nervous system. For the purposes of this review, we will focus on tests performed by neurologists, although many additional tests of autonomic function exist across various specialties. The autonomic function tests performed by neurologists typically encompass three major domains of physiological measurement: (1) measurement of parasympathetic cholinergic cardiovagal function, (2) measurement of sympathetic adrenergic vasomotor function, and (3) measurement of sympathetic cholinergic sudomotor function. These tests of autonomic physiology help to localize the lesion within the autonomic nervous system and hence will aid in the differential diagnosis. Autonomic function testing is a recommended part of polyneuropathy screening in individuals with symptoms of autonomic dysfunction. A combination of different autonomic function tests provides a higher sensitivity and specificity than a single measure. In addition to a diagnosis of sympathetic adrenergic, sympathetic cholinergic, or parasympathetic dysfunction, tilt table testing can confirm the presence of orthostatic or delayed orthostatic hypotension, postural tachycardia syndrome, and neurally mediated syncope. Recently, the expansion of the skin biopsy technique to include evaluation of somatic nociceptive C-fibers and autonomic sudomotor, pilomotor, and vasomotor fibers has improved our ability to define a structural autonomic abnormality. Although the methods to quantify the density of autonomic fibers within skin biopsies are still in evolution, substantial progress has been made in understanding the structural and functional correlates of disease.

Keywords

Heart rate variability · Valsalva maneuver
Tilting table · Acting standing test
Sympathetic skin response · Quantitative sudomotor axon reflex testing (QSART)
Thermoregulatory sweat test · Skin biopsy
Intraepidermal nerve fiber density · Sweat gland nerve fiber density · Pilomotor nerve

A. R. Abuzinadah
Neurology Division, Internal Medicine Department, King Abdulaziz University, Jeddah, Saudi Arabia

C. H. Gibbons (✉)
Neurology Department, Beth Israel Deaconess Medical Center, Harvard Medical School, Boston, MA, USA
e-mail: cgibbons@bidmc.harvard.edu

S.-T. Hsieh et al. (eds.), *Small Fiber Neuropathy and Related Syndromes: Pain and Neurodegeneration*,
https://doi.org/10.1007/978-981-13-3546-4_5

5.1 Introduction

The autonomic nervous system extends from the forebrain, the brainstem, and the spinal cord into the extensive network of ganglia and final synapsing throughout the organ systems of the body. Although testing autonomic function across nearly every organ system is available, we will focus on those tests primarily performed by neurologists. However, tests of gastric motility, urodynamic and pupillary function are common autonomic tests that are frequently performed in other specialties. Autonomic function testing by neurologists encompasses different categories of techniques which serve different purposes. The two major categories that are commonly measured in clinical practice are the cardiovascular assessment of parasympathetic and sympathetic adrenergic function and the assessment of sudomotor (sympathetic cholinergic) function. The cardiovascular testing can be subdivided into sympathetic adrenergic and parasympathetic function.

In general, the autonomic nervous system can be evaluated through measures of autonomic structure or autonomic function. Autonomic fibers, in particular, small fiber neuropathy involving the autonomic system, can be assessed through structural measures, such as by skin biopsy analysis of autonomic nerve fiber density or by more invasive nerve biopsy (although nerve biopsies are performed very rarely). Functional tests of the autonomic nervous system are more widely utilized and can provide information about the functional extent of the dysautonomia to help dictate disease management. In this chapter we will discuss the neurophysiological assessment of the autonomic nervous system as well as the histological evaluation of autonomic nerve structure.

5.2 Cardiovascular Autonomic Reflex Testing

5.2.1 Heart Rate Variability with Deep Breathing

The heart rate change with breathing is referred to as the respiratory sinus arrhythmia [1]. The afferent and efferent limbs of this reflex are mediated through the vagus nerve [2]. This reflex is triggered mainly through pulmonary stretch receptors [2]. Measurement of the heart rate variability during paced breathing provides a quantitative assessment of parasympathetic function [3, 4].

5.2.1.1 Testing Method

In a supine position, the patient is instructed to take deep, slow cycles of gradual inspiration and expiration at a rate of six breathes per minute for a total of 90 s (nine inspiratory-expiratory cycles). Six cycles per minute were found to induce the maximum change in heart rate response with deep breathing. In each inspiration the patient should gradually inhale over 5 s, while in each expiration, the patient should gradually exhale over 5 s. It is important that the patient does not pause or gasp. The heart rate is continuously recorded during this test through an electrocardiogram. The metric of this test is called E-I ratio which is the ratio of the shortest P-R intervals during inspiration to the longest P-R intervals during expiration [1, 5]. In addition, the average variation in respiratory rate during each cycle is used as a measurement to assess the cardiovagal function (known as the maximum to minimum heart rate variability). The difference between the average six fastest heart rates minus the average of the six slowest heart rates is determined. Age-adjusted normative data are available [1, 5]. The E-I ratio is more dependent on baseline heart rate than the maximum to minimum heart rate change, but both provide valuable quantitative measures of parasympathetic function. Hyperventilation may inhibit the sinus arrhythmia; thus prolonged slow breathing is not recommended, and some authors suggest using the initial respiratory cycle instead of the averaged value [6].

5.2.1.2 Interpretation of Results

A normal E-I ratio and maximum to minimum change in respiratory rate indicate normal cardiovagal (parasympathetic) function. However, the normative data for individuals above the age of 60 are not that robust [5]. Normal values have been reported, and the usual maximum to minimum variation ranges between 7 and 27

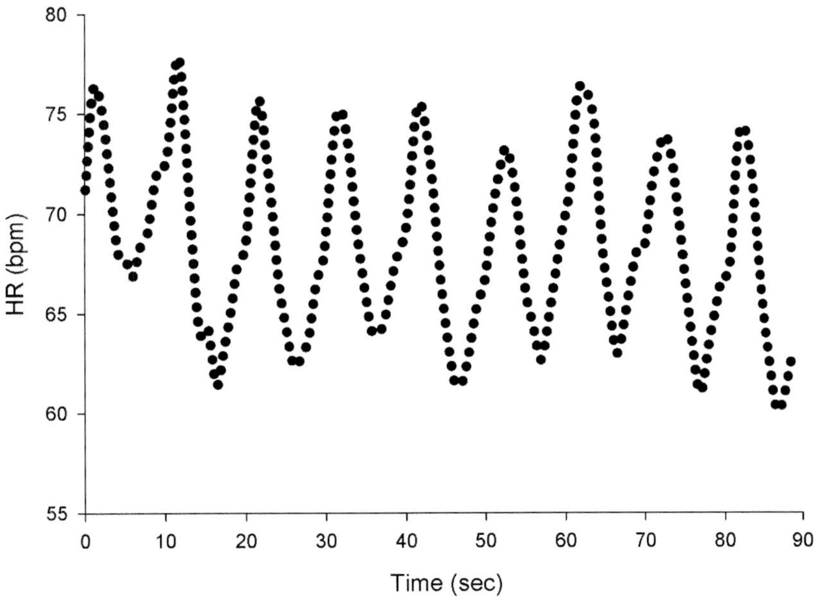

Fig. 5.1 The heart rate response to deep breathing. The heart rate varies with respiration, also known as the sinus arrhythmia. There is a typical tachycardia with inspiration and bradycardia during expiration. In this example, the heart rate varies between 63 and 75 beats per minute during respiration

beats per minute as shown in Fig. 5.1 [1, 7]. Many medications may affect the respiratory sinus arrhythmia or resting heart rate such as anticholinergic [8], nicotine [9], beta-blockers, and norepinephrine reuptake and serotonin-selective reuptake inhibitors and should be accounted for during testing [10, 11].

5.2.2 Valsalva Maneuver

The Valsalva maneuver assesses both sympathetic adrenergic and parasympathetic function simultaneously. The test is performed by exhaling against resistance for 15 s. This results in a dynamic change in both blood pressure and heart rate. First, the Valsalva maneuver assesses cardiovagal (parasympathetic) function through the change in heart rate, known as the Valsalva ratio. Second, it assesses sympathetic adrenergic function through measurement of the beat-to-beat blood pressure changes during the Valsalva [5, 12].

5.2.2.1 Testing Method

The patient should be comfortably resting in the supine position for several minutes to establish a baseline blood pressure. Then the subject takes a maximally deep breath in and exhales into a tube with measurable pneumatic resistance for 15 s. The patient should be instructed to blow out hard enough to generate a constant pressure of 40 mmHg. Pressure below 20 mmHg is insufficient and above 60 mmHg is not reliable [13]. During this maneuver, the electrocardiogram and beat-to-beat blood pressure should be continuously monitored [14]. A summary display of a Valsalva maneuver is shown in Fig. 5.2.

5.2.2.2 Interpretation of Results

Parasympathetic function is assessed through the Valsalva ratio which is calculated by dividing the maximum heart rate during expiration over the minimal heart rate occurring within 30 s after the maximum heart rate [1, 10, 15]. This number is age- and gender-dependent and should be compared to normative values. Abnormalities in the Valsalva ratio may indicate parasympathetic dysfunction but only if sufficient expiratory force was generated to result in an adequate decrease in blood pressure. However, the tachycardia seen in late phase II of the Valsalva maneuver is mediated, in part, through sympathetic activities; thus a normal Valsalva ratio does not exclude the possibility of parasympathetic impairment [14, 16].

In addition to parasympathetic function, the Valsalva maneuver offers an evaluation of

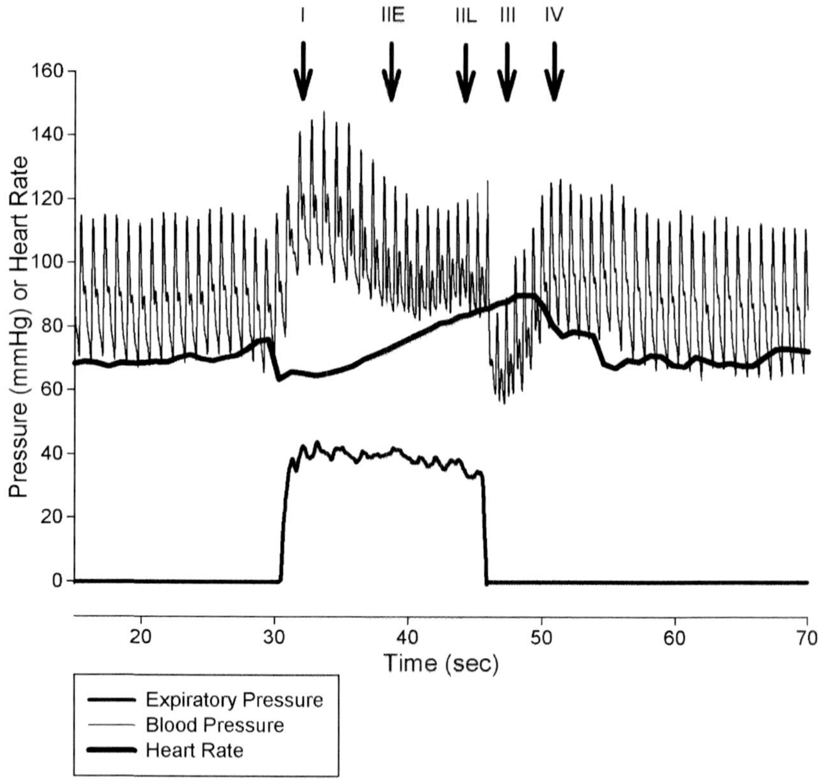

Fig. 5.2 Phases of the Valsalva maneuver. The Valsalva maneuver is shown in Fig. 5.2. The expiratory pressure should be approximately 40 mmHg for 15 s. The corresponding changes in heart rate and blood pressure are noted in the same figure. The blood pressure and heart rate components of phase I, phase 2 early (IIE), phase 2 late (IIL), phase III, and phase IV are outlined described in detail in Table 5.1

Table 5.1 Valsalva maneuver

Phase	Blood pressure changes	Mechanism
I	Increase	Mechanical compression of the aorta due to increased intrathoracic pressure
II early	Decrease	Reduced venous return due to compression of the vena cava which reduce stroke volume
II late	Increase (recovery)	Baroreceptor-mediated sympathetic vasoconstriction
III	Decrease	Mechanical decompression of the aorta due to decrease in intrathoracic pressure, with exhalation
IV	Overshoot	Persistence of vasoconstriction from phase II late

sympathetic adrenergic function through beat-to-beat blood pressure measurement. There are four phases for this test explained in Table 5.1 and Fig. 5.2. Reduced phase II (late) of the Valsalva maneuver suggests an impairment in the sympathetic adrenergic function [15]. In severe autonomic failure, there may be an exaggerated early phase II decrease in blood pressure. An absent phase VI indicates a failure in the adrenergic innervation of the heart [5, 12]. Baroreflex sensitivity (BRS) can also be measured using the Valsalva maneuver using both the vagal response (BRSv) and adrenergic response (BRSa). Another index to assess the adrenergic function is the time it takes for phase III blood pressure to return to baseline which is called blood pressure recovery time (PRT). It is found to increase progressively with the severity of autonomic failure [17].

Phase II (late) can be selectively blocked by phentolamine, while phase VI can be blocked by propranolol. This indicates that phase II (late) is primarily mediated through α-adrenergic function, while phase VI is primarily mediated through β-adrenergic function [18]. It also indicates that caution is necessary when interpreting results if the patient is taking medications that alter alpha or beta activity. There are several distinct patterns described with Valsalva maneuvers. A "V" pattern is a prolonged decrement in systolic blood pressure with no recovery in late phase II and is correlated with neurogenic orthostatic hypotension (causing the blood pressure to appear as a "V"). The "M" pattern indicates a prominent increment in systolic blood pressure during late phase II as well as rapid decrement in phase IV overshoot and is correlated with inappropriate sinus tachycardia. The "N" pattern indicates an overshoot in phase IV that is prolonged and is correlated with postural tachycardia syndrome [19].

5.2.3 Head-Up Tilt Table Testing

The head-up tilt test resembles the physiology of standing but eliminates the active muscle contraction of the legs that occurs with standing and also does so in a more controlled environment (the patient is strapped to a table so cannot be hurt if fainting occurs). When moving to the upright tilted position, there is a reduction in cardiac output and a decline in blood pressure that is detected by the baroreceptors and results in sympathetic activation which causes peripheral vasoconstriction as well as parasympathetic withdrawal causing tachycardia [16, 20].

5.2.3.1 Testing Method

Tilt table testing requires several ongoing measurements: a continuous single or multi-lead electrocardiogram, continuous beat-to-beat blood pressure monitoring, and oscillometric blood pressure through an arm cuff. The patient stays in a supine rested position for a period of time (the recommended time frame is 20 min). The table is then tilted to 60–80 degrees (laboratory angles vary by generally provide similar data), and continuous recordings continue until the end of the test. The tilt angle allows the patient to relax by reducing the effect of the leg muscle pump [14, 20–22]. In general, administration of pharmacologic agents during tilt table testing is not recommended for autonomic testing because it increases the false-positive rate. It is often used to induce neurally mediated syncope but does not provide any value in measuring autonomic function.

5.2.3.2 Interpretation of Results

Three major abnormalities can be detected during tilt table testing, and it is important to be specific about the findings (i.e., a "positive" tilt table test does not provide any valuable information). The first abnormality is a fall in blood pressure (systolic $\geq$ 20 mmHg or diastolic $\geq$ 10 mmHg) within 3 min of head-up tilt [23]. A fall in blood pressure of this magnitude is defined as orthostatic hypotension (which in conjunction with abnormal blood pressure findings on the Valsalva maneuver can indicate neurogenic orthostatic hypotension—because it is due to a failure of the sympathetic adrenergic system). If there is an equivalent fall in blood pressure beyond 3 min of testing, it is defined as delayed orthostatic hypotension [24, 25]. There are many causes of orthostatic hypotension, and medication effects need to be considered in the differential diagnosis. The second abnormality detected by tilt table testing is an increase in heart rate by $\geq$30 beats without a drop in blood pressure. This indicates a postural tachycardia; if the tachycardia is chronic and is associated with postural symptoms, it is consistent with postural tachycardia syndrome (POTS). POTS is a heterogenous disorder with many potential etiologies including medication effects, deconditioning, denervation, and underlying medical conditions. The third abnormality commonly seen during tilt table testing is neurally mediated syncope (also known as vasovagal syncope). Typically there is a simultaneous drop in blood pressure and heart rate [14, 16].

5.2.4 Active Standing Test

Although standing is a daily activity for most of us, standing for a period of time without moving can be a significant challenge for someone with autonomic dysfunction. This is physiologically similar to the tilt table test but does add the component of the muscle pump to the test.

5.2.4.1 Testing Method

Active standing requires the same setup as the tilt table test but does not require a tilt table. Standard equipment for autonoimc testing includes a continuous single or multi-lead electrocardiogram, continuous beat-to-beat blood pressure monitoring, and oscillometric blood pressure through an arm cuff. The electrocardiogram and blood pressure are monitored as described above in the tilt table testing section. In general, the patient is monitored in a relaxed supine position for several minutes, and then the patient is instructed to stand up quickly and remain standing for 5 min.

5.2.4.2 Interpretation of Results

Similar to other tests of autonomic function, active standing provides information about both sympathetic adrenergic and parasympathetic function. The physiologic responses to standing are divided into the immediate response (the first 30 s of standing) and the stabilized response (beyond 30 s). During the immediate response, there is a reduction in vascular resistance and drop in blood pressure that is maximal at 5–10 s post-standing associated with a peak in heart rate at 8–15 s, and then the heart rate declines. During the stabilized response, the heart rate remains at 15–30% above the baseline, while diastolic blood pressure is 10–15% below the baseline [21, 26]. The cardiac vagal (parasympathetic) function is measured by the 30:15 ratio: the R-R interval at approximately the 30th beat (the slowest heart rate at that time) divided over the R-R interval at approximately the 15th beat (the fastest heart rate at that time). The 30:15 ratio is not considered as reliable a parasympathetic measure as the response obtained by deep breathing but can support a diagnosis of parasympathetic dysfunction. In addition, active standing can make a diagnosis of orthostatic hypotension, delayed orthostatic hypotension, postural tachycardia, and neurally mediated syncope using the same criteria for tilt table testing [23].

5.3 Sudomotor Function Testing

5.3.1 Sympathetic Skin Response (SSR)

The sympathetic skin response is a simple means to measure electrodermal activity across regions of the body and is a surrogate measure of sudomotor function. This has been utilized for many years and has modest sensitivity and specificity.

5.3.1.1 Testing Methods

A set of recording electrodes are placed over the dorsal and palmar aspects of the hands and the feet as these sites contain the highest density of eccrine sweat glands. A reference electrode is placed over the dorsum of the hand and feet. The SSR stimulation can be either through electrical stimulation or sudden deep inhalation. The SSR is a low frequency response so the low frequency setting should be 0.1 Hz or lower. Electrical stimulation of the median and tibial nerves is usually performed with at 17 mAmp with a pulse duration of 0.1 ms. Inhalation stimulation is performed by asking the patient to take a sudden deep breath followed by forceful expiration on [27, 28].

5.3.1.2 Interpretation of Results

There are a few important points to consider during interpretation of the results. First, the SSR will habituate with repetitive measures and may be reduced or disappear if performed too many times, which may give a false-negative result [29]. Second, the test is often abnormal in individuals over 50 years of age [5]. Although normal values for latency and amplitude have been described, a common way to interpret this test is to report a visible response as "normal" and an absent response as "abnormal" [27, 28, 30, 31]. Abnormalities in the SSR can be seen in diseases that damage the peripheral nervous system, such as diabetic neuropathy [32]. However, central nervous system disorders such as Parkinson's disease and stroke can be associated with an abnormal SSR [5]. SSR was also found to correlate with neuropathy affecting unmyelinated axon but did not correlate with dysautonomia [28].

5.3.2 Quantitative Sudomotor Axon Reflex Testing (QSART)

This procedure measures the integrity of the local sudomotor axon reflex. When an axon terminal is stimulated by the iontophoresis of acetylcholine, an antidromic axon potential is generated that travels to an axon branch point where it will travel orthodromically and triggers a release of acetylcholine from the neighboring sweat glands. Therefore, the stimulation of sweat glands through the iontophoresis of acetylcholine causes sweating in the areas of stimulation (the direct response) and axon reflex sweating in neighboring regions (the indirect response). The fibers involved in this reflex are C-fibers.

5.3.2.1 Testing Methods

A multicompartmental sweat cell is placed over four standard sites on (1) the medial forearm, (2) the proximal leg distal to fibular head, (3) the distal leg proximal to medial malleoli, and (4) over the extensor digitorum brevis muscle of the foot. The sweat cell has three separate concentric circular compartments. The outer compartment contains acetylcholine. The middle compartment is a "spacer" compartment and usually just contains air. The inner compartment is for recording of humidity. An electrode is attached to the outer compartment and used to deliver a constant 2 mA current which will cause iontophoresis of acetylcholine into the skin. The inner compartment contains a circulating nitrogen gas or dehumidified air that returns to a sudorometer (humidity detection). The change of the gas humidity is used to calculate the rate of sweating. The iontophoretic stimulus is applied for 5 min, with sweat recording continues for 5 min after iontophoresis, for a total of 10 min [1, 5, 7].

5.3.2.2 Interpretation of Results

The recorded humidity as well as the latency until sweat production from the time of stimulus defines the response. Normative values have been published; see Fig. 5.3 for additional details [5, 7]. Several patterns of abnormalities can be seen with QSART. The major pattern is proximal to distal reduction in the sweat responses, which are usually suggestive of length-dependent small fiber neuropathy. QSART is abnormal in around 80% of small fiber neuropathies [33–35]. Combining this test with other modalities for small fiber assessment may increase the yield of confirming the diagnosis [36]. The test retest reliability of QSART was moderate indicating that it may not be a good tool to follow up patients [37]. This test assesses the integrity of the peripheral part of the nervous system; therefore, a central cause of sweating dysfunction will not be detected.

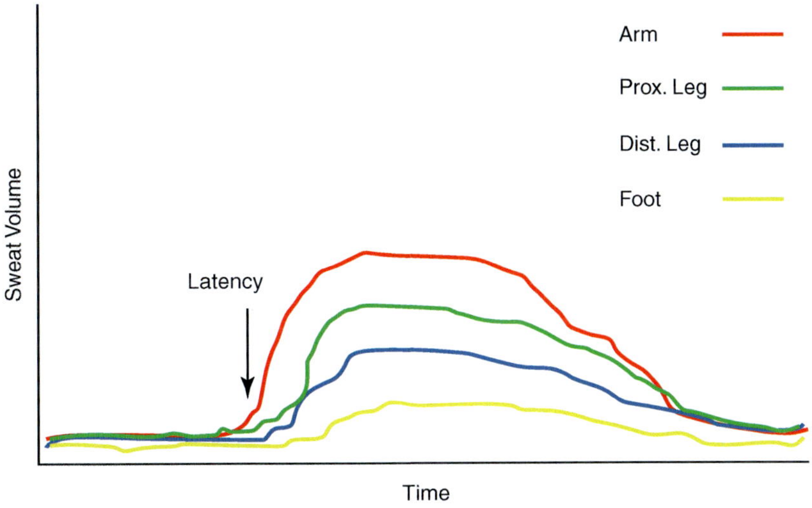

Fig. 5.3 A normal quantitative sudomotor axon reflex testing (QSART) response. A normal QSART showing sweat output as a waveform from the four standard sites. The total volume of sweat (the Y axis) and latency is within normal values. The volume of sweat production is typically measured by an area under the curve analysis from stimulation to 10 min for each lead

5.3.3 Thermoregulatory Sweat Testing (TST)

The thermoregulatory sweat test measures the integrity of the thermoregulatory system from the hypothalamus to the sweat glands. The patient is placed within a "sauna-like" environment to raise the core body temperature. When the core body temperature increases by 1 degree, the hypothalamus detects the change, and significant thermal sweating occurs, which can be measured by a topical indicator.

5.3.3.1 Testing Methods

The patient lies supine in a heat- and humidity-controlled chamber with an air temperature of 43 °C to 50 °C and humidity of 35–40%. The goal is to raise core temperature to 38 °C or 1 °C above baseline if the starting temperature was above 37.1 °C. Skin temperature should be maintained between 39 °C and 40 °C. The patient core and skin temperature must be monitored closely to prevent dehydration, thermal injury, or hyperthermia. Sweat production is monitored by the application of an indicator powder (a mixture of alizarin red, cornstarch, and sodium carbonate). This powder turns dark purple when in contact with moisture (i.e., sweat). Iodinated cornstarch is an alternative indicator dye. The test duration is 60 min maximum or less if generalized sweating occurred [1, 7, 38].

5.3.3.2 Interpretation of Results

Many different patterns have been reported with TST, generally conforming to the distribution of a neuropathy or local nerve injury. In the presence of a distal small fiber neuropathy, there is distal anhidrosis on TST which is characterized by reduced sweating in the distal limbs, as seen in Fig. 5.4. A TST is abnormal in 72–74% of small fiber neuropathy cases [33, 35]. Other patterns of abnormalities such as regional anhidrosis and global anhidrosis can be seen in non-length-dependent neuropathies or autonomic failure syndromes. Regional and global anhidroses are commonly seen in multiple system atrophy [39]. Segmental anhidrosis (large continuous zones of anhidrosis with demarcated

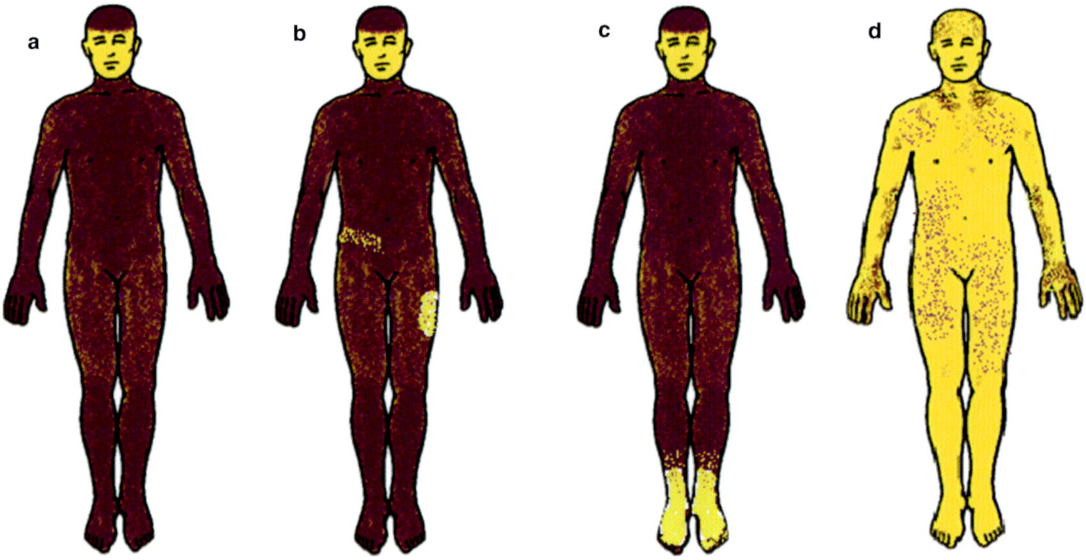

Fig. 5.4 Thermoregulatory sweat testing. The thermoregulatory sweat test provides a topographical representation of sudomotor dysfunction, from the central to the peripheral nervous system (but does not determine if it is a central or peripheral lesion). In (**a**), there is a normal sweat response (the indicator powder is not placed around the face or eyes to prevent complications). In (**b**), a right-sided T12 radiculopathy and a left lateral cutaneous nerve injury can be identified. In (**c**), there is a length-dependent neuropathy. In (**d**), there is global anhidrosis due to multiple system atrophy

borders) is likely to be associated with Ross syndrome, pure autonomic failure, and/or diabetic or immune-mediated sympathectomy [1, 38, 40]. The TST may also be interpreted through the percent of body surface anhidrosis (TST%) which roughly correlates with autonomic neuropathy manifestations [38].

5.4 Skin Biopsy Analysis of Nerve Fiber Density

5.4.1 Intraepidermal Nerve Fiber Density (IENFD)

Over the past 20 years, the utility of skin biopsy as means to diagnose small fiber neuropathy has expanded dramatically (Fig. 5.5). Initially considered a research tool, it has become the gold standard approach to the diagnosis of small fiber neuropathy. National and international groups have established normative databases and standard methodologies for diagnosis of small fiber neuropathy [41, 42]. Additional details about the use of skin biopsies for analysis of small fiber sensory neuropathy are discussed in Chap. 2.

5.4.1.1 Testing Methods

The standard punch biopsy utilizes a 3 mm punch tool at a depth of 3–4 mm for analysis of IENFD [43]. The same locations and methodology can be used to acquire tissue for study of autonomic innervation. The skin biopsies are typically taken from one to three standardized sites which are 10 cm above the lateral malleolus, 10 cm above the lateral aspect of the knee, and 20 cm below the anterior iliac spine [43, 44]. Nonstandard sites are also used for a variety of reasons but typically in the setting of an asymmetric neuropathy where a control sample is obtained from the contralateral (unaffected) body site. The specimens are typically fixed in Zamboni's solution (2% paraformaldehyde and picric acid), although other fixatives have been utilized [43]. The specimens are stained with antibodies against protein gene product 9.5 (PGP 9.5), a nonselective neuronal stain [42]. Additional stains for selective identification of autonomic nerve fibers can be considered but are generally only used with immunofluorescent imaging, not for the use with light microscopy.

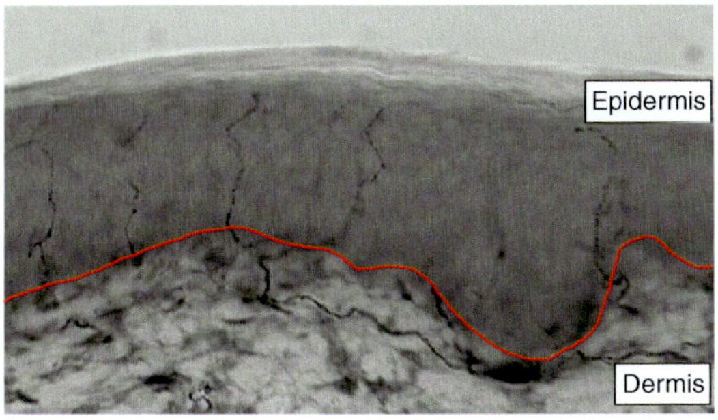

Fig. 5.5 Skin biopsy showing intraepidermal nerve fibers. This skin biopsy demonstrates the epidermal and dermal layers with staining for PGP 9.5, a pan-axonal marker. The basement membrane is highlighted in *red*. The dark lines are nerve fibers; when they cross the dermal-epidermal junction, they are counted. As they extend into the epidermis, they are referred to as intraepidermal nerve fibers

Sudomotor Nerve Fibers

For analysis of sympathetic cholinergic nerve fibers, immunostaining with PGP 9.5 will provide adequate detail for many studies of sudomotor nerve fibers (which are therefore identified by location around sweat glands, not by fiber type) [45, 46] (Fig. 5.6). Various methods to enhance background contrast have been developed to improve the ability to define the sweat gland region of interest [45, 47]. Some investigators have used Congo red staining to outline the background of the sweat gland [46, 47], while others have utilized the "unsharp mask filter" photographic technique to enhance the sweat gland image [45]. More specific information about fiber type can be gathered by using antibodies against vasoactive intestinal peptide (VIP), which selectively stains for sympathetic cholinergic nerve fibers. Quantitation of sudomotor nerve fibers is more complex than that for analysis of IENFD. The more random distribution of fibers and the nonlinear and overlapping aspects of the nerves create significant challenges to analysis [45]. A variety of approaches have been proposed. Quantification of sweat gland nerve fiber density (SGNFD) can be obtained using an unbi-

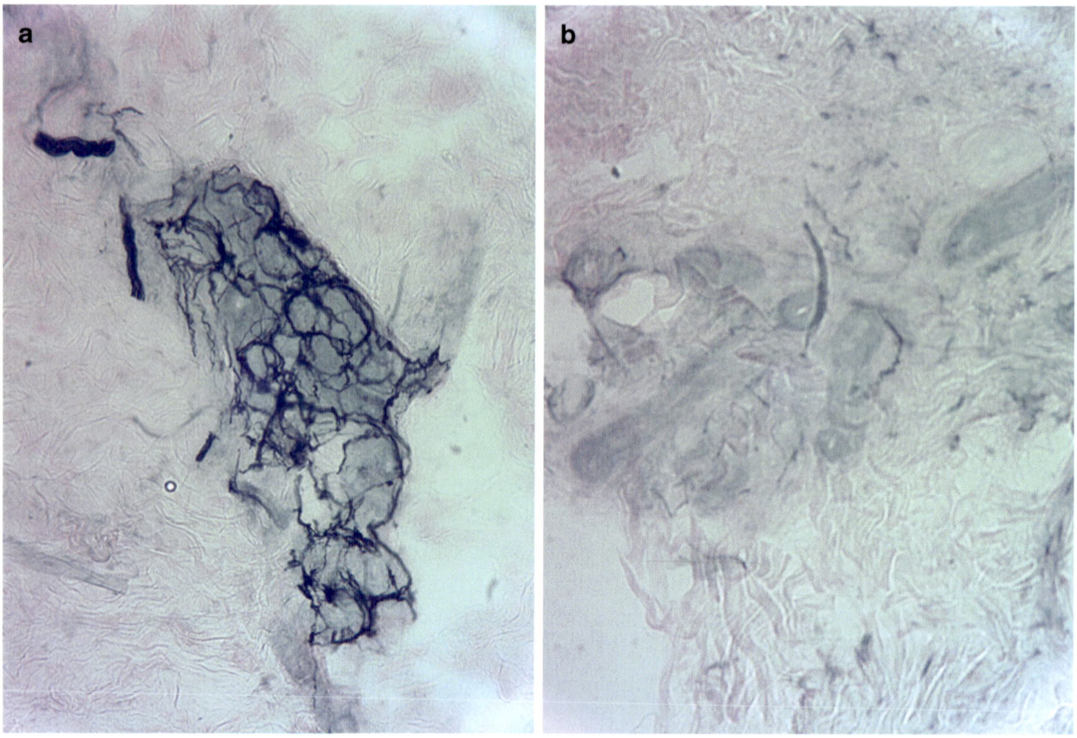

Fig. 5.6 Sweat gland innervation. Sweat glands are located within the deeper dermal tissue. Nerve fibers are stained with protein gene product 9.5 (PGP 9.5) (a pan-axonal marker), and a normal appearing sweat gland with robust innervation is shown in (**a**). In (**b**), there is a completely denervated sweat gland appearing within the tissue

ased stereologic approach [45–48]. The density of fibers is estimated based on stereology. Another approach utilizes an area analysis, where the SGNFD is estimated based on the area of the sweat gland that is occupied, sometimes referred to as the sweat gland innervation index [47, 48]. A third approach is a semiquantitative analysis of nerve density provided by the neuropathologist [48, 49]. The unbiased stereologic approach tends to provide more reliable estimates of SGNFD but is much more labor-intensive than the area analysis [48]. The semiquantitative approach has much greater inter and intra-reader variability [48].

Pilomotor Nerve Fibers

The pilomotor nerves are identified through the use of the pan-axonal marker PGP 9.5 and easily identified because of their location within pilomotor muscles. Methods to quantitate pilomotor nerve fibers have been reported and are fairly straightforward compared to sudomotor nerve fibers [50, 51]. The number of nerve fibers crossing through a given pilomotor muscle at several cross-sectional points is measured. The results are reported as the number of fibers per millimeter of muscle width [50, 51]. If more selective imaging of pilomotor fiber types is required, the use of antibodies against tyrosine hydroxylase or dopamine beta-hydroxylase can provide selective markers for sympathetic adrenergic nerve fibers. The vast majority of pilomotor fibers are sympathetic adrenergic.

5.4.1.2 Interpretation of Results

Quantitation of sudomotor and pilomotor nerves has been correlated with their physiological counterparts (sweat function and piloerection) in a study looked at effects of capsaicin application to induce loss of cutaneous nerves and their related function [51]. There was degeneration followed by regeneration of sudomotor nerve fiber density (SGNFD) as well as the correlative functional loss and recovery seen by QSART. The authors found that the degeneration and regeneration of SGNFD paralleled the loss and restoration of the sweat gland function which supported the validity of SGNFD in assessment of sudomotor function. Similarly they measured the degeneration and regeneration of pilomotor nerve fiber density as well as the functional loss and restoration through a silicone imprint of the chemical stimulation of pilomotor muscles [51, 52]. They also found that the degeneration and regeneration of pilomotor density parallel the loss of and restore the pilomotor function which supports the validity of pilomotor nerve density in assessment of pilomotor function [51]. In autonomic neuropathy, the sensitivity and specificity of sudomotor evaluation were 91% and 87%, respectively, and for pilomotor evaluation were 100% and 87%, respectively [53]. However, these results were derived from a case controlled study, and individuals had clear evidence of neuropathy compared to healthy control subjects, so it is not yet clear how sensitive and specific these tests will be in routine clinical practice. Other investigators have also identified clear differences in autonomic nerve density between individuals with and without diabetes [46, 50]. Assessment of sudomotor and pilomotor fibers has also been investigated in Ross syndrome and Holmes-Adie syndrome, where the sudomotor fibers were severely reduced or absent in Ross syndrome, while in Holmes-Adie syndrome it is relatively preserved [54]. Additional investigation into the use of sudomotor and pilomotor innervation is needed, but the techniques hold significant promise as another potential set of tools for the neurologist to use in the diagnosis and management of autonomic neuropathies [55].

5.5 Conclusion

The assessments of autonomic nerves, in particular the autonomic nerves of the cardiovascular and sudomotor systems, have become established components of neurological examinations. In addition to functional evaluations of the entire autonomic pathways, the examination of postganglionic pathology provides evidence of anatomical localization, such as QSART to measure local sudomotor function and skin biopsy of sudomotor, pilomotor, and vasomotor innervation to investigate structural changes to sympathetic cholinergic, sympathetic adrenergic, and vasomotor innervation, respectively. Taken together, the combination of these tests constitutes an integrated view of dysautonomia in small fiber neuropathy.

References

1. Daube JR, Rubin DI, editors. Clinical neurophysiology. New York: Oxford University Press; 2009.
2. Katona PG, Jih F. Respiratory sinus arrhythmia: noninvasive measure of parasympathetic cardiac control. J Appl Physiol. 1975;39:801–5.
3. Bennett T, Fentem PH, Fitton D, Hampton JR, Hosking DJ, Riggott PA. Assessment of vagal control of the heart in diabetes. Measures of R-R interval variation under different conditions. Br Heart J. 1977;39:25–8.
4. Pfeifer MA, Cook D, Brodsky J, Tice D, Reenan A, Swedine S, et al. Quantitative evaluation of cardiac parasympathetic activity in normal and diabetic man. Diabetes. 1982;31:339–45.
5. Low PA, Benarroch EE, editors. Clinical autonomic disorders. New York: Wolters Kluwer; 2009.
6. Bennett T, Farquhar IK, Hosking DJ, Hampton JR. Assessment of methods for estimating autonomic nervous control of the heart in patients with diabetes mellitus. Diabetes. 1978;27:1167–74.
7. Novak P. Quantitative autonomic testing. J Vis Exp. 2011;53:2502.
8. Lacroix D, Logier R, Kacet S, Hazard JR, Dagano J, Lekieffre J. Effects of consecutive administration of central and peripheral anticholinergic agents on respiratory sinus arrhythmia in normal subjects. J Auton Nerv Syst. 1992;39:211–7.
9. Wang J, Wang X, Irnaten M, Venkatesan P, Evans C, Baxi S, et al. Endogenous acetylcholine and nicotine activation enhances GABAergic and glycinergic inputs to cardiac vagal neurons. J Neurophysiol. 2003;89:2473–81.
10. Low PA, Opfer-Gehrking TL. Differential effects of amitriptyline on sudomotor, cardiovagal, and adrenergic function in human subjects. Muscle Nerve. 1992;15:1340–4.
11. Davidson J, Watkins L, Owens M, Krulewicz S, Connor K, Carpenter D, et al. Effects of paroxetine and venlafaxine XR on heart rate variability in depression. J Clin Psychopharmacol. 2005;25:480–4.

12. Pstras L, Thomaseth K, Waniewski J, Balzani I, Bellavere F. The Valsalva manoeuvre: physiology and clinical examples. Acta Physiol (Oxf). 2016;217:103–19.
13. Korner PI, Tonkin AM, Uther JB. Reflex and mechanical circulatory effects of graded Valsalva maneuvers in normal man. J Appl Physiol. 1976;40:434–40.
14. Freeman R. Assessment of cardiovascular autonomic function. Clin Neurophysiol. 2006;117:716–30.
15. Jones PK, Gibbons CH. The role of autonomic testing in syncope. Auton Neurosci. 2014;184:40–5.
16. Benarroch E, editor. Autonomic neurology. New York: Oxford University Press; 2014.
17. Vogel ER, Sandroni P, Low PA. Blood pressure recovery from Valsalva maneuver in patients with autonomic failure. Neurology. 2005;65:1533–7.
18. Sandroni P, Benarroch EE, Low PA. Pharmacological dissection of components of the Valsalva maneuver in adrenergic failure. J Appl Physiol (1985). 1991;71:1563–7.
19. Palamarchuk IS, Baker J, Kimpinski K. The utility of Valsalva maneuver in the diagnoses of orthostatic disorders. Am J Physiol Regul Integr Comp Physiol. 2016;310:R243–52.
20. Saal DP, Thijs RD, van Dijk JG. Tilt table testing in neurology and clinical neurophysiology. Clin Neurophysiol. 2016;127:1022–30.
21. Wieling W, Krediet CT, van Dijk N, Linzer M, Tschakovsky ME. Initial orthostatic hypotension: review of a forgotten condition. Clin Sci (Lond). 2007;112:157–65.
22. Fitzpatrick AP, Theodorakis G, Vardas P, Sutton R. Methodology of head-up tilt testing in patients with unexplained syncope. J Am Coll Cardiol. 1991;17:125–30.
23. Freeman R, Wieling W, Axelrod FB, Benditt DG, Benarroch E, Biaggioni I, et al. Consensus statement on the definition of orthostatic hypotension, neurally mediated syncope and the postural tachycardia syndrome. Auton Neurosci. 2011;161:46–8.
24. Gibbons CH, Freeman R. Delayed orthostatic hypotension: a frequent cause of orthostatic intolerance. Neurology. 2006;67:28–32.
25. Gibbons CH, Freeman R. Clinical implications of delayed orthostatic hypotension: a 10-year follow-up study. Neurology. 2015;85:1362–7.
26. Smith JJ, Porth CM, Erickson M. Hemodynamic response to the upright posture. J Clin Pharmacol. 1994;34:375–86.
27. Kucera P, Goldenberg Z, Kurca E. Sympathetic skin response: review of the method and its clinical use. Bratisl Lek Listy. 2004;105:108–16.
28. Shahani BT, Halperin JJ, Boulu P, Cohen J. Sympathetic skin response – a method of assessing unmyelinated axon dysfunction in peripheral neuropathies. J Neurol Neurosurg Psychiatry. 1984;47:536–42.
29. Seto-Poon M, Madronio M, Kirkness JP, Amis TC, Byth K, Lim CL. Decrement of the skin conductance response to repeated volitional inspiration. Clin Neurophysiol. 2005;116:1172–80.
30. Drory VE, Korczyn AD. Sympathetic skin response: age effect. Neurology. 1993;43:1818–20.
31. Lanctin C, Magot A, Chambellan A, Tich SN, Pereon Y. Respiratory evoked potentials and occlusion elicited sympathetic skin response. Neurophysiol Clin. 2005;35:119–25.
32. Huang YN, Jia ZR, Shi X, Sun XR. Value of sympathetic skin response test in the early diagnosis of diabetic neuropathy. Chin Med J (Engl). 2004;117:1317–20.
33. Low VA, Sandroni P, Fealey RD, Low PA. Detection of small-fiber neuropathy by sudomotor testing. Muscle Nerve. 2006;34:57–61.
34. Tobin K, Giuliani MJ, Lacomis D. Comparison of different modalities for detection of small fiber neuropathy. Clin Neurophysiol. 1999;110:1909–12.
35. Stewart JD, Low PA, Fealey RD. Distal small fiber neuropathy: results of tests of sweating and autonomic cardiovascular reflexes. Muscle Nerve. 1992;15:661–5.
36. Kamel JT, Vogrin SJ, Knight-Sadler RJ, Willems NK, Seiderer L, Cook MJ, et al. Combining cutaneous silent periods with quantitative sudomotor axon reflex testing in the assessment of diabetic small fiber neuropathy. Clin Neurophysiol. 2015;126:1047–53.
37. Berger MJ, Kimpinski K. Test-retest reliability of quantitative sudomotor axon reflex testing. J Clin Neurophysiol. 2013;30:308–12.
38. Fealey RD, Low PA, Thomas JE. Thermoregulatory sweating abnormalities in diabetes mellitus. Mayo Clin Proc. 1989;64:617–28.
39. Coon EA, Fealey RD, Sletten DM, Mandrekar JN, Benarroch EE, Sandroni P, et al. Anhidrosis in multiple system atrophy involves pre- and postganglionic sudomotor dysfunction. Mov Disord. 2017;32:397–404.
40. Cohen J, Low P, Fealey R, Sheps S, Jiang NS. Somatic and autonomic function in progressive autonomic failure and multiple system atrophy. Ann Neurol. 1987;22:692–9.
41. Lauria G, Bakkers M, Schmitz C, Lombardi R, Penza P, Devigili G, et al. Intraepidermal nerve fiber density at the distal leg: a worldwide normative reference study. J Peripher Nerv Syst. 2010;15:202–7.
42. Nolano M, Biasiotta A, Lombardi R, Provitera V, Stancanelli A, Caporaso G, et al. Epidermal innervation morphometry by immunofluorescence and bright-field microscopy. J Peripher Nerv Syst. 2015;20:387–91.
43. Lauria G, Hsieh ST, Johansson O, Kennedy WR, Leger JM, Mellgren SI, et al. European Federation of Neurological Societies/Peripheral Nerve Society Guideline on the use of skin biopsy in the diagnosis of small fiber neuropathy. Report of a joint task force of the European Federation of Neurological Societies and the Peripheral Nerve Society. Eur J Neurol. 2010;17:903–12, e44–9.

44. McArthur JC, Stocks EA, Hauer P, Cornblath DR, Griffin JW. Epidermal nerve fiber density: normative reference range and diagnostic efficiency. Arch Neurol. 1998;55:1513–20.
45. Gibbons CH, Illigens BM, Wang N, Freeman R. Quantification of sweat gland innervation: a clinical-pathologic correlation. Neurology. 2009;72:1479–86.
46. Luo KR, Chao CC, Hsieh PC, Lue JH, Hsieh ST. Effect of glycemic control on sudomotor denervation in type 2 diabetes. Diabetes Care. 2012;35:612–6.
47. Luo KR, Chao CC, Chen YT, Huang CM, Yang NC, Kan HW, et al. Quantitation of sudomotor innervation in skin biopsies of patients with diabetic neuropathy. J Neuropathol Exp Neurol. 2011;70:930–8.
48. Gibbons CH, Illigens BM, Wang N, Freeman R. Quantification of sudomotor innervation: a comparison of three methods. Muscle Nerve. 2010;42:112–9.
49. Nolano M, Provitera V, Estraneo A, Selim MM, Caporaso G, Stancanelli A, et al. Sensory deficit in Parkinson's disease: evidence of a cutaneous denervation. Brain. 2008;131:1903–11.
50. Nolano M, Provitera V, Caporaso G, Stancanelli A, Vitale DF, Santoro L. Quantification of pilomotor nerves: a new tool to evaluate autonomic involvement in diabetes. Neurology. 2010;75:1089–97.
51. Gibbons CH, Wang N, Freeman R. Capsaicin induces degeneration of cutaneous autonomic nerve fibers. Ann Neurol. 2010;68:888–98.
52. Siepmann T, Illigens B, Gibbons C, Freeman R. The quantitative pilomotor axon-reflex test (QPART)-A technique to assess autonomic nerve fiber function (P05.197). Neurology. 2012;78(Meeting Abstracts 1):P05.197.
53. Donadio V, Incensi A, Giannoccaro MP, Cortelli P, Di Stasi V, Pizza F, et al. Peripheral autonomic neuropathy: diagnostic contribution of skin biopsy. J Neuropathol Exp Neurol. 2012;71:1000–8.
54. Nolano M, Provitera V, Perretti A, Stancanelli A, Saltalamacchia AM, Donadio V, et al. Ross syndrome: a rare or a misknown disorder of thermoregulation? A skin innervation study on 12 subjects. Brain. 2006;129:2119–31.
55. McArthur JC, Griffin JW. Another tool for the neurologist's toolbox. Ann Neurol. 2005;57:163–7.

Part II

Small Fiber Neuropathy in Peripheral Nerve Disorders

Diabetes-Related Neuropathies

6

Christopher H. Gibbons

Abstract

With an estimate of 350 million people in the world having diabetes by the year 2030, diabetes will be an enormous cause of neuropathy. Approximately 50% of those people developing diabetes will develop a clinically significant neuropathy. In addition, the increase in prediabetes prevalence, and the associated increase in neuropathy risk in that cohort, suggests an epidemic of neuropathy that is going to be a challenge for the healthcare system to handle. The majority of individuals with diabetic neuropathy have type 2 diabetes (~95%), although large numbers of individuals with prediabetes are at risk for any associated distal neuropathy as well. Typically individuals with type 1 diabetes do not develop neuropathy until at least 5 years after diagnosis, while those individuals with type 2 diabetes may have neuropathy at the time of diagnosis. Neuropathies can vary in clinical presentation from a distal symmetric neuropathy, mononeuropathy, cranial neuropathy, plexus neuropathy, or even autonomic neuropathy. In this chapter, the clinical presentation, diagnosis, and management of the various diabetes-related neuropathies are reviewed.

C. H. Gibbons (✉)
Department of Neurology, Beth Israel Deaconess Medical Center, Harvard Medical School, Boston, MA, USA
e-mail: cgibbons@bidmc.harvard.edu

Keywords

Prediabetes · Distal symmetric polyneuropathy
Mixed distal symmetric polyneuropathy
Distal symmetric small fiber neuropathy
Cardiac autonomic neuropathy · Autonomic neuropathy · Diabetes Control and Complications Trial (DCCT) · Treatment-induced neuropathy of diabetes · Diabetic radiculoplexus neuropathy · Mononeuropathy
Cranial neuropathy · Screening guideline
American Diabetes Association

6.1 Introduction

One of the most common, concerning, and expensive complications of diabetes is neuropathy. Neuropathy can be associated with significant pain causing an associated increase in medication use, limitations to activities of daily living, and reduction in quality of life. Neuropathy can also result in a loss of sensation, including "protective" sensation that is necessary to prevent tissue injury. Neuropathy can also contribute to muscle weakness, loss of autonomic control of end-organ function, or some combination of these problems simultaneously [1].

Diabetes is a major worldwide cause of neuropathy, with estimates of 350 million people with diabetes by the year 2030, approximately 50% of those developing clinically significant

S.-T. Hsieh et al. (eds.), *Small Fiber Neuropathy and Related Syndromes: Pain and Neurodegeneration*,
https://doi.org/10.1007/978-981-13-3546-4_6

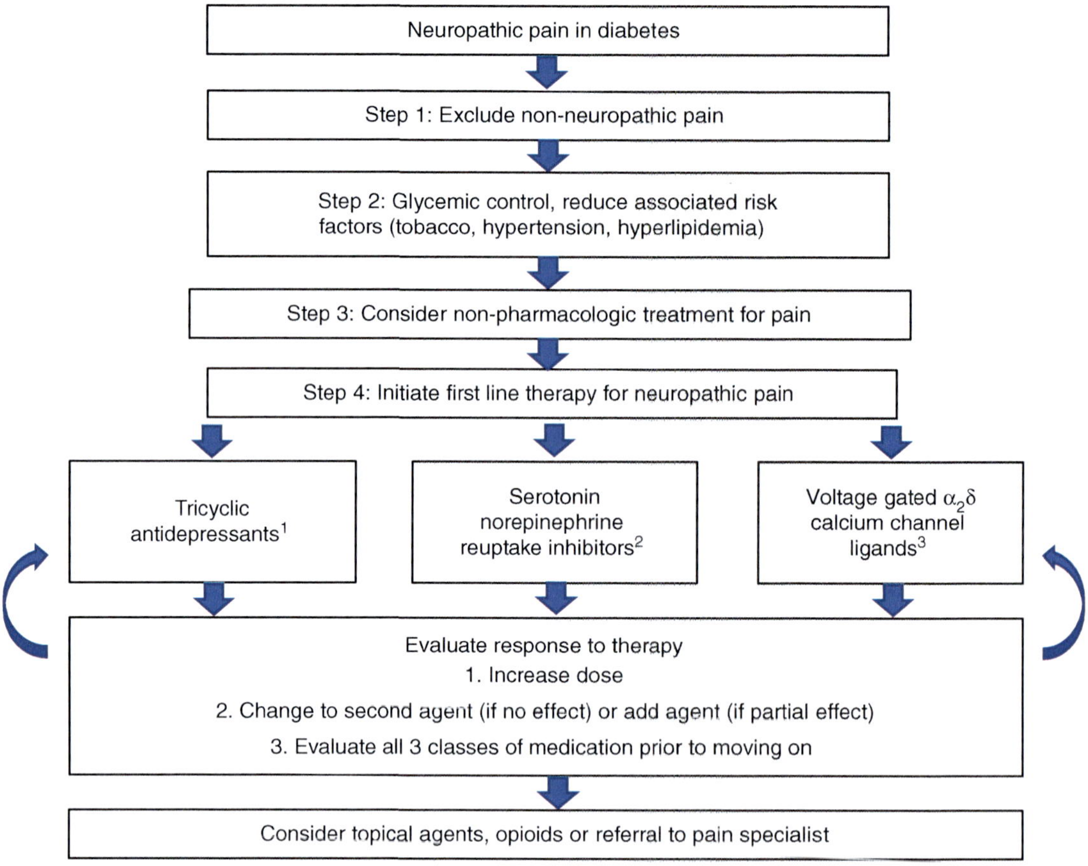

Fig. 6.1 Treatment of neuropathic pain. (1) Tricyclic antidepressants: those commonly used for neuropathic pain include amitriptyline, nortriptyline, and desipramine. Doses begin as low as 10 mg and can be titrated to clinical effect (generally <100 mg). At doses ≥100 mg, cardiac side effects become more prominent. An ECG should be performed in anyone with a history of cardiac disease or over the age of 40 years. Anticholinergic effects are prominent and should be used cautiously in the elderly. (2) Serotonin norepinephrine reuptake inhibitors: doses of 20–60 mg per day for duloxetine and 37.5–225 mg/day for venlafaxine. Side effects include nausea, somnolence, weight gain or loss, sexual side effects, serotonin syndrome. (3) $\alpha 2\delta$ calcium channel ligands: doses of 300–3600 mg per day for gabapentin and 50–600 mg per day for pregabalin. Side effects include somnolence, dizziness, peripheral edema, weight gain, and dizziness

guidelines for the treatment of neuropathic pain exist [24–26]. Figure 6.1 highlights a basic guide for the treatment of neuropathic pain. A couple of key concepts are important to consider when initiating treatment. First, the use of a single agent to control pain is often preferable to polypharmacy, as many patients with diabetes will already be taking a number of different medications and the risk of drug-drug interactions will increase [27]. Thus, attempting to achieve the maximal tolerated dose before initiating a second therapy is often helpful. Second, a plan of action needs to be clearly communicated to the patient that attempts will be made to reduce pain to a manageable level, but this is not the same as no pain at all. If no pain is the expected outcome, treatment failure is more likely. Third, it is important to reassess pain severity at every visit. If neuropathic pain has diminished, it is critical to attempt a reduction in dosing of pain medication. Otherwise there can be a slow and insidious accumulation of unnecessary medication.

6.3.2 Distal Symmetric Small Fiber Neuropathy

The presentation of a distal symmetric small fiber neuropathy in diabetes is very similar to the mixed distal polyneuropathy of diabetes but predominantly affects small unmyelinated nerve fibers without large fiber involvement [28]. In general, a diabetic small fiber neuropathy will typically progress to a mixed distal symmetric polyneuropathy over time [12, 29]. Because small unmyelinated nerve fibers are often damaged prior to large myelinated nerve fibers, many patients with prediabetes may present with a distal symmetric small fiber neuropathy, and many patients with type 2 diabetes may already have this at the time of diagnosis [30]. The reason that small fiber neuropathy is discussed as a separate entity from mixed neuropathy is the difference in history, examination, neurophysiology, and neuropathological testing results that define these disorders [31].

Although pain, thermal sensation, and autonomic function are all mediated through the small unmyelinated nerve fibers, a small fiber neuropathy typically refers to the sensory fibers, while autonomic fibers are often reported separately under "autonomic neuropathy" [9, 20]. One typical exception to the reporting of autonomic neuropathies is the sympathetic cholinergic (or sudomotor) nerve fibers that are typically involved in a distal small fiber and distal mixed polyneuropathy [32]. Pain is often the presenting symptom in patients with a small fiber neuropathy. Symptoms can include burning and stabbing pain in a distal pattern with accompanying contact allodynia or hyperalgesia in many individuals [20]. However, small fiber neuropathy may also present in a painless fashion, and the decrease in thermal or pain sensitivity can result in painless injury [20]. If peripheral sympathetic cholinergic fibers are involved, there is likely to be a distal gradient of reduced sweating [33, 34]. Paradoxically, the distal sweat loss often causes a compensatory proximal hyperhidrosis; thus patients may present for complaints of increased sweat production on the head, neck, and truncal regions.

6.3.2.1 Diagnosis of Distal Symmetric Small Fiber Neuropathy

Small fiber neuropathy is typically a clinical diagnosis with confirmation made by ancillary testing. A loss of thermal and/or pain sensitivity that is length dependent is the typical examination finding. A painful small fiber neuropathy is suggested by a history of burning or stabbing pain in a length-dependent fashion. In many cases the need for ancillary testing is not required in an individual with diabetes. However, in some situations the history and/or examination may not support a clinical diagnosis of small fiber neuropathy, and therefore additional supportive testing may be required.

At the present time, the gold standard for the diagnosis of small fiber neuropathy is a 3-mm punch skin biopsy with analysis of intraepidermal nerve fiber density [35]. The standardization of the technique and the use of international normative data by age and gender provide recommendations on the use of skin biopsy in patients with suspected small fiber neuropathy [35, 36]. In clinical practice, a standard 3-mm punch skin biopsy, taken 10 cm above the lateral malleolus and stained with an antibody to protein gene product 9.5 allows for a diagnosis of small fiber neuropathy when nerve counts are below a specific threshold [37]. It is important to recognize that a skin biopsy is not required for a diagnosis of small fiber neuropathy but may provide clarity in situations of uncertainty.

Other tests that can aid in the diagnosis of small nerve fiber neuropathy include quantitative sensory testing (QST) and quantitative sudomotor axon reflex testing (QSART). QSART is a test of peripheral sympathetic cholinergic (sudomotor or sweat) function by measuring the response from a local area of nerve stimulation [38]. QSART is both a sensitive and specific marker for the presence of a distal sudomotor neuropathy, although it is critical to standardize the technique by controlling for temperature and humidity, avoiding confounding medications, and also using age- and gender-based normative values [32, 38]. QST is generally considered a computerized method to test sensation in order to provide quantitative results. This does require the person to respond to a standardized sensory stimulus, and therefore cannot be performed on comatose, confused, or otherwise cognitively impaired individuals [39–42]. Testing with QST often involves both thermal and thermal-pain detection thresholds (warm, cold, heat-pain, and cold-pain). Although relatively sensitive and specific, QST has not generally achieved widespread clinical use because of its variability with environment and dependence on the cooperation of subjects. It has been used mainly in research setting and clinical trials [43, 44].

6.3.2.2 Management of Distal Symmetric Small Fiber Neuropathy

The management of small fiber neuropathy in diabetes is the same as that of mixed distal symmetric polyneuropathy. The long-term goal of treatment is preventing the progression of disease. Goals of care include a focus glycemic control, appropriate treatment of hyperlipidemia, normalizing blood pressure, cessation of smoking, and avoiding excessive alcohol intake. One of the primary short-term goals of treatment may be the management of neuropathic pain. In many cases neuropathic pain is the primary complaint and must therefore be considered a goal of care. A treatment paradigm for neuropathic pain is highlighted in Fig. 6.1.

6.4 Diabetic Autonomic Neuropathy

The autonomic nervous system has widespread connections through the body, allowing for protean manifestations of clinical involvement in diabetes. As a consequence, many details about the development of autonomic neuropathy may vary between studies of different organ systems, and recommendations may differ as well. Therefore, this chapter will attempt to highlight some of the major factors involved in the development of autonomic neuropathy, the clinical symptoms and signs, diagnosis, and finally management of disease. Autonomic neuropathy is often considered one of the most debilitating complications of diabetes. Different terminology in the literature may refer to this as diabetic autonomic neuropathy (DAN) or diabetic cardiac autonomic neuropathy (CAN). The frequency with which autonomic neuropathy is reported will depend entirely on what organ system is studied and what tests are used to diagnose the problem [45–49]. Damage to the autonomic nervous system may result in cardiovascular, genitourinary, gastrointestinal, pupillary, sudomotor, and neuroendocrine dysfunction [45–49]. As with most other diabetic neuropathies, the incidence of autonomic neuropathy correlates with the duration of diabetes and with glycemic control [9, 49]. Gastrointestinal autonomic neuropathy may have widespread organ involvement and can have symptoms that range from delayed gastric emptying and gastroparesis, gastroesophageal reflux disease, chronic diarrhea, or constipation [50]. Gastroparesis may present with symptoms of nausea, vomiting, food intolerance, early satiety, bloating, or abdominal pain [51–53]. Unfortunately, the onset of gastroparesis often leads to a general worsening of glycemic control because of irregular food and medication absorption. Gastroparesis therefore is a very challenging complication for those individuals that take insulin for glycemic control, because the variable absorption of food results in unpredictable periods of hypoglycemia and hyperglycemia. Other manifestations of autonomic neuropathy involving the GI tract range from heartburn, regurgitation, bronchospasm, laryngitis, or chronic cough related to reflux [54]. Some individuals develop a diarrhea that is typically nocturnal and profuse and can be associated with fecal incontinence [54]. Some individuals with significant gastrointestinal dysfunction may also progress to chronic constipation [54].

Cardiovascular autonomic involvement in diabetes is considered a very serious complication because of the associated increase in both morbidity and mortality. The first clinical sign of cardiac autonomic neuropathy may be a resting tachycardia. Over time the autonomic neuropathy may progress to clinically significant exercise intolerance, orthostatic hypotension, syncope, intraoperative cardiovascular instability (and risk of arrhythmias), and, finally, silent myocardial infarction [9, 49]. There is significant data noting the increased risk of mortality in individuals with cardiac autonomic neuropathy [47, 55]. In the Action to Control Cardiovascular Risk in Type 2 Diabetes (ACCORD) trial, it is noted that individuals with cardiovascular autonomic neuropathy have a two to three times higher risk of mortality compared to those individuals with diabetes that did not have cardiac autonomic neuropathy [56]. Orthostatic hypotension, another cardinal manifestation of cardiovascular autonomic neuropathy, is defined as a fall in systolic blood pressure of ≥20 mmHg or diastolic fall in blood pressure of ≥10 mmHg [57].

Making a diagnosis of orthostatic hypotension is sometimes challenging because of the widespread use of medications such as antidepressants, antihypertensives, and bladder antispasmodics that can cause orthostatic hypotension. It is important to differentiate between diabetic autonomic neuropathy and medication effect because of the significantly lower mortality rates of orthostatic hypotension in the setting of medication side effects [58].

The most frequently reported symptom of genitourinary autonomic neuropathy is sexual dysfunction. In men, retrograde ejaculation and erectile dysfunction may be seen in up to 50% of men with diabetes [59, 60]. In women, genitourinary autonomic dysfunction may manifest as diminished vaginal lubrication and painful intercourse (dyspareunia) [60]. Autonomic dysfunction of the bladder due to diabetes may present with urgency and frequency and in some cases may progress to diminished capacity to sense bladder fullness, resulting in urinary retention and overflow incontinence [61].

Sweating is under the control of the thermoregulatory system, and although complex central and peripheral pathways are involved in homeostatic maintenance, the peripheral sympathetic cholinergic system is the system most frequently involved in patients with diabetes. A length-dependent neuropathy results in a loss of sweat function in the same distribution as sensory loss in somatic neuropathies [33]. Sudomotor nerve dysfunction results in decreased function of sweat glands, resulting in dry, cracked skin that may provide a portal to infection [62]. As the sudomotor neuropathy worsens, thermoregulatory capacity is impaired due to inadequate capacity to cool the system. As this occurs, a proximal hyperhidrosis is a common compensatory mechanism and also a common patient complaint [63, 64].

One major patient concern is a loss of vision. Although vision loss in diabetes is typically associated with retinopathy, autonomic nervous system damage from diabetes can result in pupillary dysfunction. Of particular concern to patients is impaired pupillary dark adaptation, which will directly impair nocturnal driving thereby creating a substantial loss of independence [65].

One of the most feared complications of diabetes is the hypoglycemia associated with autonomic failure [66, 67]. With repeated cycles of hypoglycemia, the neuroendocrine response to future episodes of hypoglycemia is diminished resulting in less release of epinephrine and glucagon. The diminished neuroendocrine response results in impaired glucose counter regulatory mechanisms and further hypoglycemia unawareness, typically resulting in more profound and more frequent hypoglycemia.

6.4.1 Diagnosis

Diabetic autonomic neuropathy is diagnosed by a combination of symptoms and signs of a particular end-organ dysfunction that is due to autonomic nerve damage. This is supplemented with selected confirmatory testing [68, 69]. For example, a diagnosis of diabetic gastroparesis is made when someone presents with chronic nausea and vomiting of food eaten earlier, often largely undigested. Gastroparesis is then confirmed by a gastric emptying study [53]. If cardiovascular autonomic neuropathy develops, one of the earliest clinical signs is a resting tachycardia [32, 49]. There may be no associated clinical symptoms. A diagnosis of early diabetic autonomic neuropathy can still be made through a battery of autonomic function tests. Autonomic testing typically includes measures of cardiovascular parasympathetic function (the heart rate response to deep breathing, the heart rate response to a Valsalva maneuver, the heart rate response to standing) and measures of cardiovascular sympathetic adrenergic function (the beat-to-beat blood pressure response to a Valsalva maneuver, the blood pressure response to standing or tilt table testing) and the tests of peripheral sympathetic cholinergic function (the sudomotor response to quantitative sudomotor axon reflex testing, thermoregulatory sweat testing, or other sweat tests) [70, 71].

6.4.2 Management

It is often very challenging to manage the symptoms associated with diabetic autonomic neuropathy. There is no cure, and treatment revolves around prevention of disease progression through management of diabetes and associated risk factors. Symptomatic management will vary depending on the end-organ involvement. Although a brief overview is provided, readers are referred to several excellent publications for more in-depth information on management [53, 54, 72–75].

The major symptoms of diabetic autonomic dysfunction that necessitate management include orthostatic intolerance, gastrointestinal dysfunction, sudomotor abnormalities, and urologic and sexual dysfunction. A primary educational objective is to establish realistic treatment goals with patients; expectations should target symptom improvement but that the disease is unlikely to be cured, and long-term therapy may be required.

Orthostatic symptoms, such as lightheadedness, dizziness, and pre-syncope, are typically associated with orthostatic hypotension. More complex cases with significant orthostatic blood pressure drops often require a combination of pharmacologic and non-pharmacologic therapies. The initial therapy is non-pharmacologic and includes patient education about positional changes, heat exertion worsening the orthostatic hypotension, adequate fluid and salt intake to improve symptoms, and compression stockings (which need to be considered cautiously in the setting of a diabetic peripheral neuropathy—as they can cause pressure ulcers in individuals unable to detect painful pressure). Pharmacologic treatments may be required and could include a volume-expanding agent (such as fludrocortisone) or a sympathomimetic agent (such as midodrine, droxidopa, or ephedrine). However, all agents need to be considered very cautiously in someone with diabetic autonomic neuropathy as this is a high comorbid risk of cardiac disease. The combination of pharmacologic and non-pharmacologic therapy can improve symptoms of orthostatic intolerance in many patients. There are also many other comorbid complications of diabetes such as renal failure, cardiac disease, and retinopathy that may complicate pharmacologic treatment of orthostatic hypotension.

Treatment of other autonomic neuropathies, such as gastrointestinal disorders, should address the particular symptom involved: gastroparesis, constipation, or diarrhea. Treatment of sexual and urogenital dysfunction often requires a combination of pharmacologic intervention and lifestyle modifications for peak effectiveness [53, 76].

6.5 Treatment-Induced Neuropathy of Diabetes

Treatment-induced neuropathy of diabetes (TIND), also referred to as "insulin neuritis, acute painful diabetic neuropathy, or diabetic neuropathic cachexia" in the literature, was initially described in 1933 [77]. This is a neuropathy precipitated by a rapid improvement in glycemic control in patients with chronic hyperglycemia. More recently this has been referred to as treatment-induced neuropathy of diabetes because the change in glycemic control can be precipitated by oral hypoglycemic mediations, diet control, or insulin [78].

The most typical clinical presentation of TIND occurs 2–8 weeks after a significant improvement in glycemic control (significantly defined for this problem as a decrease in the $HbA1_C$ by 3 percentage points or more within 3 months) [79]. In one study of patients with diabetic neuropathy, approximately >10% of all cases had their neuropathy precipitated by an abrupt improvement in glycemic control, consistent with TIND [79]. In TIND, the neuropathy is primarily a painful length-dependent small fiber neuropathy with autonomic features. On rare occasions patients may present with a more mixed large and small fiber neuropathy. The neuropathic pain is typically severe and described as both burning and shooting in nature, that is, length dependent. There is a correlation between the magnitude of glycemic change and the cutaneous territory affected by the neuropathy. The associated autonomic features of TIND may result in orthostatic hypotension, gastroparesis, nausea, sexual dysfunction, and sudomotor dysfunction [78, 80]. TIND is considered a diffuse microvascular disease because there are frequently both retinal and renal complications that occur simultaneously. The vast majority of patients with TIND also have clinically significant worsening of their retinopathy (or develop retinopathy) within 6 months of the change in glycemic control [80]. In the original Diabetes Control and Complications Trial (DCCT), this was described as early worsening retinopathy [81]. Another microvascular tissue bed that is typically involved is the kidney. In patients that develop TIND, there is a significant increase in microalbuminuria [80]. In more extreme cases, there is a significant progression of disease to renal failure [82].

6.5.1 Diagnosis

TIND is often very challenging to detect, primarily because most clinicians do not consider it in the differential diagnosis for an acute neuropathy in diabetes. TIND is associated with a rapid improvement in glycemic control that occurs 2–8 weeks prior to the onset of symptoms [79]. In most cases, the $HbA1_C$ will decrease by 3 points or more in 3 months. There is a correlation between the magnitude of the decrease in the $HbA1_C$ and symptom severity [79]. In many cases of TIND, the clinical presentation is typically a small fiber sensory or autonomic neuropathy. Therefore traditional neurophysiologic testing with electromyography and nerve conduction studies may not identify any nerve damage. Both autonomic testing and skin biopsy analysis of intraepidermal nerve fiber density are helpful in confirming the diagnosis [78].

6.5.2 Management

A critical component of TIND management is education of the patient about the relationship between significant fluctuations in glycemic control and adverse outcomes. In individuals that develop TIND, there is often a very significant painful small fiber neuropathy that occurs. Patients with TIND that relax their glycemic control may have some improvement in neuropathic pain which creates a significant negative feedback loop that results in sustained hyperglycemia [77, 83]. In this circumstance many of the individuals are then at risk for developing TIND more than once, and recurrent episodes tend to carry far more comorbid complications [82]. Therefore, individuals with TIND are counseled on maintaining a stable $HbA1_C$ and avoiding any major, and rapid, changes to glycemic control.

Management of neuropathic pain is common in individuals with TIND [78]. In many cases, the pain is so severe that polypharmacy is required. Although autonomic dysfunction is common, in most cases, it is only of mild to moderate severity, and symptoms of pain are the primary concern. However, the use of tricyclic antidepressants as treatment of the painful neuropathy may worsen a resting tachycardia or exacerbate orthostatic hypotension, so it should be used with caution.

6.6 Diabetic Radiculoplexus Neuropathy

The classic description of "diabetic amyotrophy" is the acute to subacute onset of lumbosacral or cervical nerve involvement in individuals over 50 years of age, more commonly in men. The radiculoplexus neuropathies are much more common in individuals with type 2 diabetes and have been reported prior to the diagnoses of diabetes [84, 85].

Diabetic lumbosacral radiculoplexus neuropathy (DLRPN) typically presents with severe unilateral pain in the thigh, hip, and back that extends to the entire leg and may move to the contralateral limb over a period of months (or an equivalent distribution in the arm if cervical plexus is involved). The pain is severe and is typically what brings an individual with DLRPN to medical attention. The pain is followed by a pattern of progressive muscle weakness in the same limb. Examination findings will evolve over time with the initial presentation having minor sensory change but then progressing to diffuse muscle weakness, muscle atrophy, and loss of tendon reflexes. Symptoms may progress over 18 months but will eventually stabilize. Most patients will experience some degree of improvement in both pain and weakness, but it is often incomplete [85–87]. The majority of cases present with lumbosacral involvement, but approximately 1/3 of cases present as a cervical radiculoplexus neuropathy [85]. In some situations, the disease progresses from one limb to a contralateral limb or from the lower body to the upper body after the initial injury has stabilized or is starting to improve [88]. The pathology from nerve and muscle biopsies reveals that DLRPN is a microvasculitis resulting in ischemic injury to local nerves that is the cause of the disorder [85, 89].

6.6.1 Diagnosis

The diagnosis is typically a clinical one, with some supporting neurophysiology as needed. The presence of severe pain in one limb, followed by motor weakness in someone with diabetes, should always raise a suspicion for DLRPN. Neurophysiology will show multifocal involvement of nerves, roots, and nerve plexus with abnormalities detected in compound muscle action potential amplitudes. Imaging of the involved neural plexus by MRI may reveal nerve root enhancement, while testing cerebral spinal fluid may demonstrate an elevated protein with normal cell count [85].

6.6.2 Management

Although care is generally targeted toward pain control and physical therapy for the associated weakness, a number of studies have suggested the use of immunosuppressive therapies in the role of treating the underlying microvasculitis. The use of both methylprednisolone and immune globulin has been attempted in clinical trials and, although promising, has not achieved sufficient evidence for routine clinical use (although they do suggest that pain resolves more quickly) [88, 90, 91].

6.7 Mononeuropathies

Mononeuropathies, such as carpal tunnel syndrome, ulnar neuropathy at the elbow, and tibial or peroneal neuropathies in the leg, are more common in individuals with diabetes than in the general population [92, 93]. Although most of the treatments for these disorders mimic the treatment in those individuals without neuropathy, there are a few key issues in the diagnosis and management of these entrapment neuropathies that require clarification.

6.7.1 Diagnosis

The diagnosis is typically a combination of the clinical history, examination findings, and electrodiagnostic studies. However, in individuals with an underlying diabetic neuropathy, the ability to detect a mononeuropathy by electrodiagnostic study is more challenging and can create diagnostic confusion [93, 94]. Experienced electromyographers with side-to-side limb comparisons may be necessary, but, as always, clinical judgment must guide management decisions.

6.7.2 Management

In general, conservative management of upper extremity mononeuropathies is the initial treatment and includes splinting and avoidance of nerve injury. If unsuccessful, then corticosteroid injection or nerve decompression can be considered [95–98]. In cases of lower extremity nerve compression syndromes, the recommended guidance is less clear. Decompression of lower extremity mononeuropathies is not always recommended, and there is no role for splinting of the legs [99]. Education about avoidance of pressure on the common peroneal nerve (i.e., avoidance of leg crossing) is important in preventing progression.

6.8 Cranial Neuropathies

A diabetic sixth nerve palsy is one of the classic clinical syndromes. Diabetes is one of the more common causes of cranial nerve palsy and may present as a oculomotor or trochlear nerve palsy, in addition to the more common sixth nerve palsy. In a patient with a third nerve (oculomotor) palsy, the clinical presentation include the relatively acute onset of pain behind or around the eye, followed by weakness or paralysis of the oculomotor muscles, with ipsilateral ptosis. In individuals presenting with either a fourth or sixth nerve palsy, there is a more restricted eye movement abnormality. Microvascular ischemia is the cause of the diabetic cranial neuropathies [15, 92, 100]. There are many case reports of facial (seventh) nerve palsies in patients with diabetes, but it is not clear at this time if the facial nerve palsy is related pathophysiologically to diabetes in the way the other cranial neuropathies are [101].

6.8.1 Diagnosis

The diagnosis is made by history and examination. There is no need for ancillary testing to confirm a diagnosis, although imaging of the brain would be considered appropriate in ruling out other potential causes of cranial neuropathy.

6.8.2 Management

In most cases of diabetic cranial neuropathy, there is a gradual improvement and, in some cases, completely resolve. Clinical care is a generally supportive management of the diplopia with a focus on diabetes control.

6.9 Screening for Diabetic Neuropathy

The American Diabetes Association has established screening guidelines for all individuals with diabetes. The most recent recommendations, published in 2016, recommend that all patients with type 2 diabetes should be assessed for diabetic peripheral neuropathy starting at the time of diagnosis and at least annually thereafter. For individuals with type 1 diabetes, the recommendations are evaluation 5 years after the time of diagnosis of diabetes and then at least annually thereafter [102]. The guidelines on screening for distal symmetric polyneuropathy include a careful history with a focus on sensory changes or weakness and a neurologic examination focusing on sensory, strength, and reflex testing. In patients with diabetes who already have a distal symmetric polyneuropathy or other microvascular complications (renal, retinal), the then symptoms and signs of autonomic neuropathy should also be investigated.

6.10 Conclusion

The initial management of any diabetic neuropathy will involve a focus on the general guidelines for diabetes control. Glucose control is of particular importance, particularly in those individuals with type 1 diabetes [9]. The mechanisms behind neuropathy development in type 2 diabetes are more likely to be multifactorial, so treatment of other associated risk factors for neuropathy development is critical and includes lipid control, blood pressure management, and smoking cessation [9]. Regular physical activity is an important, and often overlooked, aspect of diabetes care.

Many patients with neuropathy will have associated neuropathic pain. The choice of treatment will involve a balanced discussion on pain relief while simultaneously attempting to minimize side effects. All patients with distal symmetric neuropathies should be informed about foot care safety and fall risks. Patients with distal neuropathy should examine their feet every day and contact their physician immediately if there is any sign of injury or infection.

References

1. Callaghan BC, Cheng H, Stables CL, Smith AL, Feldman EL. Diabetic neuropathy: clinical manifestations and current treatments. Lancet Neurol. 2012;11:521–34.
2. CDC, National diabetes fact sheet: national estimates and general information on diabetes and prediabetes in the United States, 2011, CDC, Editor. 2011, Department of Health and Human Services, Centers for Disease Control and Prevention.
3. DCCT Research Group. Factors in development of diabetic neuropathy. Baseline analysis of neuropathy in feasibility phase of Diabetes Control and Complications Trial (DCCT). The DCCT Research Group. Diabetes. 1988;37:476–81.
4. Diabetes Control and Complications Trial Research Group, Nathan DM, Genuth S, Lachin J, Cleary P, Crofford O, et al. The effect of intensive treatment of diabetes on the development and progression of long-term complications in insulin-dependent diabetes mellitus. N Engl J Med. 1993;329:977–86.

5. Pop-Busui R, Herman WH, Feldman EL, Low PA, Martin CL, Cleary PA, et al. DCCT and EDIC studies in type 1 diabetes: lessons for diabetic neuropathy regarding metabolic memory and natural history. Curr Diab Rep. 2010;10:276–82.
6. Ismail-Beigi F, Craven T, Banerji MA, Basile J, Calles J, Cohen RM, et al. Effect of intensive treatment of hyperglycaemia on microvascular outcomes in type 2 diabetes: an analysis of the ACCORD randomised trial. Lancet. 2010;376:419–30.
7. Callaghan B, Feldman E. The metabolic syndrome and neuropathy: therapeutic challenges and opportunities. Ann Neurol. 2013;74:397–403.
8. Tesfaye S, Selvarajah D. Advances in the epidemiology, pathogenesis and management of diabetic peripheral neuropathy. Diabetes Metab Res Rev. 2012;28(Suppl 1):8–14.
9. Pop-Busui R, Boulton AJ, Feldman EL, Bril V, Freeman R, Malik RA, et al. Diabetic neuropathy: a position statement by the american diabetes association. Diabetes Care. 2017;40:136–54.
10. Callaghan BC, Little AA, Feldman EL, Hughes RA. Enhanced glucose control for preventing and treating diabetic neuropathy. Cochrane Database Syst Rev. 2012;6:CD007543.
11. Callaghan BC, Gao L, Li Y, Zhou X, Reynolds E, Banerjee M, et al. Diabetes and obesity are the main metabolic drivers of peripheral neuropathy. Ann Clin Transl Neurol. 2018;5:397–405.
12. Sumner CJ, Sheth S, Griffin JW, Cornblath DR, Polydefkis M. The spectrum of neuropathy in diabetes and impaired glucose tolerance. Neurology. 2003;60:108–11.
13. Cortez M, Singleton JR, Smith AG. Glucose intolerance, metabolic syndrome, and neuropathy. Handb Clin Neurol. 2014;126:109–22.
14. Peltier A, Goutman SA, Callaghan BC. Painful diabetic neuropathy. BMJ. 2014;348:g1799.
15. Said G. Diabetic neuropathy – a review. Nat Clin Pract Neurol. 2007;3:331–40.
16. Clements RS Jr, Bell DS. Complications of diabetes. Prevalence, detection, current treatment, and prognosis. Am J Med. 1985;79:2–7.
17. Zochodne DW. Diabetic polyneuropathy: an update. Curr Opin Neurol. 2008;21:527–33.
18. Boulton AJ, Vinik AI, Arezzo JC, Bril V, Feldman EL, Freeman R, et al. Diabetic neuropathies: a statement by the American Diabetes Association. Diabetes Care. 2005;28:956–62.
19. Ziegler D. Treatment of diabetic neuropathy and neuropathic pain: how far have we come? Diabetes Care. 2008;31(Suppl 2):S255–61.
20. Tesfaye S, Vileikyte L, Rayman G, Sindrup SH, Perkins BA, Baconja M, et al. Painful diabetic peripheral neuropathy: consensus recommendations on diagnosis, assessment and management. Diabetes Metab Res Rev. 2011;27:629–38.
21. Kumar S, Fernando DJ, Veves A, Knowles EA, Young MJ, Boulton AJ. Semmes-Weinstein monofilaments: a simple, effective and inexpensive screening device for identifying diabetic patients at risk of foot ulceration. Diabetes Res Clin Pract. 1991;13:63–7.
22. Callaghan B, McCammon R, Kerber K, Xu X, Langa KM, Feldman E. Tests and expenditures in the initial evaluation of peripheral neuropathy. Arch Intern Med. 2012;172:127–32.
23. Callaghan BC, Burke JF, Rodgers A, McCammon R, Langa KM, Feldman EL, et al. Expenditures in the elderly with peripheral neuropathy: where should we focus cost-control efforts? Neurol Clin Pract. 2013;3:421–30.
24. Dworkin RH, O'Connor AB, Audette J, Baron R, Gourlay GK, Haanpää ML, et al. Recommendations for the pharmacological management of neuropathic pain: an overview and literature update. Mayo Clin Proc. 2010;85:S3–S14.
25. Bohlega S, Alsaadi T, Amir A, Hosny H, Karawagh AM, Moulin D, et al. Guidelines for the pharmacological treatment of peripheral neuropathic pain: expert panel recommendations for the middle East region. J Int Med Res. 2010;38:295–317.
26. Attal N, Cruccu G, Baron R, Haanpää M, Hansson P, Jensen TS, et al. EFNS guidelines on the pharmacological treatment of neuropathic pain: 2010 revision. Eur J Neurol. 2010;17:1113–e88.
27. Hovaguimian A, Gibbons CH. Clinical approach to the treatment of painful diabetic neuropathy. Ther Adv Endocrinol Metab. 2011;2:27–38.
28. Gibbons CH. Small fiber neuropathies. Continuum (Minneap Minn). 2014;20(5 Peripheral Nervous System Disorders):1398–412.
29. Dyck PJ, Norell JE, Tritschler H, Schuette K, Samigullin R, Ziegler D, et al. Challenges in design of multicenter trials: end points assessed longitudinally for change and monotonicity. Diabetes Care. 2007;30:2619–25.
30. Sommer C. Painful neuropathies. Curr Opin Neurol. 2003;16:623–8.
31. Zilliox LA, Ruby SK, Singh S, Zhan M, Russell JW. Clinical neuropathy scales in neuropathy associated with impaired glucose tolerance. J Diabetes Complicat. 2015;29:372–7.
32. Low VA, Sandroni P, Fealey RD, Low PA. Detection of small-fiber neuropathy by sudomotor testing. Muscle Nerve. 2006;34:57–61.
33. Fealey RD, Low PA, Thomas JE. Thermoregulatory sweating abnormalities in diabetes mellitus. Mayo Clin Proc. 1989;64:617–28.
34. Stewart JD. Sweating abnormalities and other autonomic disorders in diabetes mellitus. Mayo Clin Proc. 1989;64:712–5.
35. Lauria G, Hsieh ST, Johansson O, Kennedy WR, Leger JM, Mellgren SI, et al. European Federation of Neurological Societies/Peripheral Nerve Society Guideline on the use of skin biopsy in the diagnosis of small fiber neuropathy. Report of a joint task force of the European Federation of Neurological Societies and the Peripheral Nerve Society. Eur J Neurol. 2010;17:903–12, e44–9.

36. Lauria G, Bakkers M, Schmitz C, Lombardi R, Penza P, Devigili G, et al. Intraepidermal nerve fiber density at the distal leg: a worldwide normative reference study. J Peripher Nerv Syst. 2010;15:202–7.
37. Boruchow SA, Gibbons CH. Utility of skin biopsy in management of small fiber neuropathy. Muscle Nerve. 2013;48:877–82.
38. Illigens BM, Gibbons CH. Sweat testing to evaluate autonomic function. Clin Auton Res. 2009;19(2):79–87.
39. Freeman R, Chase KP, Risk MR. Quantitative sensory testing cannot differentiate simulated sensory loss from sensory neuropathy. Neurology. 2003;60:465–70.
40. Gruener G, Dyck PJ. Quantitative sensory testing: methodology, applications, and future directions. [Review] [45 refs]. J Clin Neurophysiol. 1994;11:568–83.
41. Dyck PJ, Zimmerman IR, Johnson DM, Gillen D, Hokanson JL, Karnes JL, et al. A standard test of heat-pain responses using CASE IV. J Neurol Sci. 1996;136:54–63.
42. Dyck PJ, Larson TS, O'Brien PC, Velosa JA. Patterns of quantitative sensation testing of hypoesthesia and hyperalgesia are predictive of diabetic polyneuropathy: a study of three cohorts. Nerve growth factor study group. Diabetes Care. 2000;23:510–7.
43. Brown MJ, Bird SJ, Watling S, Kaleta H, Hayes L, Eckert S, et al. Natural progression of diabetic peripheral neuropathy in the zenarestat study population. Diabetes Care. 2004;27:1153–9.
44. Bril V, Hirose T, Tomioka S, Buchanan R, Ranirestat Study Group. Ranirestat for the management of diabetic sensorimotor polyneuropathy. Diabetes Care. 2009;32:1256–60.
45. Low PA, Benrud-Larson LM, Sletten DM, Opfer-Gehrking TL, Weigand SD, O'Brien PC, et al. Autonomic Symptoms and Diabetic Neuropathy: a population-based study. Diabetes Care. 2004;27:2942–7.
46. Duby JJ, Campbell RK, Setter SM, White JR, Rasmussen KA. Diabetic neuropathy: an intensive review. Am J Health Syst Pharm. 2004;61:160–73.
47. Maser RE, Lenhard MJ. Cardiovascular autonomic neuropathy due to diabetes mellitus: clinical manifestations, consequences, and treatment. J Clin Endocrinol Metab. 2005;90:5896–903.
48. Astrup AS, Tarnow L, Rossing P, Hansen BV, Hilsted J, Parving HH. Cardiac autonomic neuropathy predicts cardiovascular morbidity and mortality in type 1 diabetic patients with diabetic nephropathy. Diabetes Care. 2006;29:334–9.
49. Spallone V, Ziegler D, Freeman R, Bernardi L, Frontoni S, Pop-Busui R, et al. Cardiovascular autonomic neuropathy in diabetes: clinical impact, assessment, diagnosis, and management. Diabetes Metab Res Rev. 2011;27:639–53.
50. Chandrasekharan B, Srinivasan S. Diabetes and the enteric nervous system. Neurogastroenterol Motil. 2007;19:951–60.
51. Enck P, Frieling T. Pathophysiology of diabetic gastroparesis. Diabetes. 1997;46(Suppl 2):S77–81.
52. Kong MF, Horowitz M, Jones KL, Wishart JM, Harding PE. Natural history of diabetic gastroparesis. Diabetes Care. 1999;22:503–7.
53. Horvath VJ, Izbéki F, Lengyel C, Kempler P, Várkonyi T. Diabetic gastroparesis: functional/morphologic background, diagnosis, and treatment options. Curr Diab Rep. 2014;14:527.
54. Sellin JH, Chang EB. Therapy Insight: gastrointestinal complications of diabetes – pathophysiology and management. Nat Clin Pract Gastroenterol Hepatol. 2008;5:162–71.
55. Dimitropoulos G, Tahrani AA, Stevens MJ. Cardiac autonomic neuropathy in patients with diabetes mellitus. World J Diabetes. 2014;5:17–39.
56. Nilsson PM. ACCORD and Risk-Factor Control in Type 2 Diabetes. N Engl J Med. 2010;362:1628–30.
57. Freeman R, Wieling W, Axelrod FB, Benditt DG, Benarroch E, Biaggioni I, et al. Consensus statement on the definition of orthostatic hypotension, neurally mediated syncope and the postural tachycardia syndrome. Auton Neurosci. 2011;161:46–8.
58. Gibbons CH, Freeman R. Clinical implications of delayed orthostatic hypotension: a 10-year follow-up study. Neurology. 2015;85:1362–7.
59. Vinik AI, Maser RE, Mitchell BD, Freeman R. Diabetic autonomic neuropathy. Diabetes Care. 2003;26:1553–79.
60. Fedele D. Therapy Insight: sexual and bladder dysfunction associated with diabetes mellitus. Nat Clin Pract Urol. 2005;2:282–90.
61. Van Den Eeden SK, Sarma AV, Rutledge BN, Cleary PA, Kusek JW, Nyberg LM, et al. Effect of intensive glycemic control and diabetes complications on lower urinary tract symptoms in men with type 1 diabetes: Diabetes Control and Complications Trial/Epidemiology of Diabetes Interventions and Complications (DCCT/EDIC) study. Diabetes Care. 2009;32:664–70.
62. Luo KR, Chao CC, Hsieh PC, Lue JH, Hsieh ST. Effect of glycemic control on sudomotor denervation in type 2 diabetes. Diabetes Care. 2012;35:612–6.
63. Low PA. Autonomic neuropathies. Curr Opin Neurol. 2002;15:605–9.
64. Cheshire WP, Freeman R. Disorders of sweating. Semin Neurol. 2003;23:399–406.
65. Lanting P, Bos JE, Aartsen J, Schuman L, Reichert-Thoen J, Heimans JJ. Assessment of pupillary light reflex latency and darkness adapted pupil size in control subjects and in diabetic patients with and without cardiovascular autonomic neuropathy. J Neurol Neurosurg Psychiatry. 1990;53:912–4.
66. Cryer PE. Mechanisms of hypoglycemia-associated autonomic failure in diabetes. N Engl J Med. 2013;369:362–72.
67. Cryer PE. Hypoglycemia-associated autonomic failure in diabetes. Handb Clin Neurol. 2013;117:295–307.

68. Gibbons CH, Freeman R, Veves A. Diabetic neuropathy: a cross-sectional study of the relationships among tests of neurophysiology. Diabetes Care. 2010;33:2629–34.
69. Buijs RM, Swaab DF. Autonomic nervous system. In Handbook of clinical neurology, volume 117, 3rd series. Edinburgh: Elsevier; 2013.
70. Illigens BM, Gibbons CH. Sweat testing to evaluate autonomic function. Clin Auton Res. 2009;19:79–87.
71. Freeman R. Assessment of cardiovascular autonomic function. Clin Neurophysiol. 2006;117:716–30.
72. Tesfaye S, Boulton AJ, Dyck PJ, Freeman R, Horowitz M, Kempler P, et al. Diabetic neuropathies: update on definitions, diagnostic criteria, estimation of severity, and treatments. Diabetes Care. 2010;33:2285–93.
73. Low PA, Benarroch EE. Clinical autonomic disorders. 3rd ed. Philadelphia: Lippincott Williams & Wilkins; 2008. xii, 780 p., 16 p. of plates.
74. Freeman R. Autonomic peripheral neuropathy. Lancet. 2005;365:1259–70.
75. Freeman R. Autonomic peripheral neuropathy. Neurol Clin. 2007;25:277–301.
76. Basu A, Ryder RE. New treatment options for erectile dysfunction in patients with diabetes mellitus. Drugs. 2004;64:2667–88.
77. Caravati CM. Insulin neuritis: a case report. Va Med Monthly. 1933;59:745–6.
78. Gibbons CH, Freeman R. Treatment-induced diabetic neuropathy: a reversible painful autonomic neuropathy. Ann Neurol. 2010;67:534–41.
79. Gibbons CH, Freeman R. Treatment-induced neuropathy of diabetes: an acute, iatrogenic complication of diabetes. Brain. 2015;138:43–52.
80. Gibbons CH. Treatment induced neuropathy of diabetes: an autonomic dysrecognition syndrome. Clin Auton Res. 2016;26:1.
81. Group DR. Early worsening of diabetic retinopathy in the Diabetes Control and Complications Trial. Arch Ophthalmol. 1998;116:874–86.
82. Gibbons CH. Treatment induced neuropathy of diabetes-long term implications in type 1 diabetes. J Diabetes Complicat. 2017;31:715–20.
83. Tesfaye S, Malik R, Harris N, Jakubowski JJ, Mody C, Rennie IG, et al. Arterio-venous shunting and proliferating new vessels in acute painful neuropathy of rapid glycaemic control (insulin neuritis). Diabetologia. 1996;39:329–35.
84. Sander HW, Chokroverty S. Diabetic amyotrophy: current concepts. [Review] [55 refs]. Semin Neurol. 1996;16:173–8.
85. Dyck PJ, Windebank AJ. Diabetic and nondiabetic lumbosacral radiculoplexus neuropathies: new insights into pathophysiology and treatment. Muscle Nerve. 2002;25:477–91.
86. Callaghan BC, Price RS, Chen KS, Feldman EL. The importance of rare subtypes in diagnosis and treatment of peripheral neuropathy: a review. JAMA Neurol. 2015;72:1510–8.
87. Younger DS. Diabetic lumbosacral radiculoplexus neuropathy: a postmortem studied patient and review of the literature. J Neurol. 2011;258:1364–7.
88. Thaisetthawatkul P, Dyck PJ. Treatment of diabetic and nondiabetic lumbosacral radiculoplexus neuropathy. Curr Treat Options Neurol. 2010;12:95–9.
89. Llewelyn JG, Thomas PK, King RH. Epineurial microvasculitis in proximal diabetic neuropathy. J Neurol. 1998;245:159–65.
90. Tamburin S, Zanette G. Intravenous immunoglobulin for the treatment of diabetic lumbosacral radiculoplexus neuropathy. Pain Med. 2009;10:1476–80.
91. Tamburin S, Magrinelli F, Favaro F, Briani C, Zanette G. Long-term response of neuropathic pain to intravenous immunoglobulin in relapsing diabetic lumbosacral radiculoplexus neuropathy. A case report. Pain Pract. 2014;14:E85–90.
92. Fraser DM, Campbell IW, Ewing DJ, Clarke BF. Mononeuropathy in diabetes mellitus. Diabetes. 1979;28:96–101.
93. Albers JW, Brown MB, Sima AA, Greene DA. Frequency of median mononeuropathy in patients with mild diabetic neuropathy in the early diabetes intervention trial (EDIT). Tolrestat Study Group for Edit (Early Diabetes Intervention Trial). Muscle Nerve. 1996;19:140–6.
94. Hansson S. Segmental median nerve conduction measurements discriminate carpal tunnel syndrome from diabetic polyneuropathy. Muscle Nerve. 1995;18:445–53.
95. Vinik A, Mehrabyan A, Colen L, Boulton A. Focal entrapment neuropathies in diabetes. Diabetes Care. 2004;27:1783–8.
96. Malkoc M, Korkmaz Ö, Oltulu I, Seker A, Say F, Bulbul AM. Surgical treatment results of carpal tunnel syndrome in patients with and without type 2 diabetes mellitus. Srp Arh Celok Lek. 2014;142:675–9.
97. Gulabi D, Cecen G, Guclu B, Cecen A. Carpal tunnel release in patients with diabetes result in poorer outcome in long-term study. Eur J Orthop Surg Traumatol. 2014;24:1181–4.
98. Thomsen NO, Anderson GS, Björk J, Dahlin LB. Neurophysiological recovery 5 years after carpal tunnel release in patients with diabetes. Muscle Nerve. 2017;56:E59–64.
99. Cornblath DR, Vinik A, Feldman E, Freeman R, Boulton AJ. Surgical decompression for diabetic sensorimotor polyneuropathy. Diabetes Care. 2007;30:421–2.
100. Ross AT. Recurrent cranial nerve palsies in diabetes mellitus. Neurology. 1962;12:180–5.
101. Pecket P, Schattner A. Concurrent Bell's palsy and diabetes mellitus: a diabetic mononeuropathy? J Neurol Neurosurg Psychiatry. 1982;45:652–5.
102. Chamberlain JJ, Rhinehart AS, Shaefer CF Jr, Neuman A. Diagnosis and management of diabetes: synopsis of the 2016 American Diabetes Association standards of medical care in diabetes. Ann Intern Med. 2016;164:542–52.

7 Genetic Small Fiber Sensory Neuropathy and Channelopathy

Rosario Privitera and Praveen Anand

Abstract

Genetic neuropathies comprise a range of conditions which can include neuropathy as a predominant feature of the disease or as part of a multisystem disease. Small fiber sensory neuropathy and symptoms have been described in association with sodium channel mutations such as in the *SCN9A*, *SCN10A*, and *SCN11A* genes, which encode $Na_v1.7$, $Na_v1.8$, and $Na_v1.9$ sodium channels, respectively. Erythromelalgia is an autosomal dominant painful condition characterized by redness of the skin and intermittent burning sensation of extremities, triggered by heat or exercise, and was shown to be related to mutations of $Na_v1.7$. Paroxysmal extreme pain disorder (PEPD, first described as familial rectal pain) is an inherited syndrome with paroxysms of rectal, ocular, or submandibular pain with flushing, also linked to gain-of-function mutations of $Na_v1.7$. In addition, *SCN9A* gene variants have been found in subjects of an idiopathic small fiber neuropathy (SFN) cohort. Predominant involvement of small nerve fibers has been documented in patients with Fabry disease, familial amyloid polyneuropathy (FAP), and Tangier's disease. This chapter focuses on small fiber neuropathies and channelopathies with sensory symptoms, particularly pain, as the sole or primary feature.

Keywords

Hereditary sensory and autonomic neuropathy (HSAN) · Sodium channel mutations Erythromelalgia · Paroxysmal extreme pain disorder · Fabry disease · Familial amyloid polyneuropathy · Tangier's disease

7.1 Introduction

Inherited neuropathies are a group of neurologic disorders with heterogeneous clinical presentations and genetic causes. Dyck et al. showed in a large prospective study that in 42% of subjects with undiagnosed neuropathy, the cause of the neuropathy may be attributed to inherited disorders [1].

Small nerve fibers can be affected in purely sensory neuropathies or mixed sensory, autonomic and motor neuropathies, or part of multi-system hereditary neuropathies. Hereditary

R. Privitera · P. Anand (✉)
Peripheral Neuropathy Unit, Centre for Clinical Translation, Division of Brain Sciences, Imperial College London, Hammersmith Hospital, London, UK
e-mail: p.anand@imperial.ac.uk

S.-T. Hsieh et al. (eds.), *Small Fiber Neuropathy and Related Syndromes: Pain and Neurodegeneration*, https://doi.org/10.1007/978-981-13-3546-4_7

sensory and autonomic neuropathies (HSN/HSAN) are clinically and genetically heterogeneous disorders showing autosomal dominant or recessive inheritance pattern, with hallmark features of prominent sensory and also autonomic involvement (e.g., anhidrosis or excessive sweating, fever, blood pressure fluctuations or postural hypotension, and gastrointestinal disturbance) [2]. The sensory and autonomic neuropathies may have a juvenile to adult onset, but for some the onset is at birth, and usually associated with more complex neurological abnormalities [3–7]. Involvement of small nerve fibers has been documented in some types of Charcot-Marie-Tooth (CMT) disease and considered to be a variant phenotype of hereditary sensory neuropathy (HSN) [8, 9], as in Charcot-Marie-Tooth neuropathy type 2 (CMT2), with nerve conduction velocities in the normal or near-normal range [10]. A more complex picture is observed in multisystem hereditary neuropathies involving the central nervous system and including sensory neuropathy, inherited with an autosomal dominant or recessive pattern [11–17].

Detailed investigations assist in identification of the cause beyond the family history [18]. Nerve conduction studies have been reported as useful to distinguish between chronic acquired and familial demyelinating neuropathy [19–21]. Quantitative sensory testing (QST) and the assessment of intraepidermal nerve fibers (IENFs) in skin biopsies may help to quantify the structural changes in small sensory nerve fibers, including in patients with inherited diseases [22, 23], especially when the family history suggests a genetic basis [24]. The assessment of small nerve fibers may result in the early diagnosis of the neuropathy, as in familial amyloid polyneuropathy [25]. Pain evoked potentials (laser evoked potentials, LEPs, or contact heat evoked potentials, CHEPs) have been suggested as helpful additional investigations, in conjunction with QST and skin biopsy [26, 27].

7.2 Sodium Channels Related to Painful Neuropathy and Channelopathy

Sodium channel blockers have been used for decades in the clinic to attenuate or abolish both acute inflammatory and chronic neuropathic pain, suggesting that peripheral input driving pain perception depends at least partly on sodium channels [28]. These sodium channels are transmembrane proteins that allow the rapid influx of sodium underlying the depolarizing upstroke of action potentials in excitable cells [29–31], which plays an important role in inflammatory pain [32–34] and chronic neuropathic pain [28, 35, 36]. $Na_v1.7$, $Na_v1.8$, and $Na_v1.9$ have been identified as the key sodium channels underlying neuropathic pain and potential targets for novel selective analgesics.

Isoforms $Na_v1.7$, $Na_v1.8$, and $Na_v1.9$ are expressed in pain-conducting pathways of the peripheral nervous system and are localized in small and medium diameters nociceptors of dorsal root ganglia (DRG) and trigeminal ganglia [35, 37–41]. $Na_v1.7$ has been considered to play a crucial role in some inherited neuropathic pain conditions [42]. Novel mutations in *SCN9A* gene encoding voltage-gated sodium channel $Na_v1.7$ have been described in patients who had a previous diagnosis of idiopathic small fiber neuropathy and exhibited both symptoms suggestive of gain-of-function of the mutant sodium channels, along with loss of IENF in skin biopsies [43]. Mutations in *SCN10A* gene, encoding $Na_v1.8$ [44], and *SCN11A* gene, encoding $Na_v1.9$ [45], have also been reported as linked to human pain conditions, with alterations of thermal pain perception thresholds. Sodium channel isoforms of $Na_v1.7$ and $Na_v1.8$ have been found to be overexpressed in acquired human painful neuropathies, e.g., in painful neuromas after trauma [46, 47].

7.3 Primary Erythromelalgia (Erythermalgia)

Primary erythromelalgia is a rare autosomal dominant condition linked to mutations in *SCN9A* gene [48–58] located on chromosome 2q [49, 59], encoding the alpha subunit of $Na_v1.7$ voltage-gated sodium channel, expressed in sensory and sympathetic neurons. Clinically, the condition is characterized by attacks or episodes of symmetrical burning pain of the feet, lower legs, and sometimes hands. Pain attacks are triggered by heat or exercise [60] and relieved by cold [61]. Additional clinical features are elevated skin temperature of affected areas, with redness of the extremities [42, 62]. Pain has been reported as early as in infants (early onset), in the second decade (delayed onset) [61, 63], and in adults (adult onset) [64, 65]. The term secondary erythromelalgia is also sometimes used for acquired conditions with a similar phenotype, in which anti-inflammatory treatments may bring swift symptom relief [42, 66].

The primary erythromelalgia mutations are linked to changes in $Na_v1.7$ channel activity, including the increased responsivity to smaller than normal depolarizations, the generation of larger than normal inward sodium currents, and faster recovery from inactivation [42, 67, 68]. Although the presence of edema, redness, warmth, and pain in human erythromelalgia patients has been considered as part of an increased neurogenic vasodilatory and inflammatory response similar to those observed in animal models [67], it has been shown that mutations of sodium channel $Na_v1.7$ can produce hyperexcitability in sensory neurons and hypoexcitability in sympathetic ganglion neurons, probably due to the contribution of the activity of $Na_v1.8$ channels available for activation at depolarized membrane potentials in sensory but not sympathetic neurons [67]. This supports the view idea that erythromelalgia is associated with dysfunction in both small sensory and autonomic fibers [69, 70]. These changes may all contribute to the clinical phenotype [71] and explain why erythromelalgia is not characterized by loss of intraepidermal nerve fiber density (IENFD) in all patients [55, 72, 73]. Skin biopsy from the affected area may be useful to show a reduction in the number of small nerve fibers [55, 73] and to differentiate the condition from erythromelalgia secondary to vasculitis [70].

The treatment of patients with erythromelalgia is very challenging [61]; some patients are responsive to treatment with the sodium channel inhibitors such as carbamazepine [74] or mexiletine [75], though most patients do not respond to significantly to pharmacotherapy.

7.4 Paroxysmal Extreme Pain Disorder (Familial Rectal Pain)

Distinct set of mutations in *SCN9A* underlie paroxysmal extreme pain disorder (PEPD, first described as familial rectal pain), a clinical condition inherited with a dominant pattern, and clinically characterized by paroxysms of pain and autonomic dysfunction [76, 77]. The clinical features of PEPD include sudden attacks of excruciating pain that affect the rectum, genitalia, face, trunk, and limbs, along with signs of autonomic dysfunction such as skin flushing and pupillary abnormalities [78]. In a 2-day-old baby, paroxysmal attacks included bradycardia, bronchospasm, and vasomotor fits, triggered by diaper changes or wiping the perineal areas [79]. Attacks may also be precipitated by yawning, crying, ingestion of cold drinks, or acidic or spicy foods but also occur spontaneously [42]. The disorder starts in the neonatal period [80], and it is most likely to be misdiagnosed as epilepsy and particularly in infancy as anoxic reflex seizures [81]. Severe pain is evident in infancy, persists throughout life, and may represent a life-threatening condition if the onset is during the neonatal period [79, 81].

Mutations of $Na_v1.7$ are linked to many of the PEPD cases reported but in some cases do not show this linkage to Nav1.7, suggesting the involvement of another target [76]. PEPD-linked *SCN9A* mutations promote prolonged action potentials and repetitive neuronal firing in response to provoking stimuli, such as stretching and exposure to cold temperatures, due to defective fast inactivation of the voltage-dependent sodium channel.

In patients with acquired or other adult-onset idiopathic conditions, rectal hypersensitivity has been linked to sensitization, sprouting, and phenotypic changes in sensory fibers. In rectal biopsies from such idiopathic cases, an increase of $Na_v1.7$ and transient receptor potential vanilloid type 1 (TRPV1) expressing nerve fibers has been reported and attributed to an increase in nerve growth factor (NGF), which acts as a regulating factor for nerve sprouting and increased expression of both $Na_v1.7$ [82] (Fig. 7.1) and TRPV1 [83]. $Na_v1.7$ channel expression was "paradoxically" decreased in rectal biopsies from patients with PEPD, suggesting that PEPD-linked gain-of-function mutations was reflected in altered (reduced) immunoreactivity of the mutated protein [82] (Fig. 7.2).

Carbamazepine and topiramate have been shown to be effective in relieving pain in some patients with PEPD, while other pain medications have been associated with a partial response [84].

7.5 Congenital Insensitivity to Pain Linked to $Na_v1.7$ Mutations

The congenital insensitivity to pain without anhidrosis (CIP without anhidrosis or HSAN V) is a very rare phenotype. This condition has been linked to an autosomal recessive condition linked to the gene *SCN9A*, with loss of function of $Na_v1.7$ [85–88].

Patients with congenital insensitivity to pain show only an impaired pain perception, but otherwise, they are mostly healthy, including sweating (unlike congenital inability to sense pain with anhidrosis (CIPA) or HSAN IV, which is attributed to a loss of function of the *TRKA* gene, the NGF high-affinity receptor). In some cases, congenital insensitivity to pain can be associated with skeletal lesions [89].

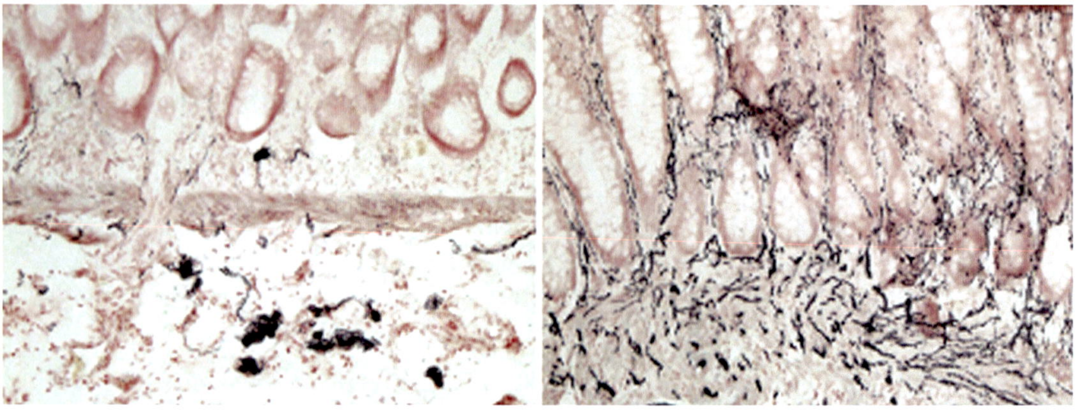

Fig. 7.1 $Na_v1.7$ expression in rectal mucosal and submucosal nerve fibers and cell bodies. $Na_v1.7$ immunoreactive rectal mucosal and submucosal nerve fibers and cell bodies in control subject (left panel) and patient with idiopathic rectal pain and hypersensitivity (right panel), magnification × 10

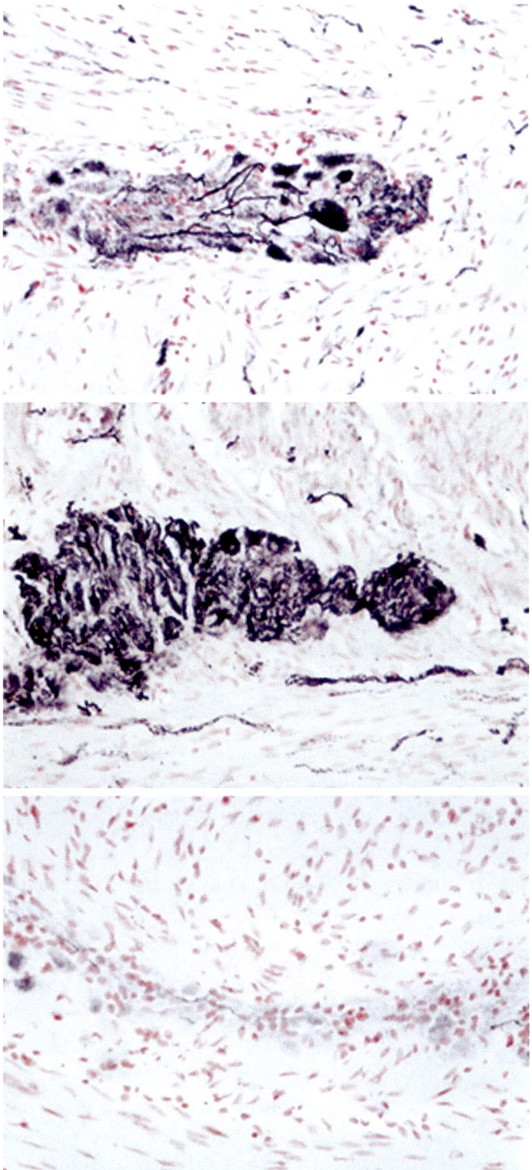

Fig. 7.2 Expression of $Na_V1.7$ in the myenteric plexus. $Na_V1.7$ immunoreactivity in the myenteric plexus of normal rectum (top), patient with idiopathic rectal hypersensitivity (middle), and patient with paroxysmal extreme pain disorder (bottom), magnification × 20

7.6 Fabry Disease

Fabry disease is a rare X-linked glycosphingolipid disorder caused by mutations in the lysosomal glycohydrolase, i.e., α-galactosidase A (α-Gal A). The enzyme defect leads to progressive lysosomal deposition of globotriaosylceramide (GL-3) in many cell types, with resulting in increased risk of developing ischemic stroke, small fiber peripheral neuropathy, cardiac dysfunction, and chronic kidney disease [90]. Despite being X-linked, females can also be affected, and both males and females may have reduced life expectancy.

Phenotypically, Fabry disease can be divided into a classical or nonclassical form. The classical form usually has its onset in childhood or adolescence [91] in men who have no residual enzyme activity and often exhibits Fabry-specific symptoms such as severe pain in the extremities, the appearance of cutaneous vascular lesions (angiokeratomas), sweating abnormalities (anhidrosis, hypohidrosis, and rarely hyperhidrosis), characteristic corneal and lenticular opacities (cornea verticillata), and proteinuria [92]. The nonclassical form of Fabry disease is characterized by late onset and variable disease course in which patients are generally less severely affected and show limited disease manifestations [93, 94].

Neuropathic pain is a disabling and frequently observed feature in the majority of patients affected by both early-onset [95, 96] and late-onset [97, 98] Fabry diseases. Pain is typically episodic and usually described as "burning" pain, triggered by physical activity and thermal stimuli and mainly affecting hands and feet [99].

Nerve conduction studies have been reported as within normal range, whereas sympathetic skin responses (SSR) have been reported as significantly altered [100, 101] or preserved [102], suggesting more prominent involvement of small nerve fibers [100, 103]. Quantitative sensory testing has shown abnormal thermal thresholds, indicating dysfunction of small sensory fibers – these were often more severely impaired for cold than warm [102], while the involvement of pinprick and light touch varied [104].

The involvement of small cutaneous fibers has been documented by skin biopsies, showing a severe loss of intraepidermal innervation at the ankle and less severe loss at the thigh, consistent with a length-dependent neuropathy [105]. Skin biopsy is considered a useful tool to provide direct evidence of small fiber neuropathy, especially in heterozygous Fabry females [106, 107] who may show reduced IENF as the only manifestation of small fiber disease present, before the onset of clinical symptoms [108].

7.7 Familial Amyloid Polyneuropathy (FAP, See Chap. 8)

Familial amyloid polyneuropathy (FAP) is described in detail (Chap. 8 of this monograph) and included in this chapter in brief. FAP is part of a multisystem life-threatening disorder which includes nerve lesions induced by deposits of amyloid fibrils, most commonly due to mutated transthyretin (*TTR*) gene [109] located on chromosome 18 [110].

Patients with FAP may present with different patterns of neuropathy, including focal neuropathies, sensorimotor polyneuropathy, autonomic neuropathy, or their combinations. The sensorimotor polyneuropathy is usually symmetric, length-dependent, lower-limb predominant, and progressive [111].

Distal paresthesias are the most frequent initial symptoms [112], and at this stage patients may complain of discomfort in their feet. Other symptoms may include numbness, pins and needles sensation, and burning pain worse at night associated with allodynia, in addition to weakness and numbness from loss of large nerve fibers [113]. Carpal tunnel syndrome is frequent at the onset and associated with polyneuropathy, including significant vibration perception deficit and a non-length-dependent progression of motor deficit [113, 114].

7.8 Tangier's Disease

Tangier's disease is a rare, autosomal recessive inherited disorder caused by a mutation in the adenosine triphosphate-binding cassette transporter 1 (*ABCA1*) gene, which results in a decrease in plasma high-density lipoprotein (HDL) levels [115].

Its clinical and pathological features are related to storage of cholesteryl esters in multiple organs [116]. Patients affected by Tangier's disease present with peripheral neuropathy, hypocholesterolemia, hypertrophic tonsils, enlarged lymph nodes, splenomegaly, and nodular skin lesions [117]. Peripheral neuropathy is present in approximately 50% of patients with Tangier's disease, which usually occurs in childhood or adolescence [118]. The peripheral neuropathy is characterized by relapsing-remitting mono- or polyneuropathy, often with dissociated loss of pain and temperature sensation. The electrophysiological changes are heterogeneous, with signs of demyelination and axonal degeneration [119].

7.9 Conclusions

Advances in the diagnosis of genetic small fiber sensory neuropathy and channelopathy have provided insight into their underlying pathophysiological mechanisms, which may lead to the development of symptomatic and disease-modifying treatments.

References

1. Dyck PJ, Oviatt KF, Lambert EH. Intensive evaluation of referred unclassified neuropathies yields improved diagnosis. Ann Neurol. 1981;10:222–6.
2. Auer-Grumbach M. Hereditary sensory and autonomic neuropathies. Handb Clin Neurol. 2013;115:893–906.
3. Klein CJ, Wu Y, Kruckeberg KE, Hebbring SJ, Anderson SA, Cunningham JM, et al. *SPTLC1* and *RAB7* mutation analysis in dominantly inherited and idiopathic sensory neuropathies. J Neurol Neurosurg Psychiatry. 2005;76:1022–4.
4. Kurth I. Hereditary sensory and autonomic neuropathy type II. In: Adam MP, Ardinger HH, Pagon RA, et al., editors. GeneReviews®. Seattle, WA: University of Washington; 1993.
5. Pinsky L, DiGeorge AM. Congenital familial sensory neuropathy with anhidrosis. J Pediatr. 1966;68:1–13.
6. Axelrod FB. Familial dysautonomia. Muscle Nerve. 2004;29:352–63.

7. Einarsdottir E, Carlsson A, Minde J, Toolanen G, Svensson O, Solders G, et al. A mutation in the nerve growth factor beta gene (*NGFB*) causes loss of pain perception. Hum Mol Genet. 2004;13:799–805.
8. Zambelis T. Small fiber neuropathy in Charcot-Marie-Tooth disease. Acta Neurol Belg. 2009;109:294–7.
9. Mersiyanova IV, Perepelov AV, Polyakov AV, Sitnikov VF, Dadali EL, Opqrin AN, et al. A new variant of Charcot-Marie-Tooth disease type 2 is probably the result of a mutation in the neurofilament-light gene. Am J Hum Genet. 2000;67:37–46.
10. Bird TD. Charcot-marie-tooth neuropathy type 2. In: Adam MP, Ardinger HH, Pagon RA, et al., editors. GeneReviews®. Seattle, WA: University of Washington; 1993.
11. Stevanin G, Dürr A, Brice A. Clinical and molecular advances in autosomal dominant cerebellar ataxias: from genotype to phenotype and physiopathology. Eur J Hum Genet. 2000;8:4–18.
12. Spring PJ, Kok C, Nicholson GA, Ing AJ, Spies JM, Bassett ML, et al. Autosomal dominant hereditary sensory neuropathy with chronic cough and gastro-oesophageal reflux: clinical features in two families linked to chromosome 3p22-p24. Brain. 2005;128:2797–810.
13. Margolis RL. The spinocerebellar ataxias: order emerges from chaos. Curr Neurol Neurosci Rep. 2002;2:447–56.
14. Harding AE. Friedreich's ataxia: a clinical and genetic study of 90 families with an analysis of early diagnostic criteria and intrafamilial clustering of clinical features. Brain. 1981;104:589–620.
15. Hewer RL. Study of fatal cases of Friedreich's ataxia. Br Med J. 1968;3:649–52.
16. Duman O, Uysal H, Skjei KL, Kizilay F, Karauzum S, Haspolat S. Sensorimotor polyneuropathy in patients with SMA type-1: electroneuromyographic findings. Muscle Nerve. 2013;48:117–21.
17. Farrar MA, Kiernan MC. The genetics of spinal muscular atrophy: progress and challenges. Neurotherapeutics. 2015;12:290–302.
18. Klein CJ, Duan X, Shy ME. Inherited neuropathies: clinical overview and update. Muscle Nerve. 2013;48:604–22.
19. Lewis RA, Sumner AJ. The electrodiagnostic distinctions between chronic familial and acquired demyelinative neuropathies. Neurology. 1982;32:592–6.
20. Lewis RA, Sumner AJ, Shy ME. Electrophysiological features of inherited demyelinating neuropathies: a reappraisal in the era of molecular diagnosis. Muscle Nerve. 2000;23:1472–87.
21. Dyck PJ, Karnes JL, Lambert EH. Longitudinal study of neuropathic deficits and nerve conduction abnormalities in hereditary motor and sensory neuropathy type 1. Neurology. 1989;39:1302–8.
22. Hilz MJ, Axelrod FB. Quantitative sensory testing of thermal and vibratory perception in familial dysautonomia. Clin Auton Res. 2000;10:177–83.
23. McDonnell A, Schulman B, Ali Z, Dib-Hajj SD, Brock F, Cobain S, et al. Inherited erythromelalgia due to mutations in *SCN9A*: natural history, clinical phenotype and somatosensory profile. Brain. 2016;139:1052–65.
24. Knowles CH, Scott SM, Wellmer A, Misra VP, Pilot MA, Williams NS, et al. Sensory and autonomic neuropathy in patients with idiopathic slow-transit constipation. Br J Surg. 1999;86:54–60.
25. González-Duarte A, Lem-Carrillo M, Cárdenas-Soto K. Description of transthyretin S50A, S52P and G47A mutations in familial amyloidosis polyneuropathy. Amyloid. 2013;20:221–5.
26. Hansen N, Kahn AK, Zeller D, Katsarava Z, Sommer C, Üçeyler N. Amplitudes of pain-related evoked potentials are useful to detect small fiber involvement in painful mixed fiber neuropathies in addition to quantitative sensory testing – an electrophysiological study. Front Neurol. 2015;6:244.
27. Valeriani M, Mariotti P, Le Pera D, Restuccia D, De Armas L, Maiese T, et al. Functional assessment of A delta and C fibers in patients with Fabry's disease. Muscle Nerve. 2004;30:708–13.
28. Wood JN, Boorman JP, Okuse K, Baker MD. Voltage-gated sodium channels and pain pathways. J Neurobiol. 2004;61:55–71.
29. Catterall WA. From ionic currents to molecular mechanisms: the structure and function of voltage-gated sodium channels. Neuron. 2000;26:13–25.
30. Cummins TR, Sheets PL, Waxman SG. The roles of sodium channels in nociception: implications for mechanisms of pain. Pain. 2007;131:243–57.
31. Beneng K, Renton T, Yilmaz Z, Yiangou Y, Anand P. Sodium channel Na_v 1.7 immunoreactivity in painful human dental pulp and burning mouth syndrome. BMC Neurosci. 2010;11:71.
32. Nassar MA, Stirling LC, Forlani G, Baker MD, Mathews EA, Dickenson AH, et al. Nociceptor-specific gene deletion reveals a major role for Na_v1.7 (PN1) in acute and inflammatory pain. Proc Natl Acad Sci U S A. 2004;101:12706–11.
33. Black JA, Liu S, Tanaka M, Cummins TR, Waxman SG. Changes in the expression of tetrodotoxin-sensitive sodium channels within dorsal root ganglia neurons in inflammatory pain. Pain. 2004;108:237–47.
34. Payne CE, Brown AR, Theile JW, Loucif AJ, Alexandrou AJ, Fuller MD, et al. A novel selective and orally bioavailable Na_v 1.8 channel blocker, PF-01247324, attenuates nociception and sensory neuron excitability. Br J Pharmacol. 2015;172:2654–70.
35. Hoeijmakers JGJ, Merkies ISJ, Gerrits MM, Waxman SG, Faber CG, et al. Genetic aspects of sodium channelopathy in small fiber neuropathy. Clin Genet. 2012;82:351–8.
36. Devor M. Sodium channels and mechanisms of neuropathic pain. J Pain. 2006;7:S3–S12.
37. Levinson SR, Luo S, Henry MA. The role of sodium channels in chronic pain. Muscle Nerve. 2012;46:155–65.
38. Coward K, Plumpton C, Facer P, Birch R, Caristedt T, Tate S, et al. Immunolocalization of SNS/PN3 and NaN/SNS2 sodium channels in human pain states. Pain. 2000;85:41–50.

39. Bucknill AT, Coward K, Plumpton C, Tate S, Bountra C, Birch R, et al. Nerve fibers in lumbar spine structures and injured spinal roots express the sensory neuron-specific sodium channels SNS/PN3 and NaN/SNS2. Spine. 2002;27:135–40.
40. Coward K, Aitken A, Powell A, Plumpton C, Birch R, Tate S, et al. Plasticity of TTX-sensitive sodium channels PN1 and brain III in injured human nerves. Neuroreport. 2001;12:495–500.
41. Coward K, Jowett A, Plumpton C, Powell A, Birch R, Tate S, et al. Sodium channel beta1 and beta2 subunits parallel SNS/PN3 alpha-subunit changes in injured human sensory neurons. Neuroreport. 2001;12:483–8.
42. Drenth JPH, Waxman SG. Mutations in sodium-channel gene *SCN9A* cause a spectrum of human genetic pain disorders. J Clin Invest. 2007;117:3603–9.
43. Faber CG, Hoeijmakers JGJ, Ahn HS Cheng X, Han C, Choi JS, et al. Gain of function $Na_v1.7$ mutations in idiopathic small fiber neuropathy. Ann Neurol. 2012;71:26–39.
44. Faber CG, Lauria G, Merkies ISJ, Cheng X, Han C, Ahn HS, et al. Gain-of-function $Na_v1.8$ mutations in painful neuropathy. Proc Natl Acad Sci U S A. 2012;109:19444–9.
45. Huang J, Han C, Estacion M, Vasylyev D, Hoeijmakers JG, Gerrits MM, et al. Gain-of-function mutations in sodium channel $Na_v1.9$ in painful neuropathy. Brain. 2014;137:1627–42.
46. England JD, Happel LT, Kline DG, Gamboni F, Thouron CL, Liu ZP, et al. Sodium channel accumulation in humans with painful neuromas. Neurology. 1996;47:272–6.
47. Black JA, Nikolajsen L, Kroner K, Jensen TS, Waxman SG, et al. Multiple sodium channel isoforms and mitogen-activated protein kinases are present in painful human neuromas. Ann Neurol. 2008;64:644–53.
48. Tanaka BS, Nguyen PT, Zhou EY, Yang Y, Yarov-Yarovoy V, Dib-Haji SD, et al. Gain-of-function mutation of a voltage-gated sodium channel $Na_v1.7$ associated with peripheral pain and impaired limb development. J Biol Chem. 2017;292:9262–72.
49. Yang Y, Wang Y, Li S, Xu Z, Li H, Ma L, et al. Mutations in *SCN9A*, encoding a sodium channel alpha subunit, in patients with primary erythermalgia. J Med Genet. 2004;41:171–4.
50. Han C, Lampert A, Rush AM, Dib-Haji SD, Wang X, Yang Y, et al. Temperature dependence of erythromelalgia mutation L858F in sodium channel $Na_v1.7$. Mol Pain. 2007;3:3.
51. Harty TP, Dib-Hajj SD, Tyrrell L, Blackman R, Hisama FM, Rose JB, et al. $Na_v1.7$ mutant A863P in erythromelalgia: effects of altered activation and steady-state inactivation on excitability of nociceptive dorsal root ganglion neurons. J Neurosci. 2006;26:12566–75.
52. Michiels JJ, te Morsche RHM, Jansen JBMJ, Drenth JPH. Autosomal dominant erythermalgia associated with a novel mutation in the voltage-gated sodium channel alpha subunit $Na_v1.7$. Arch Neurol. 2005;62:1587–90.
53. Choi J-S, Dib-Hajj SD, Waxman SG. Inherited erythermalgia: limb pain from an S4 charge-neutral Na channelopathy. Neurology. 2006;67:1563–7.
54. Lampert A, Dib-Hajj SD, Tyrrell L, Waxman SG. Size matters: erythromelalgia mutation S241T in $Na_v1.7$ alters channel gating. J Biol Chem. 2006;281:36029–35.
55. Lee MJ, Yu HS, Hsieh ST, Stephenson DA, Lu CJ, Yang CC, et al. Characterization of a familial case with primary erythromelalgia from Taiwan. J Neurol. 2007;254:210–4.
56. Zhang LL, Lin ZM, Ma ZH, Xu Z, Yang YL, Yang Y. Mutation hotspots of *SCN9A* in primary erythermalgia. Br J Dermatol. 2007;156:767–9.
57. Han C, Rush AM, Dib-Hajj SD, Li S, Xu Z, Wang Y, et al. Sporadic onset of erythermalgia: a gain-of-function mutation in $Na_v1.7$. Ann Neurol. 2006;59:553–8.
58. Wu B, Zhang Y, Tang H, Yang M, Long H, Shi G, et al. A novel *SCN9A* mutation (F826Y) in primary erythromelalgia alters the excitability of $Na_v1.7$. Curr Mol Med. 2017;17:450–7.
59. Drenth JP, Finley WH, Breedveld GJ, Testers L, Michiels JJ, Guillet G, et al. The primary erythermalgia-susceptibility gene is located on chromosome 2q31-32. Am J Hum Genet. 2001;68:1277–82.
60. Hisama FM, Dib-Hajj SD, Waxman SG. *SCN9A*-related inherited erythromelalgia. In: Adam MP, Ardinger HH, Pagon RA, et al., editors. GeneReviews®. Seattle, WA: University of Washington; 1993.
61. Low SA, Robbins W, Tawfik VL. Complex management of a patient with refractory primary erythromelalgia lacking a *SCN9A* mutation. J Pain Res. 2017;10:973–7.
62. Cook-Norris RH, Tollefson MM, Cruz-Inigo AE, Sandroni P, Davis MD, Davis DM. Pediatric erythromelalgia: a retrospective review of 32 cases evaluated at Mayo Clinic over a 37-year period. J Am Acad Dermatol. 2012;66:416–23.
63. Estacion M, Dib-Hajj SD, Benke PJ, Te Morsche RH, Eastman EM, Macala LJ, et al. $Na_v1.7$ gain-of-function mutations as a continuum: A1632E displays physiological changes associated with erythromelalgia and paroxysmal extreme pain disorder mutations and produces symptoms of both disorders. J Neurosci. 2008;28:11079–88.
64. Dib-Hajj SD, Cummins TR, Black JA, Waxman SG. Sodium channels in normal and pathological pain. Annu Rev Neurosci. 2010;33:325–47.
65. Han C, Dib-Hajj SD, Lin Z, Eastman EM, Tyrrell L, Cao X, et al. Early- and late-onset inherited erythromelalgia: genotype-phenotype correlation. Brain. 2009;132:1711–22.
66. Van Genderen PJ, Michiels JJ, Drenth JP. Hereditary erythermalgia and acquired erythromelalgia. Am J Med Genet. 1993;45:530–2.
67. Rush AM, Dib-Hajj SD, Liu S, Cummins TR, Black JA, Waxman SG. A single sodium channel mutation produces hyper- or hypoexcitability in different types of neurons. Proc Natl Acad Sci U S A. 2006;103:8245–50.

68. Cummins TR, Dib-Hajj SD, Waxman SG. Electrophysiological properties of mutant $Na_v1.7$ sodium channels in a painful inherited neuropathy. J Neurosci. 2004;24:8232–6.
69. Davis MDP, Sandroni P, Rooke TW, Low PA. Erythromelalgia: vasculopathy, neuropathy, or both? A prospective study of vascular and neurophysiologic studies in erythromelalgia. Arch Dermatol. 2003;139:1337–43.
70. Drenth JP, Vuzevski V, Van Joost T, Casteels-Van Daele M, Vermylen J, Michiels JJ. Cutaneous pathology in primary erythermalgia. Am J Dermatopathol. 1996;18:30–4.
71. Anand P, Privitera R, Yiangou Y, Donatien P, Birch R, Misra P. Trench foot or non-freezing cold injury as a painful vaso-neuropathy: clinical and skin biopsy assessments. Front Neurol. 2017;8:514.
72. Mantyh WG, Dyck PJB, Dyck PJ, Engelstad JK, Litchy WJ, Sandroni P, et al. Epidermal nerve fiber quantification in patients with erythromelalgia. JAMA Dermatol. 2017;153:162–7.
73. Davis MDP, Weenig RH, Genebriera J, Wendelschafer-Crabb G, Kennedy WR, Sandroni P. Histopathologic findings in primary erythromelalgia are nonspecific: special studies show a decrease in small nerve fiber density. J Am Acad Dermatol. 2006;55:519–22.
74. Fischer TZ, Gilmore ES, Estacion M, Eastman E, Taylor S, Melanson M, et al. A novel $Na_v1.7$ mutation producing carbamazepine-responsive erythromelalgia. Ann Neurol. 2009;65:733–41.
75. Choi J-S, Zhang L, Dib-Hajj SD, Han C, Tyrrell L, Lin Z, et al. Mexiletine-responsive erythromelalgia due to a new $Na_v1.7$ mutation showing use-dependent current fall-off. Exp Neurol. 2009;216:383–9.
76. Fertleman CR, Baker MD, Parker KA, Moffatt S, Elmslie FV, Abrahamsen B, et al. *SCN9A* mutations in paroxysmal extreme pain disorder: allelic variants underlie distinct channel defects and phenotypes. Neuron. 2006;52:767–74.
77. Hayden R, Grossman M. Rectal, ocular, and submaxillary pain; a familial autonomic disorder related to proctalgia fugaz: report of a family. AMA J Dis Child. 1959;97:479–82.
78. Fertleman CR, Ferrie CD. What's in a name-familial rectal pain syndrome becomes paroxysmal extreme pain disorder. J Neurol Neurosurg Psychiatry. 2006;77:1294–5.
79. Bednarek N, Arbuès AS, Motte J, Sabouraud P, Plouin P, Morville P. Familial rectal pain: a familial autonomic disorder as a cause of paroxysmal attacks in the newborn baby. Epileptic Disord. 2005;7:360–2.
80. Hayden R. Rectal, ocular, and submaxillary pain. AMA Am J Dis Child. 1959;97:479.
81. Fertleman CR, Ferrie CD, Aicardi J, Bednarek NA, Eeg-Olofsson O, Elmslie FV, et al. Paroxysmal extreme pain disorder (previously familial rectal pain syndrome). Neurology. 2007;69:586–95.
82. Yiangou Y, Facer P, Chessell IP, Bountra C, Chan C, Fertleman C, et al. Voltage-gated ion channel $Na_v1.7$ innervation in patients with idiopathic rectal hypersensitivity and paroxysmal extreme pain disorder (familial rectal pain). Neurosci Lett. 2007;427:77–82.
83. Chan CLH, Facer P, Davis JB, Smith JB, Egerton J, Bountra C, et al. Sensory fibres expressing capsaicin receptor TRPV1 in patients with rectal hypersensitivity and faecal urgency. Lancet. 2003;361:385–91.
84. Suter MR. What are the treatment options for paroxysmal extreme pain disorder? Pain Manag. 2015;5:229–32.
85. Cox JJ, Reimann F, Nicholas AK, Thornton G, Roberts E, Springell K, et al. An SCN9A channelopathy causes congenital inability to experience pain. Nature. 2006;444:894–8.
86. Goldberg YP, MacFarlane J, MacDonald ML, Thompson J, Dube MP, Mattice M, et al. Loss-of-function mutations in the $Na_v1.7$ gene underlie congenital indifference to pain in multiple human populations. Clin Genet. 2007;71:311–9.
87. Ahmad S, Dahllund L, Eriksson AB, Hellgren D, Karlsson U, Lund PE, et al. A stop codon mutation in *SCN9A* causes lack of pain sensation. Hum Mol Genet. 2007;16:2114–21.
88. Nilsen KB, Nicholas AK, Woods CG, Mellgren SI, Nebuchennykh M, Aasly J. Two novel *SCN9A* mutations causing insensitivity to pain. Pain. 2009;143:155–8.
89. Silverman FN, Gilden JJ. Congenital insensitivity to pain: a neurologic syndrome with bizarre skeletal lesions. Radiology. 1959;72:176–90.
90. Schiffmann R, Ries M. Fabry disease: a disorder of childhood onset. Pediatr Neurol. 2016;64:10–20.
91. Ries M, Ramaswami U, Parini R, Lindblad B, Whybra C, Willers I, et al. The early clinical phenotype of Fabry disease: a study on 35 European children and adolescents. Eur J Pediatr. 2003;162:767–72.
92. Arends M, Körver S, Hughes DA, Mehta A, Hollak CEM, Biegstraaten M. Phenotype, disease severity and pain are major determinants of quality of life in Fabry disease: results from a large multicenter cohort study. J Inherit Metab Dis. 2018;41:141–9.
93. Mehta A, Ricci R, Widmer U, Dehout F, Garcia de Lorenzo A, Kampmann C, et al. Fabry disease defined: baseline clinical manifestations of 366 patients in the Fabry Outcome Survey. Eur J Clin Invest. 2004;34:236–42.
94. Smid BE, van der Tol L, Biegstraaten M, Linthorst GE, Hollak CE, Poorthuis BJ. Plasma globotriaosylsphingosine in relation to phenotypes of Fabry disease. J Med Genet. 2015;52:262–8.
95. Laney DA, Peck DS, Atherton AM, Manwaring LP, Christensen KM, Shankar SP, et al. Fabry disease in infancy and early childhood: a systematic literature review. Genet Med. 2015;17:323–30.
96. Hoffmann B, Beck M, Sunder-Plassmann G, Borsini W, Ricci R, Mehta A, et al. Nature and prevalence of pain in Fabry disease and its response to enzyme replacement therapy-a retrospective analysis from the Fabry Outcome Survey. Clin J Pain. 2007;23:535–42.

97. MacDermot KD, Holmes A, Miners AH. Anderson-Fabry disease: clinical manifestations and impact of disease in a cohort of 98 hemizygous males. J Med Genet. 2001;38:750–60.
98. MacDermot KD, Holmes A, Miners AH. Anderson-Fabry disease: clinical manifestations and impact of disease in a cohort of 60 obligate carrier females. J Med Genet. 2001;38:769–75.
99. Üçeyler N, Ganendiran S, Kramer D, Sommer C. Characterization of pain in Fabry disease. Clin J Pain. 2014;30:915–20.
100. Gomes I, Nora DB, Becker J, Ehlers JA, Schwartz IV, Giugliani R, et al. Nerve conduction studies, electromyography and sympathetic skin response in Fabry's disease. J Neurol Sci. 2003;214:21–5.
101. Dütsch M, Marthol H, Stemper B, Brys M, Haendl T, Hilz MJ. Small fiber dysfunction predominates in Fabry neuropathy. J Clin Neurophysiol. 2002;19:575–86.
102. Luciano CA, Russell JW, Banerjee TK, Quirk JM, Scott LJ, Dambrosia JM, et al. Physiological characterization of neuropathy in Fabry's disease. Muscle Nerve. 2002;26:622–9.
103. Akpinar ÇK, Türker H, Bayrak O, Cengiz N. Electroneuromyographic features in fabry disease: a retrospective review. Arch Neuropsychiatr. 2015;52:258–62.
104. Low M, Nicholls K, Tubridy N, Hand P, Velakoulis D, Kiers L, et al. Neurology of Fabry disease. Intern Med J. 2007;37:436–47.
105. Scott LJ, Griffin JW, Luciano C, Barton NW, Banerjee T, Crawford T, et al. Quantitative analysis of epidermal innervation in Fabry disease. Neurology. 1999;52:1249–54.
106. Torvin Møller A, Winther Bach F, Feldt-Rasmussen U, Rasmussen A, Hasholt L, Lan H, et al. Functional and structural nerve fiber findings in heterozygote patients with Fabry disease. Pain. 2009;145:237–45.
107. Laaksonen SM, Röyttä M, Jääskeläinen SK, Kantola I, Penttinen M, Falck B. Neuropathic symptoms and findings in women with Fabry disease. Clin Neurophysiol. 2008;119:1365–72.
108. Liguori R, Di Stasi V, Bugiardini E, Mignani R, Burlina A, Borsini W, et al. Small fiber neuropathy in female patients with Fabry disease. Muscle Nerve. 2010;41:409–12.
109. Planté-Bordeneuve V, Said G. Familial amyloid polyneuropathy. Lancet Neurol. 2011;10:1086–97.
110. Wallace MR, Naylor SL, Kluve-Beckerman B, Long GL, McDonald L, Shows TB, et al. Localization of the human prealbumin gene to chromosome 18. Biochem Biophys Res Commun. 1985;129:753–8.
111. Suanprasert N, Berk JL, Benson MD, Dyck PJ, Klein CJ, Gollob JA, et al. Retrospective study of a TTR FAP cohort to modify NIS+7 for therapeutic trials. J Neurol Sci. 2014;344:121–8.
112. Kim DH, Zeldenrust SR, Low PA, Dyck PJ. Quantitative sensation and autonomic test abnormalities in transthyretin amyloidosis polyneuropathy. Muscle Nerve. 2009;40:363–70.
113. Shin SC, Robinson-Papp J. Amyloid neuropathies. Mt Sinai J Med. 2012;79:733–48.
114. Carr AS, Pelayo-Negro AL, Evans MR, Laurà M, Blake J, Stancanelli C, et al. A study of the neuropathy associated with transthyretin amyloidosis (ATTR) in the UK. J Neurol Neurosurg Psychiatry. 2016;87:620–7.
115. Per H, Canpolat M, Bayram AK, Ulgen E, Baran B, Kardas F, et al. Clinical, electrodiagnostic, and genetic features of Tangier disease in an adolescent girl with presentation of peripheral neuropathy. Neuropediatrics. 2015;46:420–3.
116. Bracco G, Dotti G, Levis F, David E, Saracco G, Rizzetto M, et al. Familial high-density lipoprotein deficiency (Tangier disease): the third italian case. J Inherit Metab Dis. 1988;11:155–7.
117. Hoffman HN, Fredrickson DS. Tangier disease (familial high density lipoprotein deficiency). Clinical and genetic features in two adults. Am J Med. 1965;39:582–93.
118. Puntoni M, Sbrana F, Bigazzi F, Sampietro T. Tangier disease: epidemiology, pathophysiology, and management. Am J Cardiovasc Drugs. 2012;12:303–11.
119. Zyss J, Béhin A, Couvert P, Bouhour F, Sassolas A, Kolev I, et al. Clinical and electrophysiological characteristics of neuropathy associated with Tangier disease. J Neurol. 2012;259:1222–6.

Amyloid Neuropathy

8

Chi-Chao Chao, Hung-Wei Kan, Ti-Yen Yeh, Ya-Yin Cheng, and Sung-Tsang Hsieh

Abstract

Amyloid neuropathy is a syndrome of peripheral nerve degeneration associated with deposition of amyloid attributed to various molecular compositions including (1) monoclonal protein of immunoglobulin light chain (AL amyloidosis) and (2) misfolded proteins produced by amyloid-prone genes, in particular, familial amyloid polyneuropathy (FAP) due to transthyretin (TTR) mutations. During the early stage of amyloid neuropathy, small fiber of the nociceptive type is frequently affected. Neuropathy of AL amyloidosis is a chronic hematological disorder. In addition to peripheral nerves, the kidney, liver, and heart are major organs of amyloid deposition resulting in diverse clinical presentations. In AL amyloidosis, progressive polyneuropathy is the most common neurological manifestations while mononeuropathy multiplex is also documented depending on the distribution of amyloid pathology. Since AL amyloidosis is due to B-cell dyscrasia, treatments include (1) a combination of prednisolone and chemotherapy (melphalan) and (2) target therapy of proteasome inhibitor (e.g., bortezomib). TTR-FAP is the most common genotype of FAP compared with other amyloid-producing genes (gelsolin and apolipoprotein A1). There are more than 100 mutations of TTR with variations in geographic and ethnic distributions. For example, V30M is the most common genotype and endemic in certain areas of Portugal and Japan. Small-fiber neuropathy is usually the initial presentation of FAP, including neuropathic pain, loss of protective sensation, and autonomic dysfunctions. Neurological deficits progress to affect large fibers leading to weakness of four limbs and unsteadiness, and many patients became ambulation-dependent or bedridden within a

C.-C. Chao (✉)
Department of Neurology, National Taiwan University Hospital, Taipei, Taiwan
e-mail: b1401019@ms17.hinet.net

H.-W. Kan · T.-Y. Yeh · Y.-Y. Cheng
Department of Anatomy and Cell Biology, National Taiwan University College of Medicine, Taipei, Taiwan

S.-T. Hsieh
Department of Neurology, National Taiwan University Hospital, Taipei, Taiwan

Department of Anatomy and Cell Biology, National Taiwan University College of Medicine, Taipei, Taiwan

Graduate Institute of Brain and Mind Sciences, National Taiwan University College of Medicine, Taipei, Taiwan

Graduate Institute of Clinical Medicine, National Taiwan University College of Medicine, Taipei, Taiwan

Center of Precision Medicine, National Taiwan University College of Medicine, Taipei, Taiwan

S.-T. Hsieh et al. (eds.), *Small Fiber Neuropathy and Related Syndromes: Pain and Neurodegeneration*,
https://doi.org/10.1007/978-981-13-3546-4_8

decade depending on the genotypes. In addition to liver transplantation which eliminates the major source of TTR-producing organ, new treatments have emerged that may slow the progression of neurological deficits; these include TTR stabilizers and RNA-based therapies. Hence to identify early biomarkers is the direction of research for slowing disease progression of FAP.

Keywords
Amyloidosis · β-pleated sheet · Light chain immunoglobulin · AL amyloidosis · Primary systemic amyloidosis · Plasma cell dyscrasia Chemotherapy · Familial amyloid polyneuropathy (FAP) · Familial amyloid cardiomyopathy (FAC) · Transthyretin (TTR) · Biomarkers · Liver transplantation Transthyretin stabilizer · Diflunisal Tafamidis · RNA interference (RNAi) Antisense oligonucleotide

8.1 Introduction

Amyloidosis is a syndrome of devastating diseases due to extracellular deposition of insoluble fibrillar materials resulting from protein misfolding of various precursor proteins [1]. These rigid and unbranching fibrils of ~10 nm in diameter are distinct for their antiparallel β-pleated sheet configuration [2–4] and produce green, yellow, or orange birefringence when stained with Congo red dye and observed under polarized microscopy. These Congo red staining properties remain the histological gold standard for confirming amyloids in tissue samples.

More than 30 proteins are responsible for amyloidosis with the current classification based on the amyloidogenic precursor proteins [1]. All these may have an intrinsic propensity to misfold, and the deposition may occur with a pathologic and sustained increase in the concentration of the protein (i.e., immunoglobulin light chain and serum amyloid A protein), a hereditary mutated protein of increased propensities for misfolding and aggregation (i.e., transthyretin, TTR), aging process (i.e., wild-type TTR amyloidosis and atrial natriuretic peptide amyloidosis), or for unknown mechanisms. Although these factors contribute to the amyloidogenic potential of a precursor protein alone or together, they cannot account for all the phenotypes of amyloidosis. Other undetermined environmental and genetic factors are potentially participant in amyloidogenesis [5].

Amyloid formation causes progressive cellular death, destruction of the normal tissue architecture, and organ dysfunction [6]. Almost any organ can be affected by amyloid deposits, and the target organs depend on the types of amyloid fibril proteins although they are often nonspecific. The clinical features of amyloidosis are diverse and depend on the organ systems involved. Renal, hepatic, and cardiac failures are among the most common manifestations of amyloidosis. Peripheral nerves are one common target of amyloid deposits in systemic amyloidosis. The involvement of peripheral nerves varies in degree and may become symptomatic at different time points during the progression of disease. Occasionally (and in hereditary amyloidosis mostly), peripheral neuropathy is the initial manifestation of the amyloidosis, in particular, the involvement of nociceptive-type small fibers. Of the proteins responsible for systemic amyloidosis, neuropathy is most consistently seen in patients with TTR-familial amyloid polyneuropathy (FAP) [7]. Often patients with the rare hereditary apolipoprotein A1 (ApoA1) amyloidosis have peripheral neuropathy [8]. Neuropathy is also a cardinal feature of gelsolin amyloidosis. The most common acquired amyloidotic neuropathy is due to immunoglobulin light chain amyloidosis (AL amyloidosis); peripheral neuropathy developed in approximately 15%~20% patients [9]. In the following section, we will focus on neuropathies associated with acquired and hereditary amyloidosis: (1) AL amyloidosis and (2) FAP, and the pathology and manifestations of small fibers will be highlighted.

8.2 Neuropathy of Immunoglobulin Light Chain Amyloidosis (AL Amyloidosis)

AL amyloidosis, previously termed "primary" systemic amyloidosis, is a plasma cell dyscrasia or B-cell lymphoproliferative disorder in which insoluble AL fragments are polymerized and deposited in various tissues [10]. Light chain amyloidosis is the most common type of amyloidosis. The clinical manifestations of amyloid deposits depend on the precise tissues or organs involved and can cause substantial morbidity. All tissues can be involved in AL amyloidosis; however, the commonly affected organs are the heart, kidney, liver, and gastrointestinal tract [11]. It is estimated to have a minimal incidence of 8 per million with a male predominance and is responsible for 0.58 of 1000 recorded deaths [12].

Peripheral nerves are also affected in AL amyloidosis, with the prevalence of peripheral neuropathy varying from 15% to 20% [13]. Males represented ~75% of all patients [13, 14], and the mean age at diagnosis is in the sixth and seventh decades [14, 15]. Peripheral nerve involvement related to systemic amyloidosis can be an incidental finding or the presenting symptom of the disease and is usually manifested as a symmetric polyneuropathy.

8.2.1 Clinical Manifestation

Length-dependent sensory symptoms in neuropathy of AL amyloidosis are prominent and sometimes initial manifestations, i.e., (1) symmetric distribution starting distally and progressing proximally and (2) more marked involvement of the lower limbs than the upper limbs in most patients. Carpal tunnel syndrome may develop concomitantly. Given the multifocal deposition of amyloid, peripheral neuropathy may manifest a pattern of multiplex mononeuropathy. Both small fiber (pain and thermal sensation) and large fiber (vibration and position sensation)-mediated sensations can be affected: (1) pain and temperature sensation was disproportionately affected in 52% patients, (2) vibration and position sensation was solely impaired in 16%, and (3) both were involved equally in the remaining 32%. This pattern indicates that cutaneous sensation was more frequently and more severely affected than discriminative sensation [14]. The patterns may vary depending on study populations, for example, in another study, the loss of both cutaneous and discriminative sensation could be seen in all patients diagnosed as polyneuropathy [15]. Neuropathic pain was a common component in AL amyloidosis polyneuropathy. Dysesthetic numbness was about two times more frequently than paresthesias and caused disability in some patients. Adjectives used by patients to describe their symptoms included terms such as "prickling, lightning, burning, tingling, aching, stabbing, painful numbness, and shock-like pains" [14]. The motor nerve involvement usually developed later than the sensory symptoms and also tended to be distal, symmetrical, progressive, and more prominent in the lower limbs. Autonomic dysfunction was also a common and prominent component in AL amyloidosis polyneuropathy. The most common symptom is orthostatic hypotension which impairs daily activities. Other common symptoms were hypohydrosis, urinary incontinence, sexual dysfunction, and gastrointestinal dysfunctions like diarrhea, constipation, or easy satiety [13–15].

8.2.2 Electrophysiology

The electrophysiological studies in AL amyloidosis polyneuropathy typically show markedly reduced amplitudes of sensory nerve action potentials without or with minimal prolongation of sensory distal latencies, except for the median nerve in patients with carpal tunnel syndrome. For motor nerve conduction studies, there were mildly reduced amplitudes of the compound muscle action potentials with mild slowing of motor conduction velocities and mild prolongation of distal motor latencies to a similar degree, except in the presence of a superimposed carpal tunnel syndrome [14]. In summary, the motor and sensory conduction velocity, the amplitude of the motor and sensory action potentials are decreased, and the distal motor latencies are prolonged in patients with AL amyloidosis in the median and tibial nerve, regardless whether or not there are clinically apparent neuropathic symptoms. This suggests that peripheral nerves are involved more extensively in AL amyloidosis than clinical manifestations might suggest [15].

8.2.3 Pathology

Sural nerves of AL amyloidosis polyneuropathy show amyloid deposition in the epineurium and endoneurium with or without involvement of vessels. Cellular infiltrates are usually absent or scanty [14, 16, 17]. Compatible with the electrophysiological findings, there is moderate to marked axonal degeneration without or with mild segmental demyelination. The major pathology shows a significant loss of both large myelinated and small myelinated nerve fibers although the degree of reduction was more severe in large myelinated nerve fibers than that in small myelinated ones [16]. Ultrastructural examination disclosed that myelinated and unmyelinated fibers are severely reduced in all cases [16, 18].

8.2.4 Pathogenesis

The mechanisms underlying nerve degeneration in AL amyloidosis polyneuropathy are poorly understood. However, several pathogenic processes have been implicated including progressive accumulation of insoluble β-pleated sheets of amyloid fibrils that may infiltrate and compress nerve fibers [17, 18]. Furthermore, amyloid deposit may compromise the vasa nervorum and result in nerve ischemia [9, 14]. In a study examining microcirculation, cutaneous blood flow mediated by the interactions of endothelial cell, smooth muscle, and C fiber was examined with laser Doppler flowmetry after applying acetylcholine and sodium nitroprusside into the forearm skin via iontophoresis. Endothelial cell-mediated vasodilation was decreased in AL amyloidosis with no changes in the C-fiber response or smooth muscle cell, suggesting that endothelial dysfunction might underlie the pathogenesis of the nerve degeneration of AL amyloidosis. Nerve biopsy specimens of AL amyloidosis showed existence of hydroxynonenal immunoreactivity (as a surrogate marker of lipid peroxidation) in areas corresponding to those with amyloid deposits, indicating oxidative stress responsible of pathogenesis of AL amyloidosis polyneuropathy.

8.2.5 Treatment and Prognosis

The comprehensive treatment protocols for AL amyloidosis polyneuropathy are managed in collaboration with hematooncologists and follow the principles same as in AL amyloidosis. There is substantial progress in this field to improve the survival by taking advantage of evidence-based algorithms [19]. Chemotherapy with the regimen of melphalan and predonine was introduced for AL amyloidosis in 1972 [20]. Only a minor portion of patients had responses to this protocol, and the median survival was 12–18 months [21]. The introduction of stem cell transplantation provided a new therapeutic policy. A prospective randomized study to test the effects of stem cell transplantation did not demonstrate an advantage in survival [22], and a meta-analysis also questioned the value of stem cell transplantation for amyloidosis [10, 23]. Other chemotherapy options include thalidomide, bortezomib, lenalidomide, and bendamustine. Several studies have investigated the effects of these treatment modalities, but there was ongoing debate about the optimal treatment regimens [24]. Usually the AL amyloidosis polyneuropathy has a chronic and relentless progression with development of generalized disabling neuropathy during the course of the illness. In a series of 16 patients with AL amyloidosis polyneuropathy receiving chemotherapy containing melphalan and prednisone, the monoclonal protein was decreased in 10 patients, but only 2 of them had stabilization of neuropathy [9]. As to the treatment of melphalan followed by stem cell transplantation in AL amyloidosis polyneuropathy, improvement or stabilization of neurological condition was reported in some cases [25–27].

8.3 Familial Amyloid Polyneuropathy (FAP)

FAP is a group of adult-onset genetic neuropathy inherited with an autosomal dominant mode and characterized by the systemic amyloid deposition in addition to the preferential involvement of peripheral nerves. There are three major molecular types of FAP, defined according to the precursor protein of amyloid: TTR, ApoA1, and gelsolin. FAP due to TTR mutations (TTR-FAP) is by far the most common variant worldwide. ApoA1 amyloidosis is mainly deposited in the kidney, causing renal failure; other affected organs included the liver and heart. The main feature of gelsolin-related familial amyloid neuropathy is cranial neuropathy. This review will focus on TTR-FAP.

8.3.1 TTR-Familial Amyloid Polyneuropathy

The three main phenotypes of TTR amyloidosis are FAP, familial amyloid cardiomyopathy (FAC), and familial leptomeningeal amyloidosis. TTR-FAP is a progressive and lethal neurodegenerative disorder affecting approximately 5000~10,000 patients worldwide [28]. TTR-FAP was first described by Corinho Andrade (1952) in patients from north Portugal. In the original report, TTR-FAP appeared "a peculiar form of peripheral neuropathy" for generations as "mal dos pesinhos," i.e., pain in the feet [29]. This disease was later identified in Japan, Sweden, and other countries and has been reported worldwide [7, 30, 31]. There are endemic regions where families are clustered with a high incidence rate such as in Portugal, Japan, Brazil, and Sweden. In other non-endemic places, the disease is rare and usually occurs as an isolated and sporadic neuropathy [7]. However, it is likely that the disease remains largely underdiagnosed in some areas.

8.3.2 Transthyretin

TTR was initially termed prealbumin because of its presence as a component that migrates ahead of albumin on electrophoresis. This protein was first described in the cerebrospinal fluid and later on in the plasma [32, 33] and normally circulates as a soluble protein in the blood with a tetrameric structure formed by four identical subunits. The name "transthyretin" derives from its physiological role as a transporter carrying thyroxine (T4) and retinol-binding protein [28]. The majority of TTR is made in the liver and is secreted into the blood. The choroid plexus and retinal pigment epithelium also synthesize a minor portion of TTR into the cerebrospinal fluid and eyes [34, 35]. TTR expression starts from the embryonic stage and remains throughout the adult period. However, TTR is not essential for survival since *TTR* gene knockout mice have normal fetal development and a normal lifespan, although their levels of plasma thyroid hormone and retinol were significantly depressed [36].

TTR tetramer comprises four identical subunits each of which contains 127 amino acid residues encoded by *TTR* gene. Each monomer has eight β-strands and one short helix. The β-strands are organized into two antiparallel four-stranded β-sheets, forming a wedge-shaped β-barrel [37]. Two TTR monomers form a dimmer by joining edge-to-edge through antiparallel hydrogen-bonding, and these two dimers form the native tetrameric structure of TTR via the hydrophobic interactions, creating a central channel that contains the two binding sites for T4 [38]. However, only a small proportion (<1%) of the T4-binding sites are occupied by T4. In human serum, most binding sites are unbound because of the high affinity of T4 for thyroid-binding globulin which is the major carrier of T4 [39]. The genomic structure of TTR is strongly conserved, and the amino acid sequences in the core of the molecule containing the T4-binding site are also highly conserved [40].

8.3.3 Genetics of TTR

The human *TTR* gene is localized to chromosome 18p11.1–q12.3. It spans approximately 7 kb and has four exons encoding the mature protein and its cleaved signal peptide at the first 20 amino acids: (1) exon 1, including the first 3 amino acids of the mature form; (2) exon 2, amino acids 4–47; (3) exon 3, amino acids 48–92; and (4) exon 4, amino acids 93–127 [41]. More than 100 mutations of the *TTR* gene have been reported causing systemic amyloidosis [7]. Most mutations have been documented in exon 3, and no mutation was reported in exon 1 [42]. A single nucleotide substitution in the *TTR* gene accounts for the majority of the mutations. Other mutations are the products of two nucleotide substitutions in a codon or a trinucleotide deletion in exon 4 (Val122) [43]. Given an autosomal dominant inheritance, the majority of patients with TTR amyloidosis carry one allele of pathogenic mutation producing normal and mutant TTR proteins. Cases of homozygous Val30Met mutation are reported with their serum levels of variant TTR being twofolds of those in heterozygous patients. Intriguingly there were no differences in the onset age and the progression compared with heterozygous patients [44, 45].

8.3.4 Pathogenesis of TTR-FAP

TTR tetramer circulates in peripheral blood and cerebrospinal fluid. The natively folded TTR tetramer converts into insoluble amyloid fibrils through a multiple-step and dynamic process. Amyloidogenic mutations make the native structures of TTR unstable, leading to the dissociation of tetramers into non-native monomers which have low conformational stability and subsequently aggregate into amyloid fibrils [46–49]. The wild type TTR could form amyloidogenic species under the condition of increased temperature or hydrostatic pressure [50, 51]. There is a correlation between amyloidogenesis and the thermodynamic stability of TTR variants. In comparison with wild-type TTR, amyloidogenic TTR variants have higher thermodynamic instability, promoting tetramer dissociation and monomer unfolding, and all these characters favor aggregation of amyloid fibrils [39, 52, 53]. For TTR variants, there is a correlation between amyloidogenic potential and TTR variants instability under high pressure [51]. Therefore, conditions leading to an increase in instability are probably critical for the dissociation and misfolding of native structure and the formation of amyloidogenic intermediates.

The mechanism of TTR-induced neurotoxicity in FAP is poorly understood with the assumption that amyloid aggregations can cause tissue damage by direct compression or obstruction of blood vessels resulting in local blood circulation failure. However, several lines of evidences indicate that low molecular weight oligomers and protofibrils of amyloidogenic proteins might be neurotoxic and neuronal apoptosis has been proposed as the pathogenesis of TTR-FAP [54]. Mutant TTR has been shown to be toxic to cells in culture [55, 56]. Studies on cell biology have demonstrated that the transient low molecular mass assemblies aggregated from misfolded TTR monomers were highly cytotoxic in tissue cultures [57]. The cellular toxicity of these aggregates may be triggered by induction of oxidative stress molecules, pro-inflammatory cytokines (e.g., tumor necrosis factor α (TNFα)), signaling cascades, nuclear transcription factor κB (NFκB), interleukin-1βb (IL-1βb), and macrophage colony-stimulating factor (M-CSF). Furthermore, the activation of receptor for advanced glycation end products (RAGE) may contribute to neuronal apoptosis [58]. Low molecular weight TTR aggregates induced an increase in intracellular calcium concentration via voltage-gated calcium channels, which contributed to TTR-induced cytotoxicity [59].

8.3.5 Clinical Features of TTR-FAP

FAP is the most common clinical phenotype of TTR amyloidosis. There is a wide spectrum of phenotypes associated with TTR-FAP, even in patients with the same mutation. The clinical presentations of TTR-FAP are determined by genotype, geographical variations, penetrance of the mutant genes, age at onset, and gender [60–62].

The most prevalent TTR mutation is Val30Met (V30M), most common in Portugal, Japan, and Sweden. Patients with V30M mutation are divided into two subgroups, early-onset and late-onset. In a large cohort from Portugal (n = 1223), the mean age of onset was 33.5 years (range 17–78 years), and 87% patients had onset of disease before 40 years of age. Women had a later onset than men [62]. The mean onset age of V30M FAP in northern Sweden was 56 years, much older than the patients from Portugal [63]. In Japan, the mean onset age of V30M patients from two large endemic foci (Ogawa village and Arao city) was

~40 years, much younger than those unrelated to two large endemic foci (with a mean age of 63 years) [64]. The cardinal feature of TTR-FAP is slowly progressive length-dependent sensorimotor and autonomic neuropathy [65]. The neuropathy usually begins with degeneration of small fibers, and an impairment of thermal sensation in the distal extremities is the early feature. Sensory symptoms typically include distal paresthesia progressing slowly to the proximal parts of four limbs with or without varying degrees of pain such as burning, tingling, or allodynia. They are usually symmetrical, less frequently asymmetrical with a focal or non-length-dependent distribution. Walking difficulty and weakness are less common at the onset. Motor functions tend to be relatively preserved until the sensory manifestations have advanced considerably. At a later stage, weakness occurs distally and progresses to foot drop and more proximally in the lower limbs and then in the upper limbs. Ultimately there is a flaccid paralysis of all four limbs that predominates distally with remarkable muscular atrophy [66].

Clinical examinations at early stage usually detect dissociated sensory loss, being more pronounced for pain and temperature sensations. At this stage, tendon reflexes are often preserved except for the ankle jerk reflex. Neurological deficits gradually worsen and start to affect larger sensory and motor nerve fibers over the following months to years. Proprioception and motor modalities will subsequently be damaged with impairment of vibratory sensation, joint position sense, and muscle strengths, responsible for increased disability [7]. A common site of early involvement in TTR-FAP is the median nerve at the wrist, and patients may experience symptoms of carpal tunnel syndrome preceding the symptoms of polyneuropathy [67].

Autonomic neuropathy is one important component of TTR-FAP and tends to occur relatively early in the course, especially in individuals with early-onset TTR-FAP. Gastrointestinal, cardiovascular, and genitourinary systems are usually affected. Patients may suffer from orthostatic hypotension presented with dizziness, sensation of light headedness, fatigue, and blurry vision when standing. Common gastrointestinal presentations include postprandial diarrhea, constipation, or alternative diarrhea and constipation. Gastroparesis leads to postprandial vomiting, dehydration, malnutrition, and body weight loss. In male patients, sexual dysfunction with impotence might be an early manifestation preceding sensory impairments in the distal parts of limbs. Urinary abnormalities including dysuria and retention of urine develop later; sudomotor dysfunction is less frequent. Autonomic dysfunctions could be serious problems of TTR-FAP due to its restricting daily activities [68].

Late-onset FAP, defined by clinical onset after 50 years of age, was identified decades after the early-onset pattern. The characteristics of early-onset TTR-FAP include high penetrance, clustering appearance in endemic areas, disabled autonomic dysfunctions, prominent impairments of nociception, and arrhythmia requiring pacemaker implantation, such as atrioventricular nodal block. In late-onset TTR-FAP, neuropathy shows male preponderance, and there is a 10:1 preponderance of males to females in Japan [64]. Other features of late-onset TTR-FAP include a relatively low penetrance rate, usually lack of family history (a family history in only one-third of patients), geographic scattering and unrelated to endemic foci, initial involvement of both superficial and deep sensation in a distal predominant pattern, relatively mild autonomic dysfunction, and frequent cardiac involvement [64, 69].

The non-V30M variants are much less frequent and more heterogeneous compared to V30M TTR-FAP. Polyneuropathy is generally the main clinical feature, but in some cases cardiac involvement predominates with minimal or no neurological symptoms, leading to diagnostic delays. For example, patients with Ser77Tyr TTR mutation usually have carpal tunnel syndrome as the initial symptoms in the sixth and seventh decades and later progressing to generalized neuropathy, restrictive cardiomyopathy, and intestinal malabsorption [67]. The Ala97Ser variant is common in Taiwan (Fig. 8.1) and presents as a late-onset generalized neuropathy with disabling motor, sensory, and autonomic symptoms. There is marked denervation of cutaneous and sudomotor nerves, and the degree of small-fiber degeneration is negatively correlated with disability grade [61, 70].

The manifestations of Ile107Val mutation include early proximal motor involvement with

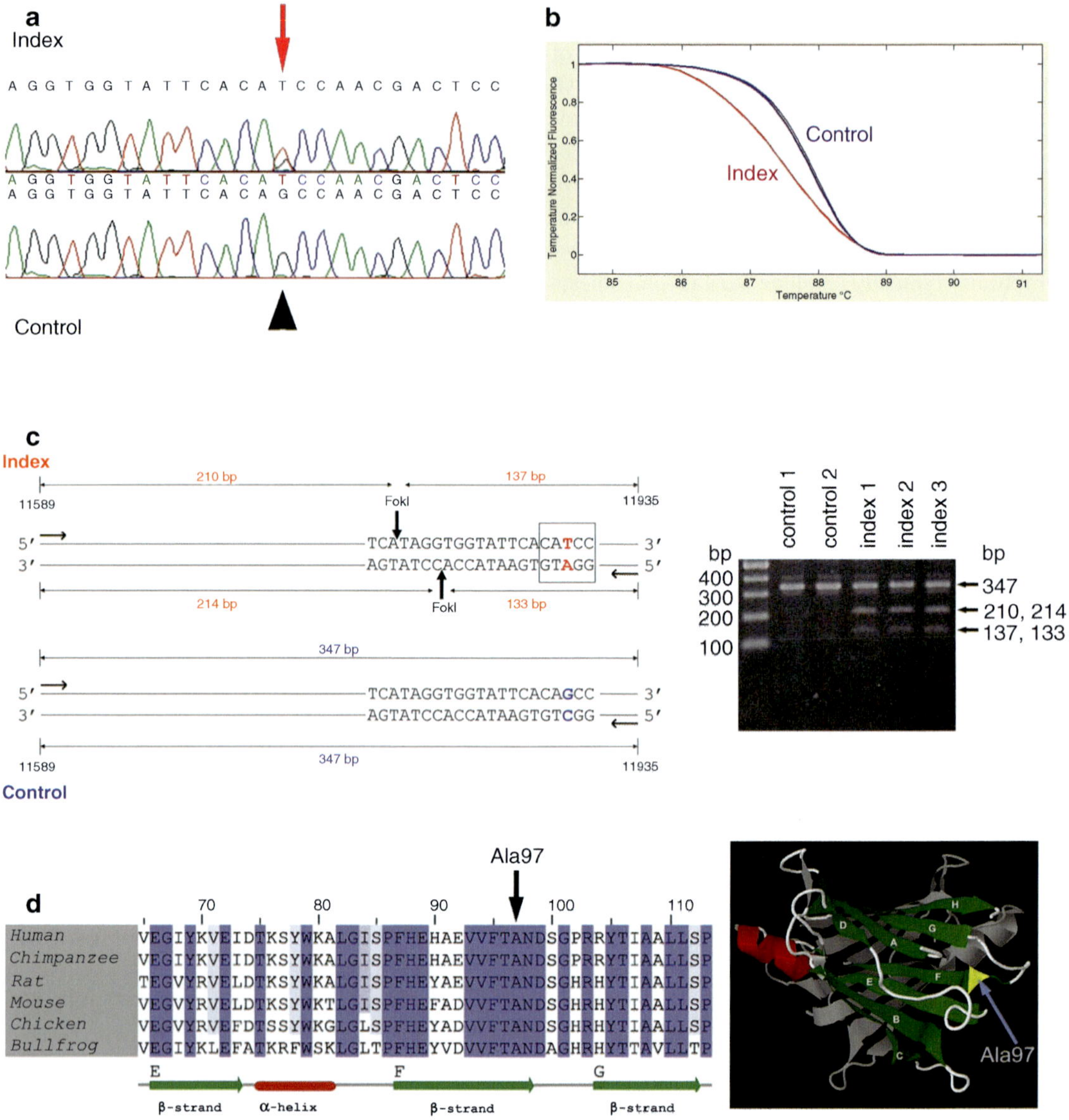

Fig. 8.1 Genetic evidence of the transthyretin (TTR) mutation. (**a**) The TTR mutation was documented with directly sequencing exon 4 of the human TTR gene from index patients and controls. The sequencing chromatography shows a sequence variant, G (black arrowhead) replaced by a T (red arrow) resulting in an amino acid change, Ala97Ser. (**b**) This mutation caused a lower melting curve with a substantial shift (red line) compared to the controls (dark blue lines). (**c**) The mutation could be further demonstrated with restriction fragment length polymorphism (RFLP). The left panel illustrates a schematic diagram of the RFLP strategy to differentiate index from control TTR alleles: no FokI cutting site in the control allele with a single 347-bp fragment. In the mutant, there are two FokI restriction fragments at ~210 bp (210, 217) and ~130 bp (137, 133). The right panel shows the PCR-RFLP after gel electrophoresis. The controls contained only one band of 347 bp, while the index patients had two additional bands corresponding to the patterns of the restriction diagram. (**d**) The sequence alignment and molecular modeling analysis of TTR in six species (human, chimpanzee, rat, mouse, chicken, and American bullfrog) indicate that Ala97 (blue arrow) is in a β-pleated sheet (β-strand, green) and conserved in all species. Each TTR monomer includes eight β-strands (A–H), and Ala97 (in yellow) lies in the F-strand and is part of the hydrophobic core. Reproduced with permission from [61]

debilitating polyneuropathy and rapid onset of gait disturbances due to tetraparesis and shorter median survival [71, 72]. Thr60Ala patients usually develop their disease late in life (a mean age of 63 years). It predominantly affects the autonomic nerves and heart with less than one-quarter of patients having peripheral neuropathy at presentation. Peripheral neuropathy is relatively infrequent during follow-up examination. Indeed, the survival in this cohort is determined by cardiac parameters [73]. Other unusually aggressive forms of TTR-FAP, fatal within 5 years, have been reported in association with Ala25Ser, Gly47Arg, Glu54Lys, and Leu55Pro [65, 74].

8.3.6 Neurophysiology of TTR-FAP

In early-onset V30M, nerve conduction studies may remain normal at early stages with pure small-fiber dysfunction. With the progression of deficits, a sensory axonal polyneuropathy develops in a length-dependent pattern. Abnormalities of motor nerve conduction studies occur later, characterized by reduced amplitude of compound muscle action potentials progressing in a similar length-dependent pattern. However in one series of V30M patients in whom the symptoms were less than 2 years, there was markedly slowing of motor and sensory conduction velocities, reduction of amplitudes of motor and sensory action potentials, and prolonged F wave latencies in the lower limbs compared to controls [75]. In late-onset TTR-FAP from non-endemic areas, usually sensorimotor polyneuropathy belonging to the type of axonal degeneration was already prominent at diagnosis, i.e., (1) absence of compound motor action potentials in 44% of the tibial nerve and (2) absence of sensory action potentials in 83% of the sural nerve [76].

Preferential involvement of small-diameter sensory and autonomic nerve fibers is the cardinal feature in TTR-FAP. In V30M FAP patients at an early stage with clinical symptoms of small-fiber neuropathy but normal findings of nerve conduction studies, thermal quantitative sensory tests showed significantly elevated thermal thresholds [77]. For the autonomic dysfunction, RR interval variability was not a sensitive tool. In contrast, the amplitude of sympathetic skin response (SSR) was markedly lower in carriers of TTR-FAP compared to control subjects with 0.53 and 0.95 for sensitivity and specificity, respectively [75].

8.3.7 Pathology of TTR-FAP

In V30M patients, postmortem examinations show amyloid deposits to a great extent in peripheral nerves including the autonomic nervous system. The deposition of amyloid due to TTR mutations was also detected in other organs except for the central nervous system or the bone marrow in which no deposits or only insignificant amounts were found [78]. Amyloid deposits are universally in (1) the endoneurium of peripheral nerves (including the proximal and distal parts) and (2) spinal and sympathetic ganglia. Amyloid distribution correlates well with peripheral nerve pathologies. The loss of small-diameter neurons was significant in spinal and sympathetic ganglia with preservation of motor neurons in the ventral horn [79]. The myelinated fiber density remains unchanged corroborating with clinical observations and electrophysiological tests. Amyloid mainly deposits around endoneurial capillaries with intact vascular walls at the early stage. With the progression of disease, the deposition of amyloids extends and damages the endoneurial vessels leading to a reduction of nerve fiber density. At the early stage, electron microscopic examinations showed degeneration of unmyelinated nerve fibers corresponding to an early loss of nociceptive sensation followed by the loss of small and then larger myelinated nerve fibers at the advanced stage [7]. There were distinct nerve pathologies between early- and late-onset V30M FAP. Sural nerve biopsies showed a reduction of small-diameter nerves in early-onset patients. In contrast, the sural nerve biopsies in the late-onset group were characterized by variable fiber size distribution and axonal sprouting with relative preservation of unmyelinated fibers. The amounts of amyloid deposition in nerves and ganglia (sensory and sympathetic) were more prominent in early-onset cases than the late-onset ones. The patterns of neuronal size distribution indicated a significant reduction of small-diameter neurons in the early-onset group and a universal loss of neurons in late-onset patients [80].

Skin biopsy is a recognized approach to demonstrate pathologic evidence of skin denervation in small-fiber sensory neuropathy. In patients of Ala97Ser mutation with late-onset disabling neuropathy, skin biopsies showed (1) depletion of intraepidermal nerve fibers and degeneration of dermal nervous plexus (Fig. 8.2) and (2) reduction of sudomotor nerves (Fig. 8.3), which

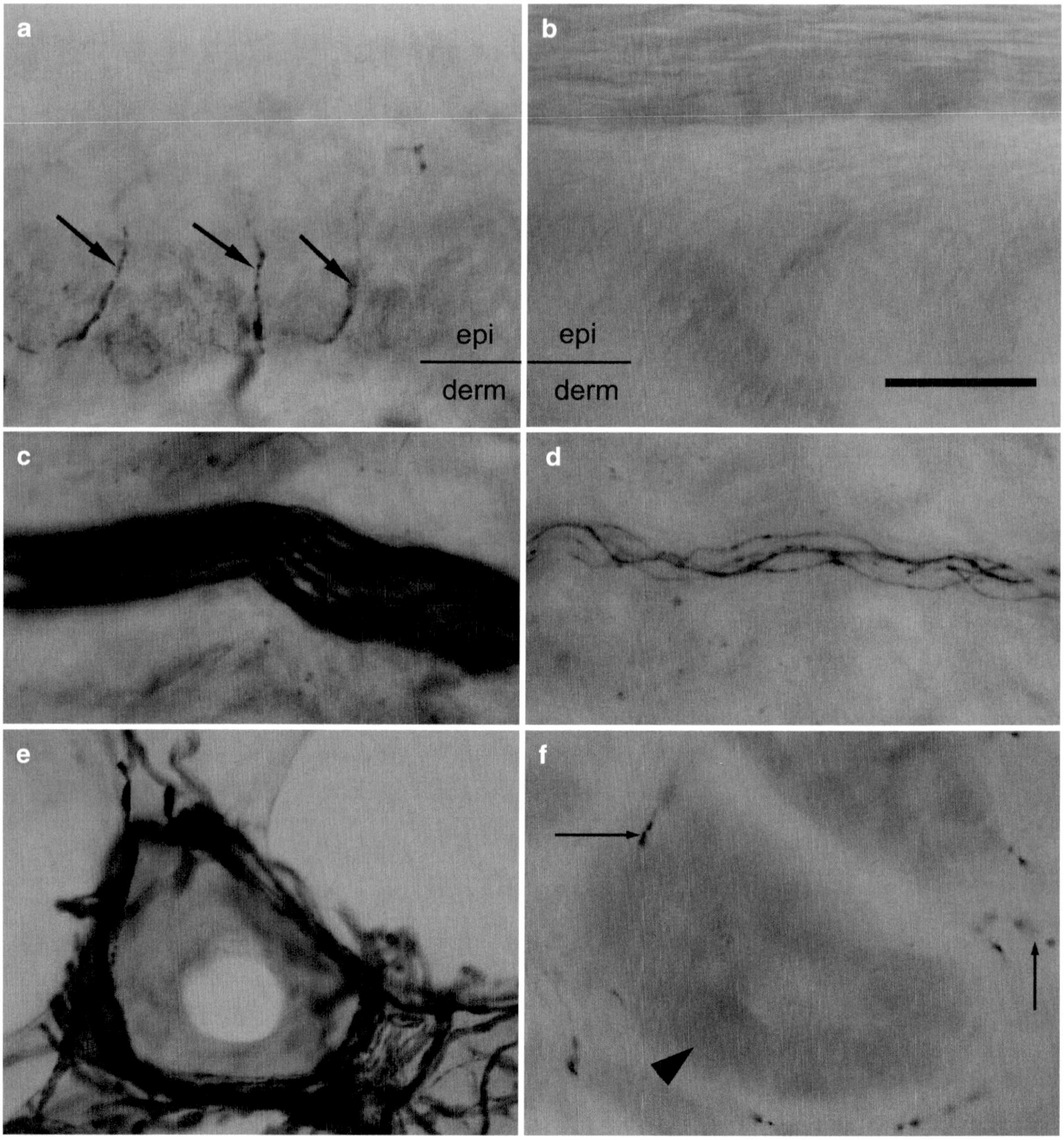

Fig. 8.2 Skin denervation in amyloid neuropathy with mutated transthyretin (Ala97Ser). Skin sections were immunostained with anti-protein gene product (PGP) 9.5 in the controls (**a**, **c**, **e**) and Ala97Ser (**b**, **d**, **f**) for nerve fibers in the epidermis (epi) and dermis (derm). (**a**) In the epidermis of the control skin, PGP 9.5 (+) intraepidermal nerve fibers (arrows) with a varicose appearance. (**b**) Intraepidermal nerve fibers were depleted in an Ala97Ser patient. (**c**) In the dermis of the control skin, nerve fibers in a nerve bundle appeared as dense immunoreactive lines. (**d**) In an Als97Ser patient, dermal nerve fibers were markedly reduced in number, and only some individual dermal nerve fibers appeared as fragmented lines. (**e**) In the control skin, nerve fibers surrounded secretory coils of sweat glands. (**f**) Innervation of sweat glands in Ala97Ser was almost completely absent and only some immunoreactive dots (*arrows*) in the surrounding of sweat coils (*arrowhead*) could be seen, which represented degenerated nerve fibers. Bar, 30 μm. Reproduced with permission from [61]

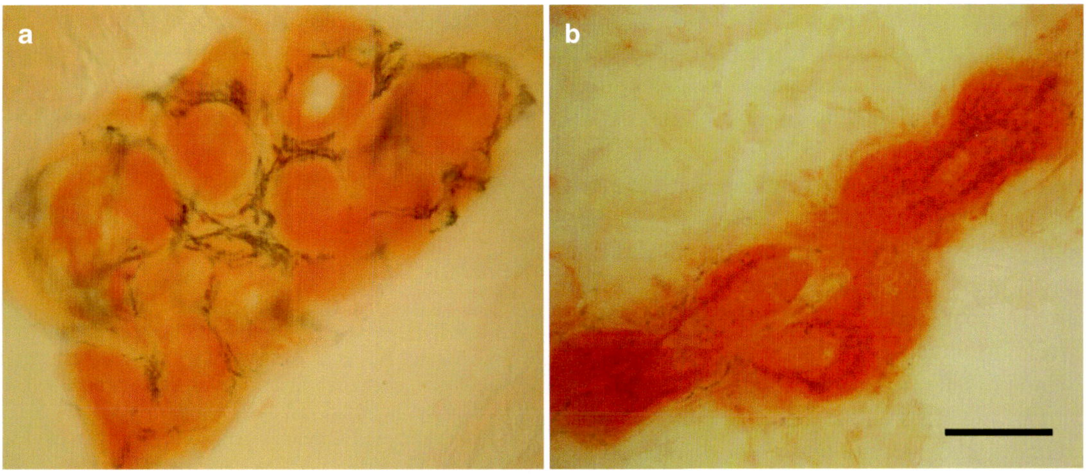

Fig. 8.3 Sudomotor innervation in familial amyloid polyneuropathy (FAP). Sudomotor nerves were stained with protein gene product 9.5 (PGP9.5), a neuronal marker, followed by counterstaining of Congo red (at pH 8.0). (**a**) Sweat glands were stained in red with surrounding PGP9.5(+) nerve fibers. (**b**) Sweat glands of FAP patients became denervated with profiles indicating nerve degeneration. Bar = 50 μm. Reproduced with permission from [70]

reflected the pathology of small-fiber sensory and autonomic nerves, paralleling clinical disability [61, 70].

8.3.8 Treatment of TTR-FAP

The therapies of TTR-FAP include symptomatic alleviation and specific treatments aiming at avoiding or slowing amyloid deposition in tissues. In addition to liver transplantation, there has been great improvement in the treatment of TTR-FAP since 2010 in particular, various pharmacological strategies and designs.

Liver transplantation inhibits the formation of amyloid deposits by depleting the major source of mutant TTR in the body. In the FAP World Transplant Registry (http://www.fapwtr.org/ram_fap.htm), there have been >2000 liver transplants as the largest number conducted in endemic areas. Retrospective studies with multivariate analysis indicated a 20-year survival of 55.3% posttransplantation with independent factors favorable for survival: (1) higher modified body mass index, (2) early-onset type (less than 50 years of age), (3) shorter duration of disease, and (4) Val30Met mutation [81].

Tafamidis, a benzoxazole derivative based on rational drug design, binds to TTR selectively with high affinity. This mode of action stabilizes the tetrameric form kinetically and slows the formation of monomers, misfolding, and aggregation of amyloid. In a double-blind and placebo-controlled study for 18 months on early-onset V30M TTR-FAP, tafamidis 20 mg once daily enhanced the stability of TTR tetramer in 98% of the tafamidis group which was associated with amelioration of neurological declines, stabilization of large- and small-fiber functions and life quality, and better nutritional profiles [82]. In the 12-month, open-label extension study, tafamidis continued to stabilize TTR tetramer in >93% of participants with the slowing of disease progression extending to 30 months in the patients continuing tafamidis. The progression of disease slowed in the patients switching from placebo to tafamidis during a 12-month active treatment [83]. Apparently the stabilizing effect of tafamidis was also present in non-V30M participants [84]. Currently in Europe, tafamidis is indicated for treating adult patients of TTR amyloidosis with stage 1 polyneuropathy to delay neurological declines [85].

Diflunisal, a generic nonsteroidal anti-inflammatory drug (NSAID), binds to the T4-binding site which could stabilize circulating TTR tetramers kinetically and reduce TTR monomers and hence amyloid formation [86, 87]. In a

2-year investigator-initiated, double-blind, and placebo-controlled study, 130 TTR-FAP patients with confirmed polyneuropathy at the clinical level took diflunisal (250 mg) or placebo twice per day. Diflunisal significantly reduced the progression of neurologic impairments and preserved quality of life according to neuropathy impairment score plus 7 nerve tests (NIS+7) and life-quality questionnaire (Short Form-36) [88]. In an observational study on TTR-FAP, the beneficial effects of diflunisal continued at 24 months and 24 months later in the context of neurological and cardiac functions. Most patients could tolerate diflunisal although periodic tests of renal functions and blood cell counts are required [89]. In a retrospective analysis on off-label use of diflunisal in TTR-FAP, discontinuation of diflunisal was reported in 57% of patients due to adverse effects, mainly gastrointestinal discomforts [85].

In addition to conventional pharmaceutical designs, two important studies in 2018 highlighted a milestone in FAP therapies by taking advantage of new biotechnological developments, i.e., RNA-based approaches at the level of transcription. Adams et al. conducted a clinical trial with a RNA interference (RNAi) therapy, patisiran, in TTR-FAP at the early stage and reported a significant reduction of disease progression assessed with the modified Neuropathy Impairment Score (mNIS+7), Norfolk Quality of Life-Diabetic Neuropathy (QOL-DN) questionnaire, 10-meter walk test, and modified body mass index (mBMI) [90]. Benson et al. tested antisense oligonucleotide therapy, inotersen, on TTR-FAP (stages 1 and 2) and showed a significant reduction of neurological disability (mNIS+7) and patient-reported outcome measure of QOL-DN [91]. There studies provide new insights and directions for treating TTR-FAP [92].

8.4 Summary

In conclusion, small-fiber neuropathy due to amyloidosis is frequently present, as an early and major manifestation of amyloidosis including AL amyloidosis and TTR-FAP. In addition to symptomatic treatment for neurological deficits, there has been considerable progress in the therapy of underlying disease, for example, chemotherapy and target therapy for chronic hematological disorders causing AL amyloidosis and TTR stabilizers or RNA-based therapies in FAP.

References

1. Sipe JD, Benson MD, Buxbaum JN, Ikeda SI, Merlini G, Saraiva MJ, et al. Amyloid fibril proteins and amyloidosis: chemical identification and clinical classification International Society of Amyloidosis 2016 Nomenclature Guidelines. Amyloid. 2016;23:209–13.
2. Sunde M, Serpell LC, Bartlam M, Fraser PE, Pepys MB, Blake CC. Common core structure of amyloid fibrils by synchrotron X-ray diffraction. J Mol Biol. 1997;273:729–39.
3. Sawaya MR, Sambashivan S, Nelson R, Ivanova MI, Sievers SA, Apostol MI, et al. Atomic structures of amyloid cross-beta spines reveal varied steric zippers. Nature. 2007;447:453–7.
4. Bonar L, Cohen AS, Skinner MM. Characterization of the amyloid fibril as a cross-beta protein. Proc Soc Exp Biol Med. 1969;131:1373–5.
5. Gillmore JD, Hawkins PN. Pathophysiology and treatment of systemic amyloidosis. Nat Rev Nephrol. 2013;9:574–86.
6. Wechalekar AD, Gillmore JD, Hawkins PN. Systemic amyloidosis. Lancet. 2016;387:2641–54.
7. Plante-Bordeneuve V, Said G. Familial amyloid polyneuropathy. Lancet Neurol. 2011;10:1086–97.
8. Testro AG, Brennan SO, Macdonell RA, Hawkins PN, Angus PW. Hereditary amyloidosis with progressive peripheral neuropathy associated with apolipoprotein AI Gly26Arg: outcome of hepatorenal transplantation. Liver Transpl. 2007;13:1028–31.
9. Rajkumar SV, Gertz MA, Kyle RA. Prognosis of patients with primary systemic amyloidosis who present with dominant neuropathy. Am J Med. 1998;104:232–7.
10. Gertz MA. Immunoglobulin light chain amyloidosis: 2016 update on diagnosis, prognosis, and treatment. Am J Hematol. 2016;91:947–56.
11. Kyle RA, Gertz MA. Primary systemic amyloidosis: clinical and laboratory features in 474 cases. Semin Hematol. 1995;32:45–59.
12. Pinney JH, Smith CJ, Taube JB, Lachmann HJ, Venner CP, Gibbs SD, et al. Systemic amyloidosis in England: an epidemiological study. Br J Haematol. 2013;161:525–32.
13. Duston MA, Skinner M, Anderson J, Cohen AS. Peripheral neuropathy as an early marker of AL amyloidosis. Arch Intern Med. 1989;149:358–60.
14. Kelly JJ Jr, Kyle RA, O'Brien PC, Dyck PJ. The natural history of peripheral neuropathy in primary systemic amyloidosis. Ann Neurol. 1979;6:1–7.

15. Matsuda M, Gono T, Morita H, Katoh N, Kodaira M, Ikeda S. Peripheral nerve involvement in primary systemic AL amyloidosis: a clinical and electrophysiological study. Eur J Neurol. 2011;18:604–10.
16. Vital C, Vital A, Bouillot-Eimer S, Brechenmacher C, Ferrer X, Lagueny A. Amyloid neuropathy: a retrospective study of 35 peripheral nerve biopsies. J Peripher Nerv Syst. 2004;9:232–41.
17. Verghese JP, Bradley WG, Nemni R, McAdam KP. Amyloid neuropathy in multiple myeloma and other plasma cell dyscrasias. A hypothesis of the pathogenesis of amyloid neuropathies. J Neurol Sci. 1983;59:237–46.
18. Sommer C, Schroder JM. Amyloid neuropathy: immunocytochemical localization of intra- and extracellular immunoglobulin light chains. Acta Neuropathol. 1989;79:190–9.
19. Gertz MA. Immunoglobulin light chain amyloidosis diagnosis and treatment algorithm 2018. Blood Cancer J. 2018;8:44.
20. Jones NF, Hilton PJ, Tighe JR, Hobbs JR. Treatment of "primary" renal amyloidosis with melphalan. Lancet. 1972;2:616–9.
21. Kyle RA, Bayrd ED. Amyloidosis: review of 236 cases. Medicine (Baltimore). 1975;54:271–99.
22. Jaccard A, Moreau P, Leblond V, Leleu X, Benboubker L, Hermine O, et al. High-dose melphalan versus melphalan plus dexamethasone for AL amyloidosis. N Engl J Med. 2007;357:1083–93.
23. Mhaskar R, Kumar A, Behera M, Kharfan-Dabaja MA, Djulbegovic B. Role of high-dose chemotherapy and autologous hematopoietic cell transplantation in primary systemic amyloidosis: a systematic review. Biol Blood Marrow Transplant. 2009;15:893–902.
24. Mahmood S, Palladini G, Sanchorawala V, Wechalekar A. Update on treatment of light chain amyloidosis. Haematologica. 2014;99:209–21.
25. Barsottini OG, Arantes A, Sigulem D, Kutner JM, Ribeiro AA, Moura LA, et al. Axoval neuropathy as initial manifestation of primary amyloidosis: report of a case submitted to bone marrow transplantation. Arq Neuropsiquiatr. 2004;62:725–9.
26. Katoh N, Matsuda M, Yoshida T, Yazaki M, Morita H, Sakashita K, et al. Primary AL amyloid polyneuropathy successfully treated with high-dose melphalan followed by autologous stem cell transplantation. Muscle Nerve. 2010;41:138–43.
27. Vucic S, Chong PS, Cros D. Atypical presentations of primary amyloid neuropathy. Muscle Nerve. 2003;28:696–702.
28. Raz A, Goodman DS. The interaction of thyroxine with human plasma prealbumin and with the prealbumin-retinol-binding protein complex. J Biol Chem. 1969;244:3230–7.
29. Andrade C. A peculiar form of peripheral neuropathy; familiar atypical generalized amyloidosis with special involvement of the peripheral nerves. Brain. 1952;75:408–27.
30. Araki S, Mawatari S, Ohta M, Nakajima A, Kuroiwa Y. Polyneuritic amyloidosis in a Japanese family. Arch Neurol. 1968;18:593–602.
31. Andersson R. Familial amyloidosis with polyneuropathy. A clinical study based on patients living in northern Sweden. Acta Med Scand Suppl. 1976;590:1–64.
32. Kabat EA, Moore DH, Landow H. An Electrophoretic study of the protein components in cerebrospinal fluid and their relationship to the serum proteins. J Clin Invest. 1942;21:571–7.
33. Fleming CE, Mar FM, Franquinho F, Sousa MM. Chapter 17: transthyretin: an enhancer of nerve regeneration. Int Rev Neurobiol. 2009;87:337–46.
34. Aleshire SL, Bradley CA, Richardson LD, Parl FF. Localization of human prealbumin in choroid plexus epithelium. J Histochem Cytochem. 1983;31:608–12.
35. Jacobsson B, Collins VP, Grimelius L, Pettersson T, Sandstedt B, Carlstrom A. Transthyretin immunoreactivity in human and porcine liver, choroid plexus, and pancreatic islets. J Histochem Cytochem. 1989;37:31–7.
36. Episkopou V, Maeda S, Nishiguchi S, Shimada K, Gaitanaris GA, Gottesman ME, et al. Disruption of the transthyretin gene results in mice with depressed levels of plasma retinol and thyroid hormone. Proc Natl Acad Sci U S A. 1993;90:2375–9.
37. Blake CC, Geisow MJ, Oatley SJ, Rerat B, Rerat C. Structure of prealbumin: secondary, tertiary and quaternary interactions determined by Fourier refinement at 1.8 A. J Mol Biol. 1978;121:339–56.
38. Blake CC, Burridge JM, Oatley SJ. X-ray analysis of thyroid hormone binding to prealbumin. Biochem Soc Trans. 1978;6:1114–8.
39. Sekijima Y. Transthyretin (ATTR) amyloidosis: clinical spectrum, molecular pathogenesis and disease-modifying treatments. J Neurol Neurosurg Psychiatry. 2015;86:1036–43.
40. Richardson SJ, Bradley AJ, Duan W, Wettenhall RE, Harms PJ, Babon JJ, et al. Evolution of marsupial and other vertebrate thyroxine-binding plasma proteins. Am J Phys. 1994;266:R1359–70.
41. Tsuzuki T, Mita S, Maeda S, Araki S, Shimada K. Structure of the human prealbumin gene. J Biol Chem. 1985;260:12224–7.
42. Ando Y, Nakamura M, Araki S. Transthyretin-related familial amyloidotic polyneuropathy. Arch Neurol. 2005;62:1057–62.
43. Uemichi T, Liepnieks JJ, Benson MD. A trinucleotide deletion in the transthyretin gene (delta V 122) in a kindred with familial amyloidotic polyneuropathy. Neurology. 1997;48:1667–70.
44. Skare J, Yazici H, Erken E, Dede H, Cohen A, Milunsky A, et al. Homozygosity for the met30 transthyretin gene in a Turkish kindred with familial amyloidotic polyneuropathy. Hum Genet. 1990;86:89–90.
45. Yoshinaga T, Nakazato M, Ikeda S, Ohnishi A. Homozygosity for the transthyretin-Met30 gene in three Japanese siblings with type I familial amyloidotic polyneuropathy. Neurology. 1992;42:2045–7.
46. Quintas A, Vaz DC, Cardoso I, Saraiva MJ, Brito RM. Tetramer dissociation and monomer partial unfolding precedes protofibril formation in

amyloidogenic transthyretin variants. J Biol Chem. 2001;276:27207–13.
47. Hammarstrom P, Wiseman RL, Powers ET, Kelly JW. Prevention of transthyretin amyloid disease by changing protein misfolding energetics. Science. 2003;299:713–6.
48. Colon W, Kelly JW. Partial denaturation of transthyretin is sufficient for amyloid fibril formation in vitro. Biochemistry. 1992;31:8654–60.
49. Schneider F, Hammarstrom P, Kelly JW. Transthyretin slowly exchanges subunits under physiological conditions: a convenient chromatographic method to study subunit exchange in oligomeric proteins. Protein Sci. 2001;10:1606–13.
50. Chung CM, Connors LH, Benson MD, Walsh MT. Biophysical analysis of normal transthyretin: implications for fibril formation in senile systemic amyloidosis. Amyloid. 2001;8:75–83.
51. Ferrao-Gonzales AD, Palmieri L, Valory M, Silva JL, Lashuel H, Kelly JW, et al. Hydration and packing are crucial to amyloidogenesis as revealed by pressure studies on transthyretin variants that either protect or worsen amyloid disease. J Mol Biol. 2003;328:963–74.
52. Sekijima Y, Hammarstrom P, Matsumura M, Shimizu Y, Iwata M, Tokuda T, et al. Energetic characteristics of the new transthyretin variant A25T may explain its atypical central nervous system pathology. Lab Invest. 2003;83:409–17.
53. Shnyrov VL, Villar E, Zhadan GG, Sanchez-Ruiz JM, Quintas A, Saraiva MJ, et al. Comparative calorimetric study of non-amyloidogenic and amyloidogenic variants of the homotetrameric protein transthyretin. Biophys Chem. 2000;88:61–7.
54. Bucciantini M, Giannoni E, Chiti F, Baroni F, Formigli L, Zurdo J, et al. Inherent toxicity of aggregates implies a common mechanism for protein misfolding diseases. Nature. 2002;416:507–11.
55. Hou X, Richardson SJ, Aguilar MI, Small DH. Binding of amyloidogenic transthyretin to the plasma membrane alters membrane fluidity and induces neurotoxicity. Biochemistry. 2005;44:11618–27.
56. Sousa MM, Cardoso I, Fernandes R, Guimaraes A, Saraiva MJ. Deposition of transthyretin in early stages of familial amyloidotic polyneuropathy: evidence for toxicity of nonfibrillar aggregates. Am J Pathol. 2001;159:1993–2000.
57. Reixach N, Deechongkit S, Jiang X, Kelly JW, Buxbaum JN. Tissue damage in the amyloidoses: transthyretin monomers and nonnative oligomers are the major cytotoxic species in tissue culture. Proc Natl Acad Sci U S A. 2004;101:2817–22.
58. Nagasaka T. Familial amyloidotic polyneuropathy and transthyretin. Subcell Biochem. 2012;65:565–607.
59. Hou X, Parkington HC, Coleman HA, Mechler A, Martin LL, Aguilar MI, et al. Transthyretin oligomers induce calcium influx via voltage-gated calcium channels. J Neurochem. 2007;100:446–57.
60. Nakazato M, Shiomi K, Miyazato M, Matsukura S. Type I familial amyloidotic polyneuropathy in Japan. Intern Med. 1992;31:1335–8.
61. Yang NC, Lee MJ, Chao CC, Chuang YT, Lin WM, Chang MF, et al. Clinical presentations and skin denervation in amyloid neuropathy due to transthyretin Ala97Ser. Neurology. 2010;75:532–8.
62. Sousa A, Coelho T, Barros J, Sequeiros J. Genetic epidemiology of familial amyloidotic polyneuropathy (FAP)-type I in Povoa do Varzim and Vila do Conde (north of Portugal). Am J Med Genet. 1995;60:512–21.
63. Holmgren G, Costa PM, Andersson C, Asplund K, Steen L, Beckman L, et al. Geographical distribution of TTR met30 carriers in northern Sweden: discrepancy between carrier frequency and prevalence rate. J Med Genet. 1994;31:351–4.
64. Misu K, Hattori N, Nagamatsu M, Ikeda S, Ando Y, Nakazato M, et al. Late-onset familial amyloid polyneuropathy type I (transthyretin Met30-associated familial amyloid polyneuropathy) unrelated to endemic focus in Japan. Clinicopathological and genetic features. Brain. 1999;122:1951–62.
65. Plante-Bordeneuve V, Kerschen P. Transthyretin familial amyloid polyneuropathy. Handb Clin Neurol. 2013;115:643–58.
66. Benson MD, Cohen AS. Generalized amyloid in a family of Swedish origin. A study of 426 family members in seven generations of a new kinship with neuropathy, nephropathy, and central nervous system involvement. Ann Intern Med. 1977;86:419–24.
67. Blanco-Jerez CR, Jimenez-Escrig A, Gobernado JM, Lopez-Calvo S, de Blas G, Redondo C, et al. Transthyretin Tyr77 familial amyloid polyneuropathy: a clinicopathological study of a large kindred. Muscle Nerve. 1998;21:1478–85.
68. Obayashi K, Ando Y. Focus on autonomic dysfunction in familial amyloidotic polyneuropathy (FAP). Amyloid. 2012;19(Suppl 1):28–9.
69. Koike H, Misu K, Ikeda S, Ando Y, Nakazato M, Ando E, et al. Type I (transthyretin Met30) familial amyloid polyneuropathy in Japan: early- vs late-onset form. Arch Neurol. 2002;59:1771–6.
70. Chao CC, Huang CM, Chiang HH, Luo KR, Kan HW, Yang NC, et al. Sudomotor innervation in transthyretin amyloid neuropathy: pathology and functional correlates. Ann Neurol. 2015;78:272–83.
71. Mariani LL, Lozeron P, Theaudin M, Mincheva Z, Signate A, Ducot B, et al. Genotype-phenotype correlation and course of transthyretin familial amyloid polyneuropathies in France. Ann Neurol. 2015;78:901–16.
72. Plante-Bordeneuve V, Ferreira A, Lalu T, Zaros C, Lacroix C, Adams D, et al. Diagnostic pitfalls in sporadic transthyretin familial amyloid polyneuropathy (TTR-FAP). Neurology. 2007;69:693–8.
73. Sattianayagam PT, Hahn AF, Whelan CJ, Gibbs SD, Pinney JH, Stangou AJ, et al. Cardiac phenotype and clinical outcome of familial amyloid polyneuropathy associated with transthyretin alanine 60 variant. Eur Heart J. 2012;33:1120–7.
74. Benson MD, Kincaid JC. The molecular biology and clinical features of amyloid neuropathy. Muscle Nerve. 2007;36:411–23.

75. Conceicao IM, Castro JF, Scotto M, de Carvalho M. Neurophysiological markers in familial amyloid polyneuropathy patients: early changes. Clin Neurophysiol. 2008;119:1082–7.
76. Koike H, Tanaka F, Hashimoto R, Tomita M, Kawagashira Y, Iijima M, et al. Natural history of transthyretin Val30Met familial amyloid polyneuropathy: analysis of late-onset cases from non-endemic areas. J Neurol Neurosurg Psychiatry. 2012;83:152–8.
77. Heldestad V, Nordh E. Quantified sensory abnormalities in early genetically verified transthyretin amyloid polyneuropathy. Muscle Nerve. 2007;35:189–95.
78. Hofer PA, Anderson R. Postmortem findings in primary familial amyloidosis with polyneuropathy. Acta Pathol Microbiol Scand A. 1975;83:309–22.
79. Sobue G, Nakao N, Murakami K, Yasuda T, Sahashi K, Mitsuma T, et al. Type I familial amyloid polyneuropathy. A pathological study of the peripheral nervous system. Brain. 1990;113:903–19.
80. Koike H, Misu K, Sugiura M, Iijima M, Mori K, Yamamoto M, et al. Pathology of early- vs late-onset TTR Met30 familial amyloid polyneuropathy. Neurology. 2004;63:129–38.
81. Ericzon BG, Wilczek HE, Larsson M, Wijayatunga P, Stangou A, Pena JR, et al. Liver transplantation for hereditary transthyretin amyloidosis: after 20 years still the best therapeutic alternative? Transplantation. 2015;99:1847–54.
82. Coelho T, Maia LF, Martins da Silva A, Waddington Cruz M, Plante-Bordeneuve V, Lozeron P, et al. Tafamidis for transthyretin familial amyloid polyneuropathy: a randomized, controlled trial. Neurology. 2012;79:785–92.
83. Coelho T, Maia LF, da Silva AM, Cruz MW, Plante-Bordeneuve V, Suhr OB, et al. Long-term effects of tafamidis for the treatment of transthyretin familial amyloid polyneuropathy. J Neurol. 2013;260:2802–14.
84. Merlini G, Plante-Bordeneuve V, Judge DP, Schmidt H, Obici L, Perlini S, et al. Effects of tafamidis on transthyretin stabilization and clinical outcomes in patients with non-Val30Met transthyretin amyloidosis. J Cardiovasc Transl Res. 2013;6:1011–20.
85. Adams D, Suhr OB, Hund E, Obici L, Tournev I, Campistol JM, et al. First European consensus for diagnosis, management, and treatment of transthyretin familial amyloid polyneuropathy. Curr Opin Neurol. 2016;29(Suppl 1):S14–26.
86. Sekijima Y, Dendle MA, Kelly JW. Orally administered diflunisal stabilizes transthyretin against dissociation required for amyloidogenesis. Amyloid. 2006;13:236–49.
87. Miller SR, Sekijima Y, Kelly JW. Native state stabilization by NSAIDs inhibits transthyretin amyloidogenesis from the most common familial disease variants. Lab Invest. 2004;84:545–52.
88. Berk JL, Suhr OB, Obici L, Sekijima Y, Zeldenrust SR, Yamashita T, et al. Repurposing diflunisal for familial amyloid polyneuropathy: a randomized clinical trial. JAMA. 2013;310:2658–67.
89. Sekijima Y, Tojo K, Morita H, Koyama J, Ikeda S. Safety and efficacy of long-term diflunisal administration in hereditary transthyretin (ATTR) amyloidosis. Amyloid. 2015;22:79–83.
90. Adams D, Gonzalez-Duarte A, O'Riordan WD, Yang CC, Ueda M, Kristen AV, et al. Patisiran, an RNAi therapeutic, for hereditary transthyretin amyloidosis. N Engl J Med. 2018;379:11–21.
91. Benson MD, Waddington-Cruz M, Berk JL, Polydefkis M, Dyck PJ, Wang AK, et al. Inotersen treatment for patients with hereditary transthyretin amyloidosis. N Engl J Med. 2018;379:22–31.
92. Buxbaum JN. Oligonucleotide drugs for transthyretin amyloidosis. N Engl J Med. 2018;379:82–5.

9 Small Fiber Pathology and Functional Impairment in Syndromes of Predominantly Large Fiber Neuropathy

Chi-Chao Chao, Chun-Liang Pan, and Sung-Tsang Hsieh

Abstract

In peripheral neuropathies with sensory ataxia or weakness of distal limbs, traditionally nerve conduction studies are the first-line tests, and these neuropathies are classified as large fiber neuropathy including (1) inflammatory neuropathies of Guillain-Barré syndrome (GBS), chronic inflammatory demyelinating polyneuropathy (CIDP), and polyneuropathy associated with organomegaly, endocrinopathy, monoclonal gammopathy, and skin hyperpigmentation (POEMS), (2) hereditary neuropathies of Charcot-Marie-Tooth (CMT) disease encompassing different genotypes, and (3) chemotherapy-induced peripheral neuropathy (CIPN). Patients with this phenotype of large fiber neuropathy frequently have symptoms and signs suggesting small fiber involvement, i.e., small fiber nerve degeneration and functional impairment. With the applications of skin biopsy for measuring intraepidermal nerve fiber density (IENFD), quantitative sensory testing for assessing thermal thresholds, and autonomic function tests for examining sympathetic and parasympathetic axis, concomitant small fiber pathology and functional impairment (small fiber pathology or syndrome) were documented in some of these neuropathies traditionally considered as pure large fiber neuropathy. Not only serving as a criterion of small fiber pathology, parameters of small fiber examinations also provide a surrogate marker of disease activities and progression reflecting generalized ambulation difficulty and correlated with patient-reported outcome measure of quality of life.

Keywords

Large fiber neuropathy · Small fiber neuropathy · Pain-evoked potential · Quality of life Guillain-Barré syndrome · Charcot-Marie-Tooth disease · Chronic inflammatory demyelinating polyneuropathy · Chemotherapy-induced peripheral neuropathy

C.-C. Chao
Department of Neurology, National Taiwan University Hospital, Taipei, Taiwan

C.-L. Pan
Graduate Institute of Molecular Medicine, National Taiwan University College of Medicine, Taipei, Taiwan

Center of Precision Medicine, National Taiwan University College of Medicine, Taipei, Taiwan

S.-T. Hsieh (✉)
Department of Neurology, National Taiwan University Hospital, Taipei, Taiwan

Center of Precision Medicine, National Taiwan University College of Medicine, Taipei, Taiwan

Department of Anatomy and Cell Biology, National Taiwan University College of Medicine, Taipei, Taiwan

Graduate Institute of Brain and Mind Sciences, National Taiwan University College of Medicine, Taipei, Taiwan
e-mail: shsieh@ntu.edu.tw

S.-T. Hsieh et al. (eds.), *Small Fiber Neuropathy and Related Syndromes: Pain and Neurodegeneration*,
https://doi.org/10.1007/978-981-13-3546-4_9

9.1 Introduction

Traditionally most neuropathies are examined with nerve conduction studies as the first-line assessment, and hence large fiber impairment is easily documented. This is obvious in most neuropathies with limb weakness and incoordination due to large fiber neuropathy involving motor and sensory nerves, for example, Guillain-Barré syndrome (GBS) and vasculitic neuropathy. However, neuropathic pain and autonomic dysfunction do exist in a certain proportion of patients with clinically overt large fiber neuropathy [1, 2]. These observations raised the possibility of superimposed small fiber nerve degeneration or functional impairment, frequently seen in diabetes, chronic kidney disease, and metabolic syndromes [3–7]. Since the definition of small fiber neuropathy excludes large fiber neuropathy, the concomitant small fiber involvement, i.e., small fiber nerve degeneration and functional impairment, is termed small fiber pathology or syndrome in this chapter. With applications of skin biopsy, quantitative sensory testing, and autonomic functional examinations, small fiber neuropathy has been readily demonstrated in traditional large fiber neuropathy, including Guillain-Barré syndrome and inflammatory disease (discussed in Chap. 10).

9.2 Small Fiber Nerve Degeneration in Inflammatory Neuropathies: Guillain-Barré Syndrome and Chronic Inflammatory Demyelinating Polyneuropathy

Typical inflammatory neuropathies include Guillain-Barré syndrome (GBS) and chronic inflammatory demyelinating polyneuropathy (CIDP) with inflammatory infiltrates as the major pathology, and the clinical manifestations mainly present with large fiber neuropathy of progressive weakness and sensory ataxia [8, 9]. Neuropathic pain over the affected limbs with weakness is common in patients with GBS, which raised the possibility of a concomitant small fiber involvement. Our group was among the first to explore this issue by examining skin innervation and quantitative sensory testing [10]. Small fiber nerve degeneration, manifested with depletion of intraepidermal nerve fibers (IENFs) which was confirmed by decreased intraepidermal nerve fiber density (IENFD) and elevated thermal thresholds, did exist in a considerable proportion of patients with classical GBS, which typically develops acute ascending weakness from lower limbs to upper limbs within 2 weeks after disease onset [10]. Several groups later confirmed this observation and documented the correlation of neuropathic pain and the degree of skin observation measured as IENFD [11–13]. In addition to thermal thresholds on quantitative sensory testing that act as psychophysical surrogate markers of small fiber impairment, pain-evoked potentials, in particular, contact heat-evoked potentials (CHEP), serve as physiological signatures of small fiber deficits. This is done by stimulating the limbs with defined temperatures and recording the evoked waveforms over the cerebral cortex [14–18]. Zhang et al. provided physiological evidence of small fiber neuropathy in GBS by demonstrating reduced amplitudes and prolonged latencies of CHEP [19]. However, CHEP parameters were similar between the patients with or without paresthesias of the limbs, which suggested that additional central mechanisms may contribute to the clinical manifestations of neuropathic pain besides peripheral mechanisms due to nerve degeneration. Further studies are required to test this hypothesis of altered activation of the limbic system as reported in diabetic neuropathy [20–22].

In addition to skin denervation in classical GBS, i.e., ascending weakness of limbs within 4 weeks of onset, there are reports of acute sensory neuropathy with several patterns: pure small fiber neuropathy or association with neuropathic deficits of large fiber or autonomic nerves, which are common manifestations of GBS [23]. The Sobue group systematically characterized a series of 21 patients with acute sensory and autonomic neuropathy and documented early degeneration of small fiber in sural nerve biopsies

[24]. Small neurons in sensory and autonomic ganglia were affected at the initial phase, followed by large neuron degeneration later, according to an autopsy case [24]. There existed a unique syndrome of acute small fiber neuropathy with erythromelalgia as the sole presentation [25]. Pediatric or adolescent patients developed acute painful limbs with vasodilation refractory to conventional antihyperalgesic treatments. These patients had rapid relief of symptoms after steroid therapy, suggesting immune-related neuropathy affecting small fibers [25, 26]. Most case reports of acute pure small fiber neuropathy are diagnosed based on symptom description and clinical examinations [26]. Tests for autonomic function or small fiber physiology were applied to provide objective assessment, for example, abnormal Valsalva test on acute small fiber neuropathy following mycoplasma infection [27]. A syndrome of acute erythromelalgia mainly in childhood and young adulthood apparently represented a variant form of acute small fiber sensory neuropathy [25, 26]. The underlying pathophysiology was explored by passive transfer experiments and examining the binding of immunoglobulin from human sera on rat dorsal ganglia neurons [28]. Accordingly, acute small fiber neuropathy can be categorized as (1) small fiber neuropathy associated with classical GBS [10–13], (2) pure sensory neuropathy with the involvement of both large and small fibers [23, 25, 29], and (3) pure small fiber neuropathy [12, 23, 26, 27]. Based on operational definition of sensory GBS, i.e., acute neuropathy with exclusive sensory symptoms and signs, Uncini and Yuki proposed that sensory GBS could be further classified into three subtypes of acute sensory demyelinating polyneuropathy, acute sensory large fiber axonopathy-ganglionopathy, and acute sensory small fiber neuropathy-ganglionopathy [23].

A further question is whether small fiber nerve degeneration or functional impairments were present in chronic inflammatory demyelinating polyneuropathy (CIDP). We demonstrated skin denervation with reduced IENFD in CIDP, suggesting the coexistence of small fiber neuropathy in CIDP, although the mechanisms or significance could be different from GBS, as IENFD was correlated with functional disability scale in GBS but not in CIDP [30]. Furthermore, the degree of functional impairment in CIDP was not as severe as in GBS, i.e., most small fiber impairment and autonomic dysfunctions were subclinical [30, 31].

Polyneuropathy, organomegaly, endocrinopathy, monoclonal gammopathy, and skin hyperpigmentation (POEMS) is a distinct syndrome with clinical manifestations of progressive weakness in four limbs and electrophysiological and pathological characteristics of diffuse demyelination and myelin pathology, such as uncompacted myelin [32, 33]. Although POEMS shares clinical features of CIDP, such as progressive weakness of limbs and neurophysiological characteristics on nerve conduction studies, the pathology of POEMS was distinct compared to that of CIDP, with more severe axonal degeneration and increased epineurial arterioles but less inflammation [32–34]. IENFD was decreased in POEMS, but the unmyelinated nerve fiber density was preserved in sural nerves of POEMS, suggesting limited involvement of the most distal parts of the nerve fibers [35]. Importantly, the reduction of IENFD was correlated with the presence of glucose intolerance, indicating the systemic effects of metabolic derangement of POEMS on the small fibers [35]. Neuropathic pain was reported in ~50% of POEMS patients and was related to the loss of myelinated nerve fibers and increased circulating pro-inflammatory cytokines [36]. Vascular endothelial growth factor (VEGF) and erythropoietin (EPO) showed opposite changes in the pathogenesis of nerve damage of POEMS: higher VEGF and lower EPO [37].

9.3 Small Fiber Nerve Degeneration in Chemotherapy-Induced Peripheral Neuropathy (CIPN)

In most patients with CIPN, traditional assessments included clinical instruments or patient-reported questionnaires of functional disability or neuropathic pain [38–40], for example, Participant Neurotoxicity Questionnaire (PNQ) with the highest patient-reported outcome measures and Total Neuropathy Score clinical (TNSc) version with the highest quality according to meta-analysis of instruments [38, 39]. A critical issue was a lack of objective and quantifiable evaluations among these scoring systems, for example, Grade 3 neuropathies defined as sensory loss or paresthesia interfering with activities of daily living and Grade 4 defined as disabling sensory loss that interferes with function according to National Cancer Institute Common Terminology Criteria for Adverse Events (CTCAE) [41]. The FACT/GOG-Neurotoxicity subscale (FACT/GOG-Ntx) consisted of 11 items of symptoms: numbness and discomfort in hands and feet, joint and muscle pain, and trouble feeling and walking, assessed with a 5-point scale, and the total score was categorized into tertiles (tertile 1, no or few symptoms; tertile 2, moderate neuropathy; and tertile 3, severe neuropathy). Clinical judgments include light touch, pain sensation, cold sensation, vibration, muscle strength, grip strength, Purdue Pegboard test, Timed Get Up and Go test, and Fullerton Advanced Balance test which documented long-term disability in cancer survivors [42]. There are several critical issues in applying these assessments, for example, the discordant between clinical assessments and patient-reported outcome measures, which affected the exploration of biomarkers for vulnerability to CIPN of paclitaxel [43]. Furthermore, the dosing regimens and the time course of CIPN among different agents certainly affected the outcome measure, e.g., oxaliplatin vs. paclitaxel [44].

With improvement in cancer therapies, CIPN has become an important and perhaps the most frequent etiology of toxic or medication-related neuropathy. Furthermore, CIPN has become a dose-limiting factor in designing the protocol for treating malignancies [42]. The major symptoms of CIPN include sensory ataxia and neuropathic pain. The former is examined with nerve conduction studies, and the latter raised the possibility of small fiber involvement. Alterations of IENFD were variable in some reports [45], but usually the frequency of small fiber impairment was higher than that of large fiber deficits according to the comparisons among nerve conduction studies, quantitative sensory testing, and skin biopsy [46]. Neuropathic pain is common among different chemotherapeutic agents; platinum-based drugs, such as oxaliplatin; and taxol-based medications, e.g., paclitaxel [40, 47, 48]. These can be documented by somatosensory phenotypes and various patient-reported outcome measures or questionnaires [38, 39, 48]. Among these, 21% of patients had neuropathic pain of burning and tingling, suggesting the involvement of small fibers [47] and the development of neuropathic pain was related to the blood concentration of nerve growth factor but independent of intraepidermal nerve fiber density [47]. In addition to cutaneous sensory nerves, sudomotor functions assessed by electrochemical skin conductance on sudoscan were applied on patients with chemotherapy [49]. In a 3-minute test of sudoscan, the electrochemical skin conductance was reduced in hands and feet after oxaliplatin treatment with corresponding changes in Total Neuropathy Score clinical version [49]. In addition to direct testing of autonomic functions, surrogate markers of perfusion technique, laser Doppler imager-FLARE (LDI_{FLARE}), were applied to patients with platinum analogue and taxanes [50]. LDI_{FLARE} had a higher abnormal rate compared to that of conventional nerve conduction studies [50].

In a study on survivors of pediatric acute lymphoblastic leukemia receiving adjuvant chemotherapy, large fiber deficits were reported in two-thirds of patients, and small fiber manifestations of pain sensitization were documented in one-third of patients according to quantitative sensory testing [51]. A long-term follow-up of patients receiving oxaliplatin indicated both acute neuropathic pain of cold and mechanical hyperalgesia and chronic neuropathic numbness and large fiber deficits even after cessation of chemotherapy [52]. In vincristine and paclitaxel neuropathy, pathology and functional impairments of both large and small fibers were reported by examining touch and thermal sensation with reduction of Meissner corpuscles and IENFs [53]. Remarkably neuropathic pain persisted for more than 4 years after cessation of vincristine and paclitaxel [53]. The combined treatment of paclitzel and carboplatin induced both systolic and diastolic orthostatic hypotension [54]. Systolic orthostatic hypotension was reversible, but diastolic orthostatic hypotension was irreversible, indicating differential involvement in the sympathetic and parasympathetic components of the autonomic nervous system [54].

9.4 Small Fiber Pathology in Hereditary Neuropathies Mainly Affecting Large Fibers

Charcot-Marie-Tooth disease (CMT disease), the most common hereditary neuropathy, is a syndrome with major involvement of large fibers including motor and proprioceptive nerves [55, 56]. Patients exhibit weakness, unsteadiness, and sensory ataxia. Currently more than 70 genes are identified to be associated with CMT disease [55, 57, 58]. Among these different genotypes, CMT1A due to duplication of peripheral myelin protein 22 (PMP22) is the most prevalent one of CMT disease [55]. Skin biopsies on fingers have been applied to investigate the loss of distal myelinated nerves and encapsulated sensory receptors of Meissner corpuscles [59, 60]. Furthermore, this approach also showed increased PMP22 transcripts in CMT1A [61]. Although this technique could not serve as a biomarker of disease progression and therapeutic responses [62], it was able to identify cutaneous transcripts together with plasma metabolome for predicting the progression of the disease assessed with various outcome measures and scores [56, 63]. In addition to symptoms of large fiber neuropathy, patients may also have sensory symptoms of superficial sensitivity, which raised the possibility of small fiber involvement. Corneal confocal microscopy detected small fiber involvement including corneal sensitivity, corneal nerve fiber density, corneal nerve branch density, corneal nerve fiber length, and corneal nerve fiber tortuosity which were significantly reduced [64]. Skin biopsies from the conventional site of the distal leg revealed significant loss of IENFD, suggesting the involvement of small fibers [65, 66]. Furthermore, patients with lower IENFD were associated with superficial pain sensitivity disorder but not with deep pain sensitivity disorder, indicating that association of structural skin denervation was associated with functional deficits [66]. Intriguing, the number of Langerhans cells was reduced in CMT1A [66], which was in contrast to increased Langerhans cell density in diabetes-associated small fiber neuropathy [67] and experimental nerve injury [68]. Since Langerhans cells were reduced in recently diagnosed diabetes mellitus and there was no correlation between Langerhans cell density and IENFD [69], the relationship between IENFs and Langerhans cell and the clinical significance require prospective studies [66].

9.5 Clinical Implications of Small Fiber Neuropathy: Functional Correlations with Disease Activities, Ambulation Disability, and Neuropathic Pain

Skin biopsy not only provides objective evidence of nociceptive nerve degeneration by measuring IENFD but also offers assessments of microvasculitis in the dermis complementary to nerve biopsy for non-systemic vasculitic neuropathy [70]. Furthermore, the biomarkers of skin biopsy including the degrees of skin innervation and microvasculitis provide clinical significance. Importantly, the IENFD and the degree of microvasculitis were correlated with clinical parameters of systemic lupus erythematosus, including circulating complements and anti-double-stranded DNA (anti-dsDNA) [71]. In addition to laboratory examinations, IENFD was correlated with functional disability. This was most pronounced in GBS and eosinophilia, in which IENFD was correlated with functional disability as examined with the grading score of independent ambulation [72]. Moreover, the IENFD at

peripheral neuropathy in a cooperative group trial (S0221): the DELCaP study. J Natl Cancer Inst. 2017;109(12). https://doi.org/10.1093/jnci/djx098.
42. Miaskowski C, Mastick J, Paul SM, et al. Chemotherapy-induced neuropathy in cancer survivors. J Pain Symptom Manag. 2017;54:204–18.
43. Park SB, Kwok JB, Asher R, et al. Clinical and genetic predictors of paclitaxel neurotoxicity based on patient- versus clinician-reported incidence and severity of neurotoxicity in the ICON7 trial. Ann Oncol. 2017;28:2733–40.
44. Pachman DR, Qin R, Seisler D, et al. Comparison of oxaliplatin and paclitaxel-induced neuropathy (Alliance A151505). Support Care Cancer. 2016;24:5059–68.
45. Koskinen MJ, Kautio AL, Haanpaa ML, et al. Intraepidermal nerve fibre density in cancer patients receiving adjuvant chemotherapy. Anticancer Res. 2011;31:4413–6.
46. Kroigard T, Schroder HD, Qvortrup C, et al. Characterization and diagnostic evaluation of chronic polyneuropathies induced by oxaliplatin and docetaxel comparing skin biopsy to quantitative sensory testing and nerve conduction studies. Eur J Neurol. 2014;21:623–9.
47. Velasco R, Navarro X, Gil-Gil M, et al. Neuropathic pain and nerve growth factor in chemotherapy-induced peripheral neuropathy: prospective clinical-pathological study. J Pain Symptom Manag. 2017;54:815–25.
48. Ventzel L, Madsen CS, Karlsson P, et al. Chronic pain and neuropathy following adjuvant chemotherapy. Pain Med. 2018;19(9):1813–24.
49. Saad M, Psimaras D, Tafani C, et al. Quick, non-invasive and quantitative assessment of small fiber neuropathy in patients receiving chemotherapy. J Neuro-Oncol. 2016;127:373–80.
50. Sharma S, Venkitaraman R, Vas PRJ, et al. Assessment of chemotherapy-induced peripheral neuropathy using the LDI (FLARE) technique: a novel technique to detect neural small fiber dysfunction. Brain Behav. 2015;5:e00354.
51. Lieber S, Blankenburg M, Apel K, et al. Small-fiber neuropathy and pain sensitization in survivors of pediatric acute lymphoblastic leukemia. Eur J Paediatr Neurol. 2018;22(3):457–69.
52. Kokotis P, Schmelz M, Kostouros E, et al. Oxaliplatin-induced neuropathy: a long-term clinical and neurophysiologic follow-up study. Clin Colorectal Cancer. 2016;15:e133–40.
53. Boyette-Davis JA, Cata JP, Driver LC, et al. Persistent chemoneuropathy in patients receiving the plant alkaloids paclitaxel and vincristine. Cancer Chemother Pharmacol. 2013;71:619–26.
54. Dermitzakis EV, Kimiskidis VK, Lazaridis G, et al. The impact of paclitaxel and carboplatin chemotherapy on the autonomous nervous system of patients with ovarian cancer. BMC Neurol. 2016;16:190.
55. Weis J, Claeys KG, Roos A, et al. Towards a functional pathology of hereditary neuropathies. Acta Neuropathol. 2017;133:493–515.
56. Fledrich R, Mannil M, Leha A, et al. Biomarkers predict outcome in Charcot-Marie-Tooth disease 1A. J Neurol Neurosurg Psychiatry. 2017;88:941–52.
57. Pisciotta C, Shy ME. Neuropathy. In: Geschwind DH, Paulson HL, Klein C, editors. Handbook of clinical neurology (Neurogenetics, Part II) 2018;148. p. 653–65.
58. Jerath NU, Shy ME. Hereditary motor and sensory neuropathies: understanding molecular pathogenesis could lead to future treatment strategies. Biochim Biophys Acta (BBA) - Mol Basis Dis. 2015;1852:667–78.
59. Li J, Bai Y, Ghandour K, et al. Skin biopsies in myelin-related neuropathies: bringing molecular pathology to the bedside. Brain. 2005;128:1168–77.
60. Manganelli F, Nolano M, Pisciotta C, et al. Charcot-Marie-Tooth disease: new insights from skin biopsy. Neurology. 2015;85:1202–8.
61. Katona I, Wu X, Feely SME, et al. PMP22 expression in dermal nerve myelin from patients with CMT1A. Brain. 2009;132:1734–40.
62. Nobbio L, on behalf of the CMT-TRIAAL Group, Visigalli D, et al. PMP22 messenger RNA levels in skin biopsies: testing the effectiveness of a Charcot-Marie-Tooth 1A biomarker. Brain. 2014;137:1614–20.
63. Soldevilla B, Cuevas-Martin C, Ibanez C, et al. Plasma metabolome and skin proteins in Charcot-Marie-Tooth 1A patients. PLoS One. 2017;12:e0178376.
64. Tavakoli M, Marshall A, Banka S, et al. Corneal confocal microscopy detects small fiber neuropathy in CMT1A patients. Muscle Nerve. 2012;46:698–704.
65. Duchesne M, Danigo A, Richard L, et al. Skin biopsy findings in patients with CMT1A: baseline data from the CLN-PXT3003-01 study provide new insights into the pathophysiology of the disorder. J Neuropathol Exp Neurol. 2018;77:274–81.
66. Nolano M, Manganelli F, Provitera V, et al. Small nerve fiber involvement in CMT1A. Neurology. 2015;84:407–14.
67. Casanova-Molla J, Morales M, Planas-Rigol E, et al. Epidermal Langerhans cells in small fiber neuropathies. Pain. 2012;153:982–9.
68. Hsieh ST, Choi S, Lin WM, et al. Epidermal denervation and its effects on keratinocytes and Langerhans cells. J Neurocytol. 1996;25:513–24.
69. Strom A, Bruggemann J, Ziegler I, et al. Pronounced reduction of cutaneous Langerhans cell density in recently diagnosed type 2 diabetes. Diabetes. 2014;63:1148–53.
70. Uceyler N, Devigili G, Toyka K, et al. Skin biopsy as an additional diagnostic tool in non-systemic vasculitic neuropathy. Acta Neuropathol. 2010;120:109–16.
71. Tseng MT, Hsieh SC, Shun CT, et al. Skin denervation and cutaneous vasculitis in systemic lupus erythematosus. Brain. 2006;129:977–85.
72. Chao CC, Hsieh ST, Shun CT, et al. Skin denervation and cutaneous vasculitis in eosinophilia-associated neuropathy. Arch Neurol. 2007;64:959–65.

73. Yang NC, Lee MJ, Chao CC, et al. Clinical presentations and skin denervation in amyloid neuropathy due to transthyretin Ala97Ser. Neurology. 2010;75:532–8.
74. Chao CC, Huang CM, Chiang HH, et al. Sudomotor innervation in transthyretin amyloid neuropathy: pathology and functional correlates. Ann Neurol. 2015;78:272–83.
75. Suenaga G, Ikeda T, Masuda T, et al. Inflammatory state exists in familial amyloid polyneuropathy that may be triggered by mutated transthyretin. Sci Rep. 2017;7:1579.
76. Martins D, Moreira J, Goncalves NP, et al. MMP-14 overexpression correlates with the neurodegenerative process in familial amyloidotic polyneuropathy. Dis Model Mech. 2017;10:1253–60.
77. Waldfogel JM, Nesbit SA, Dy SM, et al. Pharmacotherapy for diabetic peripheral neuropathy pain and quality of life. Neurology. 2017;88:1958–67.
78. Coelho T, Maia LF, Martins da Silva A, et al. Tafamidis for transthyretin familial amyloid polyneuropathy: a randomized, controlled trial. Neurology. 2012;79:785–92.
79. Bakkers M, Faber CG, Hoeijmakers Janneke GJ, et al. Small fibers, large impact: quality of life in small fiber neuropathy. Muscle Nerve. 2013;49:329–36.
80. Lin MT, Lee LJH, Chao CC, et al. Quality of life in polyneuropathy: association with biomarkers of small fiber impairment. Health Qual Life Outcomes. 2015;13:169.

10 Autoimmune and Infectious Neuropathies

Ahmad R. Abuzinadah
and Christopher H. Gibbons

Abstract

Small nerve fibers, either sensory or autonomic, are commonly involved in various immune-mediated and infectious disorders. More than 60% of patients with Sjögren's disease or systemic lupus erythematosus and a third of patients with sarcoidosis have evidence of small fiber neuropathy on history, examination, or skin biopsy. Patients with antibody-mediated disorders, such as voltage-gated potassium channels, commonly complain of symptoms suggestive of small fiber neuropathy. On physical examination, the presence of non-length-dependent pattern of sensory loss suggests an immune-mediated disorder, while patchy nerve fiber involvement suggests a vasculitic etiology. Infectious disorders such as leprosy, hepatitis C virus, and HIV virus remain common causes of small fiber neuropathy. Therapeutic interventions will focus on treatment of the underlying disease with symptomatic management of the pain and autonomic symptoms.

A. R. Abuzinadah
Neurology Division, Internal Medicine Department, King Abdulaziz University, Jeddah, Saudi Arabia

C. H. Gibbons (✉)
Neurology Department, Harvard Medical School, Beth Israel Deaconess Medical Center, Boston, MA, USA
e-mail: cgibbons@bidmc.harvard.edu

Keywords

Sjögren's syndrome · Systemic lupus erythematosus · Voltage-gated potassium channel · Contactin-associated protein-like 2 (CASPR2) · Rheumatoid arthritis · Cardiac autonomic neuropathy · Sarcoidosis · Celiac disease · Human immunodeficiency virus Hepatitis C · Cryoglobulinemia · Varicella zoster virus · Leprosy · Lyme disease

10.1 Introduction

Infectious, autoimmune, and vasculitic neuropathies are commonly associated with symptoms and signs of small fiber neuropathy. In this chapter, we will discuss the entities that may cause a small fiber neuropathy and present the available information regarding the prevalence of neuropathy and small fiber neuropathy and the associated symptoms and neuropathological findings. We will also discuss specific treatments for these disorders. Immune-mediated and infectious demyelinating neuropathies will be discussed in a separate chapter.

S.-T. Hsieh et al. (eds.), *Small Fiber Neuropathy and Related Syndromes: Pain and Neurodegeneration*,
https://doi.org/10.1007/978-981-13-3546-4_10

Table 10.1 Prevalence of neuropathy and small fiber neuropathy in autoimmune diseases

Disease	Neuropathy	Small fiber neuropathy symptoms or signs
Sjögren's disease	27%	45%
Sarcoidosis	17%	10%
Systemic lupus erythematosus	4–8%	30%
Celiac disease	2.5%	2%
Voltage gated potassium channel	–	50%
Rheumatoid arthritis	57%	Unknown

10.2 Autoimmune Diseases

A summary of the autoimmune diseases and the associated risk of developing small fiber neuropathy is listed in Table 10.1. A more detailed description of each autoimmune condition is provided below.

10.2.1 Sjögren's Syndrome

Sjögren's is an autoimmune disease characterized by progressive damage to glandular tissue with a clinical presentation of sicca syndrome or dry mouth and dry eyes. Sjögren's is one of the more common autoimmune disorders, with up to 1% of the population carrying the diagnosis [1]. A-length-dependent peripheral neuropathy with large fiber features occurs in 22–27% of patients with Sjögren's syndrome [2–4]. Length-dependent axonal sensory or sensorimotor neuropathy is the most common form of neuropathy associated with Sjögren's syndrome [4–8]. A diagnosis of Sjögren's can be made by antibody testing, although anti-SSA and anti-SSB are positive in only 30% of cases with neuropathy who have lip biopsy-proven Sjögren's syndrome and in only 10% of patients presenting with sicca symptoms and neuropathy [5, 8]. Sicca symptoms are reported in 40% of patients with neuropathy and lip biopsy-confirmed Sjögren's syndrome [5]. An isolated small fiber neuropathy has been reported in up to 35–45% of patients with Sjögren's syndrome [9, 10]. In another study, 30% of patients with Sjögren's syndrome had a mixed small and large fiber neuropathy [10]. Small fiber neuropathy in Sjögren's syndrome is frequently non-length-dependent and suggests a small fiber sensory neuronopathy rather than dying back axonal neuropathy [10]. The small fiber neuropathy of Sjögren's commonly affects the trunk, face, and proximal limbs. Symptoms can include burning, shooting, or prickling pain [11]. In 1 retrospective cohort of 62 patients with primary Sjögren's syndrome, only 2 patients had intraepidermal nerve fiber density (IENFD) <3.4 fibers/mm (the threshold for a small fiber neuropathy on skin biopsy in this study), while 27% (17 patients) had evidence of neuropathy on physical examination and 55% had evidence of neuropathy on nerve conduction study [2]. An autopsy finding of neuronal loss in the dorsal root and sympathetic ganglia has led to the theory that the mechanism underlying the painful sensory neuropathy in Sjögren's is due to a ganglioneuropathy affecting small fiber neurons [7, 11, 12].

Most studies have incorporated patients with both primary and secondary Sjögren's in the results. Patients with primary Sjögren's syndrome (PSS) may have small fiber autonomic neuropathy features as well [7]. In a cohort of 92 patients with PSS, autonomic neuropathy was the prominent feature in 3 (3%). However, the prevalence of autonomic neuropathy was likely higher than that reported because among the patients presenting with sensory neuropathy, 56% had Adie's pupil, 40% had orthostatic hypotension, 70% had hypohidrosis, and 20% had constipation [12].

Another neurologic manifestation of Sjögren's syndrome is mononeuritis multiplex, a neuropathy that can impact both small and large nerve fibers in a specific nerve territory. Mononeuritis multiplex was present in 9% of patients with PSS with neuropathy, and 63% of those showed evidence of vasculitis on sural nerve biopsy [8, 12]. In patients with PSS that presented with a sensory neuropathy, 17% had evidence of vasculitis on sural nerve biopsy [12].

A variety of treatment options have been considered, although none have established a high level of evidence. The use of intravenous immunoglobulin in treating Sjögren's syndrome-associated painful neuropathy was based on a retrospective series of studies [13–15]. In one cohort of five patients treated with IVIG (but with no control group), all patients reported improvement based a visual analogue scale with a mean reduction in pain of 74% that was maintained for 2–6 months [15].

10.2.2 Systemic Lupus Erythematosus (SLE)

Systemic lupus erythematosus (SLE) is an autoimmune disorder that impacts multiple end organs with a significant female predilection. The prevalence of neuropathy among patients with SLE has been variably reported between 5% and 27% [3, 16]. In a large retrospective cohort of 1533 patients with SLE, peripheral neuropathy occurred in 14% of patients with SLE; however, a proportion of these had another common etiology for peripheral neuropathy [17]. Of those 8% of neuropathies related to SLE without the presence of other neuropathy etiology, the phenotypic presentations included 25% with pure sensory polyneuropathy, 20% with sensory motor polyneuropathy, and 15% with mononeuritis multiplex. On electrophysiology, 70% of these patients showed an axonal neuropathy, while 20% showed a demyelinating neuropathy [17]. In another large cohort of 2097 patients followed over 25 years, neuropathy was seen in 5.9%. However, only 66.7% (3.9% of the whole cohort) of these cases were thought to be related to SLE, as other common etiologies for neuropathy were present in other patients [18]. A small fiber neuropathy was found in 17% of the neuropathy patients based on skin biopsy analysis of intraepidermal nerve fiber density [18]. However, similar to Sjögren's syndrome, the small fiber neuropathies are non-length-dependent and are suggestive of small fiber dorsal root ganglia involvement [18].

Skin biopsies showing reduced IENFD were seen in 13% of all SLE patients (both with and without clinical manifestations of neuropathy), and the mean IENFDs were lower than healthy control subjects and disease control subjects with rheumatoid arthritis [19]. Skin biopsies with analysis of IENFD were more sensitive (82%) for neuropathy than quantitative sensory testing (QST) (33%) and nerve conduction studies (31%) in patients with SLE [20]. IENFD is correlated with thermal QST thresholds as well as with SLE disease activity [20].

No specific therapies for SLE-related small fiber neuropathy are reported in the literature. However, both SLE vasculitis and SLE neuropathy have been reported to respond to IVIG [14, 21]. It should be noted that these are small, unblinded, retrospective studies that do not have sufficient evidence to support treatment recommendations. Guidelines have been published on the management of neuropsychiatric lupus, and these included patients with neuropathy. It was recommended to treat neuropsychiatric lupus in a similar fashion to SLE without neuropsychiatric manifestations with glucocorticoid and immunosuppressive therapies indicated when the symptoms did not respond to conventional therapy [22].

10.2.3 Voltage-Gated Potassium Channel

Voltage-gated potassium channels (VGKC) have an important role in regulating nociceptive afferent sensory thresholds [23]. Although VGKC do not have a direct link with small fiber neuropathy, pain occurs in 50% of patients with antibodies against VGKC, without other explanations [24]. Contactin-associated protein-like 2 (CASPR2) IgG is a subtype of the VGKC antibody that is more associated with pain, unlike the leucine-rich glioma-inactivated 1 (LGI-1) IgG antibody [24]. Pain distribution occurs more commonly in the extremities but can also present with total body pain. The character of the pain is deep in 40%, burning in 33%, and lancinating in 19% [24]. Small fiber sudomotor dysfunction was present in 29% VGKC patients with pain as documented by thermoregulatory sweat testing or quantitative sudomotor axon reflex testing. Disturbances in heat and heat-pain sensation were present in 6% of VGKC patients with pain [24].

Although there is no clearly established relationship between small fiber neuropathy and VGKC antibodies, approximately 70% of patients with VGKC and pain have reported pain that is relieved with different neuropathic pain medications, typically in combination with immunomodulatory treatment [24]. The immunomodulatory therapy includes prednisone, IVIG, methotrexate, and hydroxychloroquine with some degree of improvement (24). The data is insufficient to make any recommendations for clinical practice.

10.2.4 Rheumatoid Arthritis

Rheumatoid arthritis (RA) is another autoimmune condition that has the potential to cause neurological dysfunction in some patients. A diagnosis of peripheral neuropathy, based on nerve conduction studies, has been documented in 57% of consecutive rheumatoid arthritis patients [25]. Among these RA patients, sensory and sensory motor axonal neuropathy were most common (74%), while mononeuritis multiplex occurred in 11% and demyelinating polyneuropathy in 14% [25]. However, a few case series have been reported of a non-length-dependent small fiber neuropathy in individuals with rheumatoid arthritis [19, 26].

Autonomic dysfunction has been frequently described in association with rheumatoid arthritis [27]. In a case control study, cardiovascular autonomic neuropathy (CAN) was present in 68% of individuals with RA, while sudomotor dysfunction was present in 36% of patients with RA. This study found that the presence of pro-inflammatory cytokines and greater disease activity are more commonly associated with CAN [28]. Although extensive data about the treatment of RA exists, there is no data available about the impact of immunomodulatory treatment on neuropathy severity.

10.3 Inflammatory Diseases (Noninfectious)

10.3.1 Sarcoidosis

Sarcoidosis is a systemic granulomatous disease of unknown etiology. The nervous system is affected in approximately 5% of cases. Neuropathy occurs in 17% of the cases with neurological involvement (i.e., in those with neurosarcoidosis) [29]. The number of patients with neurological involvement may be higher than previously indicated because one study reported symptoms of burning feet in 30% of patients with sarcoidosis and a small fiber neuropathy was documented in a third of such cases [30, 31]. The symptoms of small fiber neuropathy included both burning and electric-like pain. The pain was usually patchy, migratory, and distributed in a noncontiguous pattern [30]. Another large retrospective cohort study of patients with sarcoidosis reported a non-length-dependent neuropathy in 54% [32]. A low IENFD was found in 33% of patients with small fiber neuropathy symptoms and sarcoidosis, while 14% of sarcoidosis patients without small fiber neuropathy symptoms had a reduced IENFD [31].

Multiple mononeuropathies were also found in 10.5% of sarcoid neuropathy cases in a retrospective cohort [33]. Although not specific for a small fiber neuropathy, the mononeuropathies will also damage small nerve fibers that travel with the affected nerve. Sural nerve biopsy showed a necrotizing vasculitis in 63% of patients that presented with focal or multifocal neuropathy [34]. The common histopathological findings include multinucleated giant cell (80%); granulomatous inflammation in the endoneurium, perineurium, and epineurium; as well as multifocal axonal loss which may be an indirect evidence of ischemia [3, 33, 34].

Treatments for sarcoid neuropathy are limited. In one retrospective study, 76% of patients treated with IVIG reported symptomatic improvement, while 66% of patients treated with antitumor necrosis factor reported symptomatic improvement versus 15% in the group who did not receive IVIG or anti-TNF-α [32]. These data might establish a basis for future clinical trials but do not provide sufficient information to make recommendations for clinical practice.

10.3.2 Celiac Disease

Celiac disease is an autoimmune disorder characterized by small intestinal involvement in individuals that are genetically predisposed to development of the disease. Symptoms classically begin in early childhood, but there is significant clinical heterogeneity to the disorder frequently resulting in delayed diagnosis. Neurological manifestations of celiac disease occur in ~10% of patients. Biopsy-proven neuropathy in celiac diseases occurs in 2.5% of patients with no other identifiable causes of neuropathy [35]. Similarly celiac disease occurs in 2.5% of patients with idiopathic autonomic abnormalities [36]. Neurologic manifestations include paresthesia, imbalance, and length-dependent sensory loss. However, 50% of patients with celiac disease and neuropathy are reported to have normal nerve conduction studies [35]. In a small series of eight patients with celiac disease and neuropathy, 88% had reduced IENFD on skin biopsy, while 13% had abnormal nerve conduction studies. These findings suggest that small fiber neuropathy may be more frequent than large fiber neuropathy in patients with celiac disease [37]. Although antibodies are frequently used in the diagnosis of celiac disease, only 65% of biopsy-proven celiac disease had positive antibodies against gangliosides [35]. The potential mechanism of neuropathy in celiac disease has been attributed to malabsorption and deficiency of vitamin B12 and E [3]. Therapy targeting the underlying disease (i.e., avoidance of gluten) may be helpful in reducing symptoms related to neuropathy, but no long-term data exists at this time about outcomes. No other potential therapies have sufficient evidence to support clinical recommendations for treatment of celiac neuropathy at this time. Due to the potential for malabsorption of vitamin B12 and Vitamin E, supplementation of both should be considered.

10.4 Infectious Diseases

A summary of the infectious diseases and the associated risk of developing small fiber neuropathy is reported in Table 10.2. A more detailed description of each infectious disease is provided below.

10.4.1 Human Immunodeficiency Virus

Human immunodeficiency virus (HIV) infection and some of the associated treatments for the retrovirus are both known to cause neuropathy [38]. The prevalence of neuropathy in patients with HIV is approximately 31% and reported to occur in up to 69% of patients over a long enough period of time [39]. The most common form of neuropathy is a distal symmetric polyneuropathy which occurs in 35% of patients [38]. The IENFD at the distal leg and proximal thigh is lower among HIV patients with distal sensory polyneuropathy than control subjects [40]. In addition, HIV patients with IENFD of ≤10 fibers/mm at the distal leg are more likely to progress to a symptomatic distal sensory polyneuropathy [41]. The prevalence of small fiber neuropathy among HIV patients has not been well described but may be higher than previously anticipated [42]. One study reported a high prevalence of small fiber dysfunction among HIV patients with neuropathic symptoms. In patients with HIV, quantitative sudomotor axon reflex testing was significantly lower among HIV patients, and the abnormalities in QSART

Table 10.2 Prevalence of neuropathy and small fiber neuropathy in infectious diseases

Disease	Neuropathy	Isolated small fiber neuropathy	Evidence of small fiber neuropathy
Human immunodeficiency virus (HIV)	31–69%	Common	QSART, IENFD, and pain-evoked potentials
Hepatitis C	9%	Rare[a]	Clinical
Hepatitis C + cryoglobulinemia	45%	Rare[a]	Clinical
Lyme disease	32%	Rare[a]	Clinical
Varicella zoster virus	10% (PHN)	Rare[a]	Clinical
Leprosy	All symptomatic patients	Rare[a]	Clinical

[a]But commonly seen in association as part of a large fiber or mixed neuropathy

were correlated with a visual analogue scale of pain severity [42].

A more recent study of the use of skin biopsy in patients with HIV reported a significant difference in IENFD between HIV patients and control subjects [43]. However, there were no difference in IENFD between HIV patients with and without symptoms of neuropathy and no difference between those with and without exam findings of neuropathy [43]. The study concluded that in certain disorders, such as HIV, skin biopsy analysis of IENFD may not be an effective marker of HIV-associated sensory neuropathy [43].

Treatment of HIV small fiber neuropathy follows the general guidelines of symptomatic treatment of HIV neuropathy and of HIV, accordingly. The only agent specifically studied in HIV neuropathy is the capsaicin 8% patch, which showed efficacy over placebo for the treatment of pain in HIV neuropathy [44]. Despite an absence of evidence in the treatment of pain associated with HIV neuropathy, the use of amitriptyline, gabapentin, and pregabalin is widely accepted [45].

10.4.2 Hepatitis C and Cryoglobulinemia

Hepatitis C is a virus that primarily affects the liver but also causes cryoglobulinemia, a condition where proteins become insoluble at reduced temperatures leading to clumping of protein that can potentially block blood vessels. The proteins are mostly comprised of immunoglobulins (IgM) and precipitate at temperatures below 37 degrees Celsius, therefore more frequently affecting the distal limbs due to cooler temperatures. Polyneuropathy was found in 37% of patients with hepatitis C. Neuropathy was more prevalent among patients with hepatitis C with cryoglobulinemia (45%) than patients without cryoglobulinemia (9%). There are three types of cryoglobulinemia: type I is associated with clonal hematological disorders, while types II and III are associated with hepatitis C [46, 47]. Hepatic C virus replication was not detected in the muscle and nerve among patients with hepatitis C and neuropathy indicating that the neuropathy is an immune-triggered response rather than a direct effect of the virus [48].

Distal axonal polyneuropathy is the most common form of neuropathy in hepatitis C infection (83%), while mononeuritis multiplex occurred in 10% of cases [48]. Distal axonal neuropathy was typically painful and was associated with vasculitis in nerve and muscle biopsies [48]. Some studies found that mononeuritis multiplex in hepatitis C is associated with medium size vasculitis and necrotizing arteritis, while distal axonal neuropathy is typically associated with perivascular infiltrates; these findings were not confirmed by other studies [48, 49]. Small fiber neuropathy has been reported but only rarely so the true prevalence is not known [50].

There are no guidelines specific for treatment of small fiber neuropathy due to hepatitis C and cryoglobulinemia. Treatments targeting hepatitis C by genotype have become very effective at eradicating the virus, but the effects on existing neuropathy have not been studied [51].

10.4.3 Herpes and Other Viruses

The incidence of herpes varicella zoster virus (VZV) is around 3.4 per 1000 persons, and 10% of patients developed postherpetic neuralgia (PHN) [52]. The lifetime risk is 20–30% [53]. VZV causes neuropathy mainly in a dermatomal distribution. Most commonly VZV affects the thoracic segment followed by lumbosacral and trigeminal distribution [3]. Small fiber neuropathy involvement in VZV occurs in conjunction with other reported peripheral nerve involvement such as polyneuritis and polyradiculoneuritis [54, 55]. Small fiber neuropathy symptoms such as pain and itching may be a prominent feature [54, 55]. The development of a rash that is typically vesicular usually occurs in a week or two after the sensory disturbance. However, neuropathy without zoster (the vesicular rash) can occur, and it is called zoster sine herpete [54].

If pain persists for more than 4 weeks after the rash disappears, it is described as postherpetic neuralgia (PHN) [56]. The pain occurs in the affected dermatome and it has a burning and lancinating quality [55]. Possible mechanisms of PHN include structural changes in the dorsal root ganglia following reactivation of the latent virus and associated small fiber dysfunction [55].

Therapies for treatment of the pain approved by the FDA for PHN include pregabalin, gabapentin, and lidocaine patches. For refractory cases, radiofrequency ablation of the dorsal root ganglia can be tried [57]. Vaccination against VZV reduces the risk of developing acute zoster by 51% and PHN by 67% [53].

Herpes simplex virus (HSV) has been reported to mimic varicella zoster affecting the second and third trigeminal nerve distribution [58]. In addition, cytomegalovirus has been associated with painful distal polyneuropathy but more frequently in association with another infection (such as HIV) [59].

10.5 Lyme Disease

Lyme disease, caused by bacterial infection with *Borrelia burgdorferi*, can cause a host of widespread symptoms and signs. Neuropathic complications are commonly noted in late disseminated infections. Approximately 32% of patients with chronic untreated Lyme have some form of neuropathy. Other common neurological manifestations are meningoencephalitis and cranial neuropathies [60]. Lyme disease affects mainly large fibers, and it manifests as an axonal neuropathy [61]. Peripheral neuropathy usually develops 6–12 months after the initial infection. In almost 50% of patients, the neuropathy is distal and symmetric, while in the other 50%, there is an asymmetric radicular pain [62]. There is little data to suggest that an isolated small fiber neuropathy is commonly associated with Lyme disease, although single case reports exist [63].

10.5.1 Leprosy

There are two types of leprosy infection: tuberculoid leprosy and lepromatous leprosy. The bacteria reproduce at cooler temperature (27–30 °C), which is usually the temperature of the distal limbs, nose bridge, and ear pinna [64]. In tuberculoid leprosy, there is a focal inflammatory reaction to the bacteria with involvement of the skin and nerves. However, intact cell-mediated immunity contains the tuberculoid infection. As a result, there is infiltration of the nerves by epithelioid cells forming granulomas which lead to the destruction of Schwann cells and axonal damage. The neuropathy is an immune-mediated phenomenon and tends to occur more quickly than in the lymphomatous form of leprosy [65]. Because the nerve damage is not caused by bacteria, the skin Fite scan is usually negative [65]. There is loss of sensation in the affected region (anesthetic skin). The tuberculoid form of leprosy tends to affect superficial nerves such as the ulnar, median, and common peroneal nerves causing mononeuropathies [64].

In lepromatous leprosy, the cell-mediated immunity is impaired resulting in spread of the infection. There is no infiltration of nerves by lymphocytes and no granuloma formation. As a result, the damage to the nerves is to the bacteria rather than a consequence of an immune reaction. The nerve injury is slow and less prominent and hence less likely to cause anesthetic skin. The Fite scan is usually positive in skin biopsy [65]. This lepromatous form of leprosy commonly presents with a distal sensory polyneuropathy. When the face is involved, it causes madarosis (loss of eyebrows and eyelashes) as well as leonine faces. Distal symmetric polyneuropathy might be the only presentation of leprosy [64, 65].

Treatments for leprosy typically consist of dapsone and rifampin daily for 12 months in the tuberculoid form. Dapsone, rifampin, and clofazimine are used in lepromatous leprosy for a total of 2 years of therapy until the skin Fite smear is negative [65]. There is no data on the long-term effect of the treatments on an existing neuropathy.

10.6 Summary

Neuropathy is a common complication of vasculitic, autoimmune, and infectious diseases. The neuropathies are typically mixed, in many cases involving both large and small nerve fibers. It is less common for an isolated small fiber neuropathy to occur, with the exception of HIV and Sjögren's syndrome. The presentation of the neuropathy may aid in narrowing the differential diagnosis. The presence of non-length-dependent pattern of sensory loss suggests an immune-mediated disorder, while patchy nerve fiber involvement suggests a vasculitic etiology, particularly in the setting of a mononeuritis multiplex.

References

1. Brito-Zeron P, Baldini C, Bootsma H, Bowman SJ, Jonsson R, Mariette X, et al. Sjogren syndrome. Nat Rev Dis Primers. 2016;2:16047.
2. Goransson LG, Herigstad A, Tjensvoll AB, Harboe E, Mellgren SI, Omdal R. Peripheral neuropathy in primary sjogren syndrome: a population-based study. Arch Neurol. 2006;63:1612–5.
3. Neuropathies associated with systemic disease. In: Amato AA, Russell JA, editors. Neuromuscular disorders, 2nd ed. New York, NY: McGraw-Hill; 2016.
4. Gemignani F, Marbini A, Pavesi G, Di Vittorio S, Manganelli P, Cenacchi G, et al. Peripheral neuropathy associated with primary Sjogren's syndrome. J Neurol Neurosurg Psychiatry. 1994;57:983–6.
5. Gorson KC, Ropper AH. Positive salivary gland biopsy, Sjogren syndrome, and neuropathy: clinical implications. Muscle Nerve. 2003;28:553–60.
6. Mellgren SI, Conn DL, Stevens JC, Dyck PJ. Peripheral neuropathy in primary Sjogren's syndrome. Neurology. 1989;39:390–4.
7. Mellgren SI, Goransson LG, Omdal R. Primary Sjogren's syndrome associated neuropathy. Can J Neurol Sci. 2007;34:280–7.
8. Grant IA, Hunder GG, Homburger HA, Dyck PJ. Peripheral neuropathy associated with sicca complex. Neurology. 1997;48:855–62.
9. Lopate G, Pestronk A, Al-Lozi M, Lynch T, Florence J, Miller T, et al. Peripheral neuropathy in an outpatient cohort of patients with Sjogren's syndrome. Muscle Nerve. 2006;33:672–6.
10. Chai J, Herrmann DN, Stanton M, Barbano RL, Logigian EL. Painful small-fiber neuropathy in Sjogren syndrome. Neurology. 2005;65:925–7.
11. Gorson KC, Herrmann DN, Thiagarajan R, Brannagan TH, Chin RL, Kinsella LJ, et al. Non-length dependent small fibre neuropathy/ganglionopathy. J Neurol Neurosurg Psychiatry. 2008;79:163–9.
12. Mori K, Iijima M, Koike H, Hattori N, Tanaka F, Watanabe H, et al. The wide spectrum of clinical manifestations in Sjögren's syndrome-associated neuropathy. Brain. 2005;128:2518–34.
13. Molina JA, Benito-Leon J, Bermejo F, Jimenez-Jimenez FJ, Olivan J. Intravenous immunoglobulin therapy in sensory neuropathy associated with Sjogren's syndrome. J Neurol Neurosurg Psychiatry. 1996;60:699.
14. Levy Y, Uziel Y, Zandman GG, Amital H, Sherer Y, Langevitz P, et al. Intravenous immunoglobulins in peripheral neuropathy associated with vasculitis. Ann Rheum Dis. 2003;62:1221–3.
15. Morozumi S, Kawagashira Y, Iijima M, Koike H, Hattori N, Katsuno M, et al. Intravenous immunoglobulin treatment for painful sensory neuropathy associated with Sjögren's syndrome. J Neurol Sci. 2009;279:57–61.
16. Omdal R, Henriksen OA, Mellgren SI, Husby G. Peripheral neuropathy in systemic lupus erythematosus. Neurology. 1991;41:808–11.
17. Florica B, Aghdassi E, Su J, Gladman DD, Urowitz MB, Fortin PR. Peripheral neuropathy in patients with systemic lupus erythematosus. Semin Arthritis Rheum. 2011;41:203–11.
18. Oomatia A, Fang H, Petri M, Birnbaum J. Peripheral neuropathies in systemic lupus erythematosus: clinical features, disease associations, and immunologic characteristics evaluated over a twenty-five-year study period. Arthritis Rheumatol. 2014;66:1000–9.
19. Goransson LG, Brun JG, Harboe E, Mellgren SI, Omdal R. Intraepidermal nerve fiber densities in chronic inflammatory autoimmune diseases. Arch Neurol. 2006;63:1410–3.
20. Tseng MT, Hsieh SC, Shun CT, Lee KL, Pan CL, Lin WM, et al. Skin denervation and cutaneous vasculitis in systemic lupus erythematosus. Brain. 2006;129:977–85.
21. Camara I, Sciascia S, Simoes J, Pazzola G, Salas V, Karim Y, et al. Treatment with intravenous immunoglobulins in systemic lupus erythematosus: a series of 52 patients from a single centre. Clin Exp Rheumatol. 2014;32:41–7.
22. Bertsias GK, Ioannidis JP, Aringer M, Bollen E, Bombardieri S, Bruce IN, et al. EULAR recommendations for the management of systemic lupus erythematosus with neuropsychiatric manifestations: report of a task force of the EULAR standing committee for clinical affairs. Ann Rheum Dis. 2010;69:2074–82.
23. Ocana M, Cendan CM, Cobos EJ, Entrena JM, Baeyens JM. Potassium channels and pain: present realities and future opportunities. Eur J Pharmacol. 2004;500:203–19.
24. Klein CJ, Lennon VA, Aston PA, McKeon A, Pittock SJ. Chronic pain as a manifestation of potassium channel-complex autoimmunity. Neurology. 2012;79:1136–44.
25. Agarwal V, Singh R, Wiclaf CS, Tahlan A, Ahuja CK, et al. Clinical, electrophysiological, and pathological study of neuropathy in rheumatoid arthritis. Clin Rheumatol. 2008;27:841–4.
26. Gemignani F, Giovanelli M, Vitetta F, Santilli D, Bellanova MF, Brindani F, et al. Non-length

dependent small fiber neuropathy. A prospective case series. J Peripher Nerv Syst. 2010;15:57–62.
27. Adlan AM, Lip GY, Paton JF, Kitas GD, Fisher JP. Autonomic function and rheumatoid arthritis: a systematic review. Semin Arthritis Rheum. 2014;44:283–304.
28. Syngle V, Syngle A, Garg N, Krishan P, Verma I. Predictors of autonomic neuropathy in rheumatoid arthritis. Auton Neurosci. 2016;201:54–9.
29. Fritz D, van de Beek D, Brouwer MC. Clinical features, treatment and outcome in neurosarcoidosis: systematic review and meta-analysis. BMC Neurol. 2016;16:220.
30. Tavee J, Culver D. Sarcoidosis and small-fiber neuropathy. Curr Pain Headache Rep. 2011;15:201–6.
31. Bakkers M, Merkies IS, Lauria G, Devigili G, Penza P, Lombardi R, et al. Intraepidermal nerve fiber density and its application in sarcoidosis. Neurology. 2009;73:1142–8.
32. Tavee JO, Karwa K, Ahmed Z, Thompson N, Parambil J, Culver DA. Sarcoidosis-associated small fiber neuropathy in a large cohort: clinical aspects and response to IVIG and anti-TNF alpha treatment. Respir Med. 2017;126:135–8.
33. Burns TM, Dyck PJ, Aksamit AJ, Dyck PJ. The natural history and long-term outcome of 57 limb sarcoidosis neuropathy cases. J Neurol Sci. 2006;244:77–87.
34. Said G, Lacroix C, Plante-Bordeneuve V, Le Page L, Pico F, Presles O, et al. Nerve granulomas and vasculitis in sarcoid peripheral neuropathy: a clinicopathological study of 11 patients. Brain. 2002;125:264–75.
35. Chin RL, Sander HW, Brannagan TH, Green PH, Hays AP, Alaedini A, et al. Celiac neuropathy. Neurology. 2003;60:1581–5.
36. Gibbons CH, Freeman R. Autonomic neuropathy and coeliac disease. J Neurol Neurosurg Psychiatry. 2005;76:579–81.
37. Brannagan TH 3rd, Hays AP, Chin SS, Sander HW, Chin RL, Magda P, et al. Small-fiber neuropathy/neuronopathy associated with celiac disease: skin biopsy findings. Arch Neurol. 2005;62:1574–8.
38. Kaku M, Simpson DM. HIV neuropathy. Curr Opin HIV AIDS. 2014;9(6):521–6.
39. Benevides ML, Filho SB, Debona R, Bergamaschi EN, Nunes JC. Prevalence of peripheral neuropathy and associated factors in HIV-infected patients. J Neurol Sci. 2017;375:316–20.
40. Herrmann DN, McDermott MP, Henderson D, Chen L, Akowuah K, Schifitto G, et al. Epidermal nerve fiber density, axonal swellings and QST as predictors of HIV distal sensory neuropathy. Muscle Nerve. 2004;29:420–7.
41. Herrmann DN, McDermott MP, Sowden JE, Henderson D, Messing S, Cruttenden K, et al. Is skin biopsy a predictor of transition to symptomatic HIV neuropathy? A longitudinal study. Neurology. 2006;66:857–61.
42. Boger MS, Hulgan T, Haas DW, Mitchell V, Smith AG, Singleton JR, et al. Measures of small-fiber neuropathy in HIV infection. Auton Neurosci. 2012;169:56–61.
43. Shikuma CM, Bennett K, Ananworanich J, Gerschenson M, Teeratakulpisarn N, Jadwattanakul T, et al. Distal leg epidermal nerve fiber density as a surrogate marker of HIV-associated sensory neuropathy risk: risk factors and change following initial antiretroviral therapy. J Neurovirol. 2015;21:525–34.
44. Simpson DM, Brown S, Tobias J, Group N-CS. Controlled trial of high-concentration capsaicin patch for treatment of painful HIV neuropathy. Neurology. 2008;70:2305–13.
45. Phillips TJ, Cherry CL, Cox S, Marshall SJ, Rice AS. Pharmacological treatment of painful HIV-associated sensory neuropathy: a systematic review and meta-analysis of randomised controlled trials. PLoS One. 2010;5:e14433.
46. Muchtar E, Magen H, Gertz MA. How I treat cryoglobulinemia. Blood. 2017;129:289–98.
47. Cacoub P, Saadoun D, Limal N, Leger JM, Maisonobe T. Hepatitis C virus infection and mixed cryoglobulinaemia vasculitis: a review of neurological complications. AIDS. 2005;19(Suppl 3):S128–34.
48. Authier FJ, Bassez G, Payan C, Guillevin L, Pawlotsky JM, Degos JD, et al. Detection of genomic viral RNA in nerve and muscle of patients with HCV neuropathy. Neurology. 2003;60:808–12.
49. Cacoub P, Maisonobe T, Thibault V, Gatel A, Servan J, Musset L, et al. Systemic vasculitis in patients with hepatitis C. J Rheumatol. 2001;28:109–18.
50. Tembl JI, Ferrer JM, Sevilla MT, Lago A, Mayordomo F, Vilchez JJ. Neurologic complications associated with hepatitis C virus infection. Neurology. 1999;53:861–4.
51. Saadoun D, Thibault V, Si Ahmed SN, Alric L, Mallet M, Guillaud C, et al. Sofosbuvir plus ribavirin for hepatitis C virus-associated cryoglobulinaemia vasculitis: VASCUVALDIC study. Ann Rheum Dis. 2016;75:1777–82.
52. Hadley GR, Gayle JA, Ripoll J, Jones MR, Argoff CE, Kaye RJ, et al. Post-herpetic neuralgia: a review. Curr Pain Headache Rep. 2016;20:17.
53. Johnson RW. Herpes zoster and postherpetic neuralgia: a review of the effects of vaccination. Aging Clin Exp Res. 2009;21:236–43.
54. Fox RJ, Galetta SL, Mahalingam R, Wellish M, Forghani B, Gilden DH. Acute, chronic, and recurrent varicella zoster virus neuropathy without zoster rash. Neurology. 2001;57:351–4.
55. Steiner I. Herpes virus infection of the peripheral nervous system. Handb Clin Neurol. 2013;115:543–58.
56. Portenoy RK, Duma C, Foley KM. Acute herpetic and postherpetic neuralgia: clinical review and current management. Ann Neurol. 1986;20:651–64.
57. Kim K, Jo D, Kim E. Pulsed radiofrequency to the dorsal root ganglion in acute herpes zoster and postherpetic neuralgia. Pain Physician. 2017;20:E411–8.
58. Krohel GB, Richardson JR, Farrell DF. Herpes simplex neuropathy. Neurology. 1976;26:596–7.

59. Wulff EA, Wang AK, Simpson DM. HIV-associated peripheral neuropathy: epidemiology, pathophysiology and treatment. Drugs. 2000;59:1251–60.
60. Pachner AR, Steere AC. The triad of neurologic manifestations of Lyme disease: meningitis, cranial neuritis, and radiculoneuritis. Neurology. 1985;35:47–53.
61. Kindstrand E, Nilsson BY, Hovmark A, Nennesmo I, Pirskanen R, Solders G, et al. Polyneuropathy in late Lyme borreliosis—a clinical, neurophysiological and morphological description. Acta Neurol Scand. 2000;101:47–52.
62. Logigian EL, Steere AC. Clinical and electrophysiologic findings in chronic neuropathy of Lyme disease. Neurology. 1992;42:303–11.
63. Hughes PJ, Lane RJ, Cutler SJ, Wright DJ, Abraham RM, Wade JP, et al. Small-fiber dysfunction in a Borrelia burgdorferi infection. Muscle Nerve. 1993;16:221–2.
64. Altman DJ, Amato A. Lepromatous neuropathy. J Clin Neuromuscul Dis. 1999;1:68–73.
65. Ooi WW, Srinivasan J. Leprosy and the peripheral nervous system: basic and clinical aspects. Muscle Nerve. 2004;30:393–409.

Part III

Roles of Small Fiber Pathology: Pain Syndromes and Neurodegeneration

these alterations are causative to the pain in the disorders or whether they are a secondary phenomenon.

11.2 Small Fiber Pathology in Complex Regional Pain Syndrome (CRPS)

Complex regional pain syndrome (CRPS) is a post-traumatic pain disorder characterized by severe spontaneous or stimulus-induced pain that is disproportionate to the inciting event and is associated with autonomic and motor disturbances. CRPS is categorized into CRPS I, without injury of a major nerve trunk, and CRPS II, with nerve injury. However, studies have come out indicating minor nerve damage, in particular of C-fibers, also in CRPS I. Deficits in temperature sensation in CRPS were first identified using quantitative sensory testing (QST) in 2000 [1] and later confirmed by other studies [2, 3]. QST being a psychophysical method and assessing the whole tract from the sensory receptor to the brain, it could not be decided whether this was a peripheral or a central phenomenon. Later, skin biopsy studies revealed a reduced innervation density of the epidermis at the affected limb [4, 5]. First reports came from studies looking at tissue post amputation [6, 7], where innervation was found to be reduced, but numbers of cases were very small, and interpretation of these cases at the extreme end of CRPS severity was difficult. In a small cohort, Oaklander et al., using skin punch biopsies, found epidermal innervation to be reduced to about 70% of controls and also of the patients' contralateral side. The authors speculated that a small fiber neuropathy might be pathophysiologically relevant for the development of CRPS. QST findings were not different from controls in this study [4]. Kharkar et al. studied 43 patients. Skin biopsies were analyzed by a commercial laboratory. Mean intraepidermal nerve fiber density (IENFD) was not different between patients and controls, but 21% of the patients had reduced fiber densities at a distal site of the affected limb [5]. The authors concluded that the analysis in the commercial laboratory may have missed subtle findings, or that changes related to abnormal sensory thresholds are functional rather than structural. Thus, the role of small fiber dysfunction (as shown by QST) and of reduced skin innervation in CRPS is as yet unclear. Changes secondary to edema, immobilization, and local treatment, among others, have to be taken into account.

11.3 Small Fiber Pathology and Dysfunction in Fibromyalgia Syndrome (FMS)

11.3.1 QST

Earlier studies reported no difference in warm and cold detection thresholds [8], only elevated warm detection thresholds [9], or even mildly increased sensitivities to non-painful warmth over painful sites and to non-painful cold at all painful and non-painful sites [10]. More recently, functional impairment of small nerve fibers or their ascending tracts in patients with FMS was detected in several studies using QST [11–15]. On average, patients with FMS had elevated detection thresholds for warm and cold stimuli, very similar to those in small fiber neuropathy (SFN) [11]. Detection thresholds for non-noxious mechanical stimuli were also elevated. The discrepancies between earlier and more recent studies may be explained by small sample sizes of patients and controls and also by the presence of subgroups within the FMS spectrum.

11.3.2 Evoked Potentials

Like QST, evoked potentials measure the whole sensory tract from the peripheral receptor to the somatosensory cortex. In contrast to QST, evoked potentials are not dependent on patient reaction time and less on patient cooperation. Cooperation is needed, however, since information on pain thresholds when stimulating and relaxation are prerequisites for successful recordings. Laser-evoked potentials (LEP) and electrically elicited so-called pain-related evoked potentials (PREP) are both evoked potentials upon Aδ fiber stimulation. LEP can be used to evoke C-fiber potentials, but this method is rarely used.

In FMS patients, peak-to-peak amplitudes of PREP were reduced [11], irrespective of whether the foot, the hand, and the face were stimulated. This indicates a conduction deficit in the tracts arising from cutaneous Aδ fibers. This finding was corroborated by the laboratory of de Tommaso and colleagues using LEP [16]. The direct cause of this reduction in amplitude is as yet unknown. Contributing factors may be present at any level of the sensory axis. The pathology of the peripheral skin innervation described below indicates that a deficit in peripheral signal transduction very likely contributes to the reduced PREP and LEP amplitudes.

11.3.3 Skin Biopsy Studies

Some FMS patients complain of burning pain, similar to patients with SFN. This observation led several groups of researchers to investigate skin innervation in FMS. Until present, nine studies reporting on a reduction in distal IENFD on skin punch biopsies of patients with FMS have been published [11–13, 16–21].

We compared 25 FMS patients to patients with major depression but no pain and to healthy controls [11]. The disease control group with major depression was chosen because of the often-heard judgment that FMS is a subtype of depression. On average, FMS patients had a lower IENFD and a lower density of dermal bundles with unmyelinated nerve fibers, compared to controls (see Fig. 11.1 for example). Numbers of myelinated nerve fibers in the skin were not different between groups. In patients with major depression, skin innervation was not different from that in controls. Using individual patient data, about two thirds of the FMS patients in this cohort had reduced IENFD. Oaklander et al. reported on reduced IENFD in 41% of their cohort comprising 27 FMS patients [21]. In the study by Caro and Winter, approximately two thirds of 41 patients were affected [22]. Out of 20 Italian FMS patients, six had reduced IENFD and fulfilled the criteria for SFN [18], similar to 15 of 46 Greek patients, who had reduced IENFD [19].

In some studies, distal and proximal skin samples from the leg were compared. This frequently revealed a non-length-dependent distribution of pathology [11, 17], i.e., denervation was present to a similar degree in the distal and proximal skin. Data from our laboratory, studying two additional groups of FMS patients, confirm that IENFD is reduced in many but not all FMS

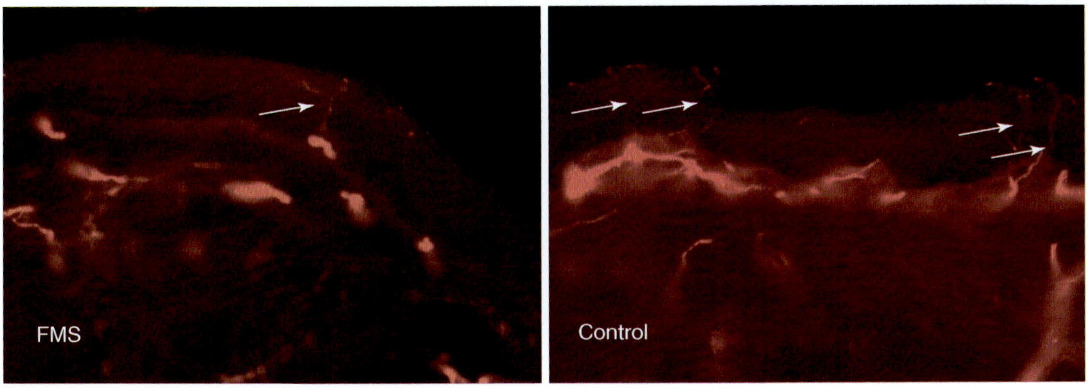

Fig. 11.1 Skin innervation in fibromyalgia syndrome. Examples of intraepidermal nerve fibers on skin punch biopsies obtained from a patient with fibromyalgia syndrome (FMS) and from a healthy control. The photomicrographs show intraepidermal nerve fibers (*arrows*) visualized using the pan-axonal marker protein gene product 9.5 (PGP9.5) in skin samples obtained from the lower leg, pseudocolor. Fiber density is normal in the healthy control and reduced in the patient with FMS

patients and that the non-length-dependent injury type is more prevalent [12, 13].

Further insight was gained from patient skin biopsies regarding possible pathophysiological processes. The number of nerve fibers immunoreactive for growth-associated protein 43 (GAP-43), a marker of nerve regeneration that is normally present in healthy skin nerves, was reduced more than what would have been expected proportionally to the density of protein gene product 9.5 (PGP9.5) reactive fibers in the skin samples from the proximal leg of FMS patients [11]. This may indicate a deficit in the natural regenerative capacity of intraepidermal nerve fibers in FMS patients, but this needs to be confirmed in larger cohorts. Patients with reduced IENFD also had diminished immunoreactivity for insulin growth factor 1 receptor [13], which among other functions is a trophic factor for C-fibers [23]. Electron microscopy revealed that C-fibers from FMS patients had smaller diameters than those from SFN patients and controls [12]. Myelinated nerve fibers were neither affected by numbers nor by diameter, supporting the notion of a specific pathological process in the small fibers [16]. This is in contrast to the finding that Meissner corpuscles, the rapidly adapting mechanoreceptors in the skin [24], were markedly reduced in FMS patients [16]. The report on reduced numbers of Meissner corpuscles will need to be independently reproduced before drawing further conclusions. In any case, the finding might correspond to elevated detection thresholds for non-noxious mechanical stimuli in QST. Autonomic skin fibers were also involved [18], in line with the complaint of many FMS patients of autonomic symptoms. A neurodegenerative process in FMS patients has been speculated upon, given that they had a more rapid reduction of IENFD with ageing [17].

Albrecht et al. studied peptidergic innervation of cutaneous arteriole–venule shunts (AVS) in the palmar glabrous skin of FMS patients [25]. They found that FMS patients had an increased innervation of the AVS compared to controls. Immunohistochemistry revealed that this increase was mainly due to vasodilatory sensory fibers and that vasoconstrictive sympathetic fibers were not different between the groups. The authors reasoned that glabrous skin AVS modulate blood flow to the skin in humans for thermoregulation but also in other tissues such as the skeletal muscle, where they allow increased blood flow during periods of high metabolic demand. They hypothesized that this increased innervation of the AVS might result in insufficient blood flow in deep tissues like skeletal muscles, which might contribute to widespread deep pain and fatigue in FMS patients [25]. Such an intriguing concept will have to be further explored in appropriate models.

11.3.4 Corneal Confocal Microscopy (CCM)

In many disorders, the quantification of corneal innervation by CCM has been used as a surrogate marker for small fiber pathology and has been compared to skin IENFD. Examples are shown in Fig. 11.2. In FMS, CCM data have been published in two studies. One study included 17 patients with FMS and 17 healthy controls [14]. The findings were a smaller diameter of stromal corneal nerves and a diminished subbasal plexus in the FMS group compared to controls. The reason and biological meaning of this finding remains to be determined. Whether an analogy can be drawn to the smaller diameter of the skin C-fibers [12] needs further research. In a second study, reduced nerve fiber length, branching, and/or density in 51% of the 39 patients with FSM were reported when comparing data with published normative values [26]. This finding underscores the concept of widespread small fiber pathology in subgroups of patients with FMS that goes beyond what has been found in the skin of the extremities.

11.3.5 Microneurography

With microneurography the activity of C- or Aδ fibers can be directly recorded in humans. In FMS, recordings from C-fibers of the peroneal nerve revealed increased spontaneous activity [27],

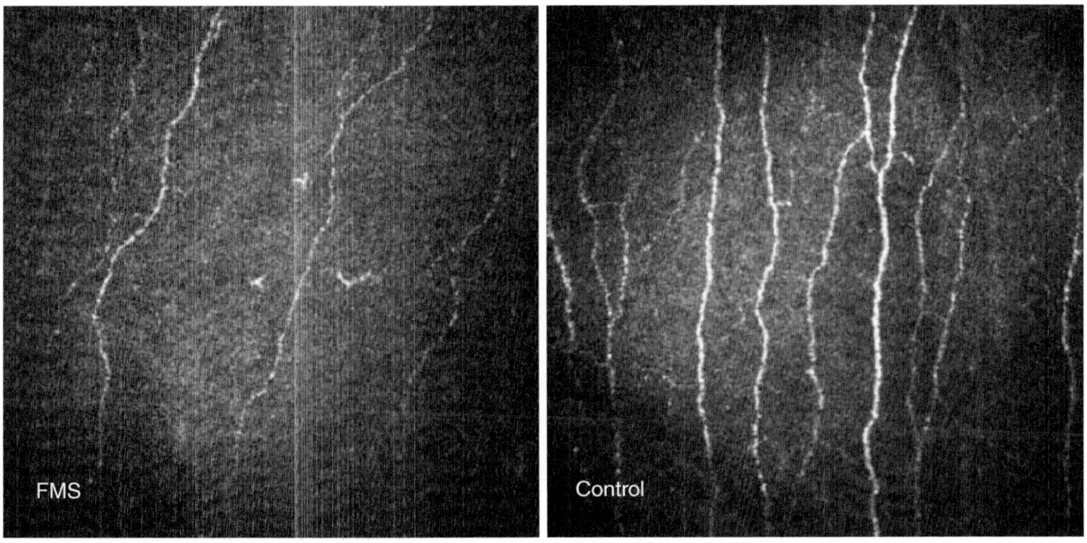

Fig. 11.2 Corneal confocal microscopy in fibromyalgia syndrome. Examples of corneal confocal microscopy (CCM) showing corneal subbasal innervation. Fiber density is normal in the healthy control and reduced in the patient with fibromyalgia syndrome (FMS)

a direct correlate of burning pain. Furthermore, nociceptors normally not responsive to mechanical stimuli ("silent nociceptors") were sensitized to mechanical stimulation. These findings, the spontaneous activity, and the mechanical sensitization were very similar to those in patients with SFN. However, one finding was particular to FMS and not present in SFN: C-fibers from FMS patients had abnormally high slowing of conduction velocity that was not observed in SFN or in controls. Underlying factors discussed as a cause for this slowing were changes in ion channel density, axon diameter, and nerve environment [27]. Intriguingly, C-fibers in the skin of FMS patients are indeed thinner than those of SFN patients and of controls [12]. Whether this is a developmental or a neurodegenerative process is as yet unknown.

11.3.6 Small Fiber Findings in FMS: Relevance and Relation to SFN

Table 11.1 summarizes the present state of findings on small fiber pathology and FMS. In the respective study cohorts, none of the findings were present in all patients investigated, indicating distinct patient subgroups. Future studies will have to show whether subgroups of FMS patients with and without small fiber pathology differ in other aspects, for example, in their response to therapy. Also, there may be a subgroup showing an overlap between SFN (with additional symptoms) and FMS (with burning pain of hands and feet).

On the other hand, IENFD, QST data, and the result of the neurological examination of patients with FMS differed from that of patients with idiopathic SFN [12], which was also true for microneurography results [27]. Thus, small fiber pathology in FMS should not be named SFN [28, 29]. Small fiber pathology indicated by reduced IENFD is present in other chronic pain syndromes, such as postherpetic neuralgia (PHN) and also in non-painful disorders like Parkinson's disease. The findings on small fiber impairment in FMS reported so far are thus not specific. Future studies will be needed to clarify if and how small nerve fiber pathology may contribute to FMS and which pathophysiology underlies the small nerve fiber dysfunction.

Whether small nerve fiber impairment contributes to the pathophysiology of FMS or if it constitutes sequelae or comorbidities of the syndrome cannot be decided at present. We can

Table 11.1 Summary of results obtained in case-control studies on small fiber impairment in fibromyalgia syndrome (FMS) patients

References	FMS	Controls	Disease controls	Methods and results
[11]	25	10	10 unipolar depression, no pain	Skin biopsy: IENFD↓ QST: thermal thresholds↑ PREP: N1-P1 amplitudes↓
[21]	27	30	None	Skin biopsy: IENFD↓
[27]	30	9	17 SFN	Microneurography: pathological C-nociceptor activity
[17]	41	47	None	Skin biopsy: IENFD↓
[19]	46	34	None	Skin biopsy: IENFD↓
[16]	199	109	None	LEP: N2-P2 amplitude↓; N2-P2 habituation↓ Skin biopsy: IENFD↓
[18]	20/0	None	None	Skin biopsy: IENFD↓
[12]	32	24	12 SFN	Skin biopsy: IENFD↓ QST: thermal thresholds↑
[14]	17	17	None	CCM: corneal nerve thickness↓
[26]	39	None	None	CCM: corneal innervation↓ QST: thermal thresholds↑
[13]	30	34	None	Skin biopsy: IENFD↓ QST: thermal thresholds↑

Abbreviations—*CCM* corneal confocal microscopy, *FMS* fibromyalgia syndrome, *IENFD* intraepidermal nerve fiber density, *LEP* laser-evoked potentials, *PREP* pain-related evoked potentials, *QST* quantitative sensory testing, *SFN* idiopathic small fiber neuropathy

only assume, based on current data, that alterations in the central and peripheral nervous system both play a pathophysiological role in subgroups of patients with FMS. Small nerve fiber impairment cannot explain all FMS symptoms, even if it may underlie one major symptom, the FMS pain, via peripheral sensitization. Given that FMS pain is mostly localized in deep tissue, the conceptual gap between impaired superficial skin innervation and deeply located pain needs to be closed.

Up to now, subgroup analysis has been hampered by the small numbers of patients studied in each laboratory. No laboratory performed all possible small fiber tests in one group of patients, so we still do not know how and if the results of different small fiber tests correlate. The heterogeneity of the methods used hinders direct comparison between laboratories. Longitudinal studies are still needed, since the cross-sectional study design does not allow conclusions about questions of state or trait and even less so about causality. We still need to find out if the small fiber changes are possibly epiphenomena of other processes, if they represent an adaptive reaction of the organism, or if they are the primary factor determining the disease. The influence of such comorbidities in FMS patients is hardly ever controlled in clinical studies but also needs to be taken into account. More disease controls are needed, for example, patients with other chronic pain syndromes or with psychiatric disorders, to define the specificity for FMS of a given finding.

At present, small nerve fiber tests cannot be used to make a diagnosis of FMS. The diagnosis of FMS is still made by descriptive criteria like the American College of Rheumatology (ACR) criteria [30, 31] and the symptom-based diagnostic criteria of the German S3 FMS guideline [32]. However, the small fiber findings in FMS for the first time show objectively measureable somatic alterations in the nervous system of patients diagnosed with FMS. This has changed the view of patients, physicians, and scientists on the syndrome. Characterization of small nerve fiber impairment in FMS may help to identify subgroups with a neuropathic component of pain, with a specific pathophysiology, and a distinct response to therapy. Further research is urgent to now understand potential pathophysiological connections that may contribute to FMS symptoms.

11.4 Small Fiber Pathology in Other Nerve Lesions

11.4.1 Postherpetic Neuralgia (PHN)

Small fiber pathology has also been reported after peripheral nerve lesion of different etiology. Postherpetic neuralgia (PHN) is one example occurring in up to 30% of patients with varicella zoster reactivation and affecting spinal nerve endings [33, 34]. Pain may already start before skin lesions are visible or develop during the acute phase of herpes zoster and persist after healing of the cutaneous efflorescences. Severe static and dynamic allodynia and dysesthesias are frequently reported by patients suffering from PHN, and analgesic treatment is often a challenge. Small nerve fibers have been investigated on a functional and morphological level in several studies. Buonocore et al. investigated 25 patients with PHN and obtained skin punch biopsies from the affected and contralateral unaffected body side. IENFD quantification revealed an overall reduction of epidermal innervation in the affected skin samples with, however, large variability among individual patients. Fiber density did not correlate with the presence or the severity of tactile allodynia [35]. We found similar results investigating 13 patients with PHN for skin innervation and QST results in the affected ipsilateral and unaffected contralateral skin area [36]. IENFD was lower in the affected skin, and sensory profiling revealed elevated warm and mechanical detection thresholds with high interindividual variance and no correlation with the presence and severity of pain. In a cross-sectional study, Truini and colleagues investigated skin biopsies obtained from the supraorbital region of patients with ophthalmic PHN [37] and found evidence for a prominent loss of unmyelinated epidermal nerve fibers in affected skin, while autonomic innervation was normal. In another study skin innervation was assessed in patients with herpes zoster at baseline and the development of IENFD was monitored ipsi- and contralaterally and at a third distant control skin area for a period of 6 months after the zoster skin lesion. The authors showed that IENFD was reduced in zoster skin by up to 40% at baseline compared to mirror and control skin. Functional testing for thermal sensation, allodynia, and capsaicin response, however, did not reveal any correlation with skin innervation in patients developing PHN [38]. Interestingly, LEP assessment of 41 patients with PHN revealed lower potential amplitudes after stimulation at the affected side compared to the unaffected side correlating with the intensity of constant pain reported by the patients [39]. Pfau et al. conducted a QST study for sensory profiling of 72 patients with truncal PHN compared with data of 162 healthy controls [40]. Thermal and tactile deficits and dynamic mechanical allodynia were reported as most prominent sensory alteration in the affected skin. These findings underscore the morphological and functional involvement of small nerve fibers in PHN; however, the vast variability of findings in individual cases is striking and reflects different pathophysiological subgroups of deafferentation (i.e., reduction in peripheral IENFD) and severe spontaneous activity of affected dorsal root ganglion neurons without changes in cutaneous innervation. The latter will also provide a neural basis for local analgesic treatment response, which is missing when skin innervation is lost.

11.4.2 Postsurgical Pain

Small nerve fiber pathology may also occur after surgery. One example is persistent inguinal postherniorrhaphy pain that develops in up to 5% of patients [40]. In a study by Bischoff et al., IENFD was reduced in the painful inguinal region compared to the contralateral side, while sensory testing revealed a subgroup of patients with thermal hyposensitivity [41]. Functional and morphological data did not correlate, and also treatment response to lidocaine patch application was not different from placebo. Similarly, 46 patients with persistent inguinal postherniorrhaphy pain were investigated using skin punch biopsy and QST [41]. Reduced IENFD at surgery site was confirmed, while baseline sensory profiles were not compared with healthy controls.

The pathophysiological role of small nerve fiber pathology in this and similar post-surgery conditions remains to be elucidated.

11.5 Other Painful Conditions with Small Fiber Pathology

Deficits in small fiber function or morphology have been reported in several pain syndromes or other disorders associated with pain. As in the conditions discussed above, it seems plausible that a C- or Aδ pathology is causative to the pain, but in most cases this has not been directly proven. For example, skin innervation is reduced in patients with the Ehlers-Danlos syndrome which is associated with generalized joint hypermobility, skin texture abnormalities, visceral and vascular dysfunctions, and pain [42]. In Morvan's syndrome, a disorder with hyperhidrosis and fasciculations, associated with malignant thymoma and autoantibodies to contactin-associated protein (CASPR)-like 2, skin innervation was reduced in three of four patients, indicating involvement of the small nerve fibers [43]. In burning mouth syndrome of unknown origin, the finding of reduced tongue innervation has led to the assumption that this might be a focal small fiber neuropathy [44, 45]. This may give a rationale for topical treatment directed at hyperactive nociceptors [46].

References

1. Birklein F, Riedl B, Sieweke N, Weber M, Neundorfer B. Neurological findings in complex regional pain syndromes—analysis of 145 cases. Acta Neurol Scand. 2000;101:262–9.
2. Üçeyler N, Eberle T, Rolke R, Birklein F, Sommer C. Differential expression patterns of cytokines in complex regional pain syndrome. Pain. 2007;132:195–205.
3. Gierthmühlen J, Maier C, Baron R, Tölle T, Treede RD, Birbaumer N, et al. Sensory signs in complex regional pain syndrome and peripheral nerve injury. Pain. 2012;153:765–74.
4. Oaklander AL, Fields HL. Is reflex sympathetic dystrophy/complex regional pain syndrome type I a small-fiber neuropathy? Ann Neurol. 2009;65:629–38.
5. Kharkar S, Venkatesh YS, Grothusen JR, Rojas L, Schwartzman RJ. Skin biopsy in complex regional pain syndrome: case series and literature review. Pain Physician. 2012;15:255–66.
6. van der Laan L, ter Laak HJ, Gabreels-Festen A, Gabreels F, Goris RJ. Complex regional pain syndrome type I (RSD): pathology of skeletal muscle and peripheral nerve. Neurology. 1998;51:20–5.
7. Albrecht PJ, Hines S, Eisenberg E, Pud D, Finlay DR, Connolly MK, et al. Pathologic alterations of cutaneous innervation and vasculature in affected limbs from patients with complex regional pain syndrome. Pain. 2006;120:244–66.
8. Hurtig IM, Raak RI, Kendall SA, Gerdle B, Wahren LK. Quantitative sensory testing in fibromyalgia patients and in healthy subjects: identification of subgroups. Clin J Pain. 2001;17:316–22.
9. Blumenstiel K, Gerhardt A, Rolke R, Bieber C, Tesarz J, Friederich HC, et al. Quantitative sensory testing profiles in chronic back pain are distinct from those in fibromyalgia. Clin J Pain. 2011;27:682–90.
10. Kosek E, Ekholm J, Hansson P. Sensory dysfunction in fibromyalgia patients with implications for pathogenic mechanisms. Pain. 1996;68:375–83.
11. Üçeyler N, Zeller D, Kahn AK, Kewenig S, Kittel-Schneider S, Schmid A, et al. Small fibre pathology in patients with fibromyalgia syndrome. Brain. 2013;136:1857–67.
12. Doppler K, Rittner HL, Deckart M, Sommer C. Reduced dermal nerve fiber diameter in skin biopsies of patients with fibromyalgia. Pain. 2015;156:2319–25.
13. Leinders M, Doppler K, Klein T, Deckart M, Rittner H, Sommer C, et al. Increased cutaneous miR-let-7d expression correlates with small nerve fiber pathology in patients with fibromyalgia syndrome. Pain. 2016;157:2493–503.
14. Ramirez M, Martinez-Martinez LA, Hernandez-Quintela E, Velazco-Casapia J, Vargas A, Martinez-Lavin M. Small fiber neuropathy in women with fibromyalgia. An in vivo assessment using corneal confocal bio-microscopy. Semin Arthritis Rheum. 2015;45:214–9.
15. da Silva LA, Kazyiama HH, Teixeira MJ, de Siqueira SR. Quantitative sensory testing in fibromyalgia and hemisensory syndrome: comparison with controls. Rheumatol Int. 2013;33:2009–17.
16. de Tommaso M, Nolano M, Iannone F, Vecchio E, Ricci K, Lorenzo M, et al. Update on laser-evoked potential findings in fibromyalgia patients in light of clinical and skin biopsy features. J Neurol. 2014;261:461–72.
17. Caro XJ, Winter EF. Evidence of abnormal epidermal nerve fiber density in fibromyalgia: clinical and immunologic implications. Arthritis Rheumatol. 2014;66:1945–54.
18. Giannoccaro MP, Donadio V, Incensi A, Avoni P, Liguori R. Small nerve fiber involvement in patients referred for fibromyalgia. Muscle Nerve. 2014;49:757–9.
19. Kosmidis ML, Koutsogeorgopoulou L, Alexopoulos H, Mamali I, Vlachoyiannopoulos PG, Voulgarelis

M, et al. Reduction of intraepidermal nerve fiber density (IENFD) in the skin biopsies of patients with fibromyalgia: a controlled study. J Neurol Sci. 2014;347:143–7.
20. Martinez-Martinez LA, Mora T, Vargas A, Fuentes-Iniestra M, Martinez-Lavin M. Sympathetic nervous system dysfunction in fibromyalgia, chronic fatigue syndrome, irritable bowel syndrome, and interstitial cystitis: a review of case-control studies. J Clin Rheumatol. 2014;20:146–50.
21. Oaklander AL, Herzog ZD, Downs HM, Klein MM. Objective evidence that small-fiber polyneuropathy underlies some illnesses currently labeled as fibromyalgia. Pain. 2013;154:2310–6.
22. Caro XJ, Winter EF. The role and importance of small fiber neuropathy in fibromyalgia pain. Curr Pain Headache Rep. 2015;19:55.
23. Gao WQ, Shinsky N, Ingle G, Beck K, Elias KA, Powell-Braxton L. IGF-I deficient mice show reduced peripheral nerve conduction velocities and decreased axonal diameters and respond to exogenous IGF-I treatment. J Neurobiol. 1999;39:142–52.
24. Hoffmann JN, Montag AG, Dominy NJ. Meissner corpuscles and somatosensory acuity: the prehensile appendages of primates and elephants. Anat Rec A Discov Mol Cell Evol Biol. 2004;281:1138–47.
25. Albrecht PJ, Hou Q, Argoff CE, Storey JR, Wymer JP, Rice FL. Excessive peptidergic sensory innervation of cutaneous arteriole-venule shunts (AVS) in the palmar glabrous skin of fibromyalgia patients: implications for widespread deep tissue pain and fatigue. Pain Med. 2013;14:895–915.
26. Oudejans L, He X, Niesters M, Dahan A, Brines M, van Velzen M. Cornea nerve fiber quantification and construction of phenotypes in patients with fibromyalgia. Sci Rep. 2016;6:23573.
27. Serra J, Collado A, Sola R, Antonelli F, Torres X, Salgueiro M, et al. Hyperexcitable C nociceptors in fibromyalgia. Ann Neurol. 2014;75:196–208.
28. Üçeyler N. Small fiber pathology-a culprit for many painful disorders? Pain. 2016;157(Suppl 1):S60–6.
29. Devigili G, Tugnoli V, Penza P, Camozzi F, Lombardi R, Melli G, et al. The diagnostic criteria for small fibre neuropathy: from symptoms to neuropathology. Brain. 2008;131:1912–25.
30. Wolfe F, Clauw DJ, Fitzcharles MA, Goldenberg DL, Katz RS, Mease P, et al. The American College of Rheumatology preliminary diagnostic criteria for fibromyalgia and measurement of symptom severity. Arthritis Care Res (Hoboken). 2010;62:600–10.
31. Wolfe F, Smythe HA, Yunus MB, Bennett RM, Bombardier C, Goldenberg DL, et al. The American College of Rheumatology 1990 criteria for the classification of fibromyalgia. Report of the Multicenter Criteria Committee. Arthritis Rheum. 1990;33:160–72.
32. Eich W, Bar KJ, Bernateck M, Burgmer M, Dexl C, Petzke F. et al. [Definition, classification, clinical diagnosis and prognosis of fibromyalgia syndrome: updated guidelines 2017 and overview of systematic review articles]. Schmerz. 2017;31:231–8.
33. Dworkin RH, Gnann JW Jr, Oaklander AL, Raja SN, Schmader KE, Whitley RJ. Diagnosis and assessment of pain associated with herpes zoster and postherpetic neuralgia. J Pain. 2008;9:S37–44.
34. Cohen JI. Clinical practice: Herpes zoster. N Engl J Med. 2013;369:255–63.
35. Buonocore M, Gatti AM, Amato G, Aloisi AM, Bonezzi C. Allodynic skin in post-herpetic neuralgia: histological correlates. J Cell Physiol. 2012;227:934–8.
36. Üçeyler N, Valet M, Kafke W, Tölle TR, Sommer C. Local and systemic cytokine expression in patients with postherpetic neuralgia. PLoS One. 2014;9:e105269.
37. Truini A, Haanpaa M, Provitera V, Biasiotta A, Stancanelli A, Caporaso G, et al. Differential myelinated and unmyelinated sensory and autonomic skin nerve fiber involvement in patients with ophthalmic postherpetic neuralgia. Front Neuroanat. 2015;9:105.
38. Petersen KL, Rice FL, Farhadi M, Reda H, Rowbotham MC. Natural history of cutaneous innervation following herpes zoster. Pain. 2010;150:75–82.
39. Truini A, Galeotti F, Haanpaa M, Zucchi R, Albanesi A, Biasiotta A, et al. Pathophysiology of pain in postherpetic neuralgia: a clinical and neurophysiological study. Pain. 2008;140:405–10.
40. Pfau DB, Krumova EK, Treede RD, Baron R, Toelle T, Birklein F, et al. Quantitative sensory testing in the German Research Network on Neuropathic Pain (DFNS): reference data for the trunk and application in patients with chronic postherpetic neuralgia. Pain. 2014;155:1002–15.
41. Bischoff JM, Ringsted TK, Petersen M, Sommer C, Uceyler N, Werner MU. A capsaicin (8%) patch in the treatment of severe persistent inguinal postherniorrhaphy pain: a randomized, double-blind, placebo-controlled trial. PLoS One. 2014;9:e109144.
42. Cazzato D, Castori M, Lombardi R, Caravello F, Bella ED, Petrucci A, et al. Small fiber neuropathy is a common feature of Ehlers-Danlos syndromes. Neurology. 2016;87:155–9.
43. Laurencin C, Andre-Obadia N, Camdessanche JP, Mauguiere F, Ong E, Vukusic S, et al. Peripheral small fiber dysfunction and neuropathic pain in patients with Morvan syndrome. Neurology. 2015;85:2076–8.
44. Lauria G, Majorana A, Borgna M, Lombardi R, Penza P, Padovani A, et al. Trigeminal small-fiber sensory neuropathy causes burning mouth syndrome. Pain. 2005;115:332–7.
45. Puhakka A, Forssell H, Soinila S, Virtanen A, Roytta M, Laine M, et al. Peripheral nervous system involvement in primary burning mouth syndrome—results of a pilot study. Oral Dis. 2016;22:338–44.
46. Jaaskelainen SK, Woda A. Burning mouth syndrome. Cephalalgia. 2017;37:627–47.

Visceral Pain and Hypersensitivity Disorders

12

Rosario Privitera and Praveen Anand

Abstract

Visceral pain and hypersensitivity disorders are common but poorly understood, and their specific treatment is even more limited than for somatic disorders. Visceral hypersensitivity is mediated by small sensory fibers, the Aδ and C afferents, and results from their lowered activation thresholds, change in phenotype, or altered regional density. Visceral and somatic hypersensitivities share some neuropathological mechanisms, as well as convergent signaling pathways, considered to be responsible in the spinal cord for referred pain and dysmotility. However, some mechanisms are distinct, both in the normal state and chronic pain conditions. The transient receptor potential vanilloid type 1 (TRPV1, heat and capsaicin) receptor plays a major role in visceral pain, and also purinergic receptor P_2X_3, which is selectively expressed by a subset of nociceptors. Neurotrophic factors, particularly nerve growth factor (NGF) and glial cell-derived neurotrophic factor (GDNF), are known to regulate nociceptor function in health and disease; overexposure of nociceptors to neurotrophic factors is a key underlying mechanism in a number of chronic visceral pain syndromes, including irritable bowel syndrome, bladder pain syndrome/interstitial cystitis, and fibromyalgia. Patients with visceral pain are more likely to have other pain conditions, and growing documentation suggests a link with genetic predisposition and environmental triggers.

Keywords

Transient receptor potential (TRP) channel
Transient receptor potential vanilloid type 1 (TRPV1) · Purinergic receptor P_2X_3 · Irritable bowel syndrome · Bladder pain syndrome/ interstitial cystitis · Fibromyalgia · Voltage-gated sodium channel (Na_v)

R. Privitera · P. Anand (✉)
Peripheral Neuropathy Unit, Centre for Clinical Translation, Division of Brain Sciences, Imperial College London, Hammersmith Hospital, London, UK
e-mail: p.anand@imperial.ac.uk

12.1 Introduction

The heightened perception of visceral sensation (hypersensitivity) has been observed in patients who have unexplained abdominal pain [1], with lower pain thresholds in comparison with healthy subjects [2–4]. Persistent neuropathic pain along with bladder and bowel hypersensitivity has been reported after surgery [5]. Visceral hypersensitivity is also considered to be an important pathophysiological mechanism in the development of functional gastrointestinal and genitourinary disorders. Understanding the pathophysiological

S.-T. Hsieh et al. (eds.), *Small Fiber Neuropathy and Related Syndromes: Pain and Neurodegeneration*,
https://doi.org/10.1007/978-981-13-3546-4_12

mechanisms underlying these common conditions and developing effective treatments have become one of the major challenges in modern medicine. Unexplained abdominal symptoms account for 40% of gastroenterological practice in the United Kingdom. Chronic unexplained symptoms have health-care resource implications, as extensive investigations may be conducted without benefit, and inappropriate treatment leading to poor patient satisfaction [6] and increased health-care cost [7]. Musculoskeletal chest pain may occur in 20% of all diagnoses, followed by reflux esophagitis, costochondritis, and angina [8]. Irritable bowel syndrome (IBS) has a prevalence between 10% and 20%, often diagnosed in young women [9, 10]. The prevalence of chronic pelvic pain has been reported ranging between 6% and 26% [11]. The prevalence of bladder pain syndrome/interstitial cystitis (BPS/IC) symptoms may range between 2% and 6%, depending on the diagnostic criteria utilized [12–14]. Urologic chronic prostatitis/chronic pelvic pain syndromes have a prevalence between 2% and 16% and are the most common urologic diseases in men below 50 years old [15]. Chronic scrotal pain without a specific etiology occurs in up to 34% of young males [16], while vulvodynia ranges from 10% to 28% in reproductive-aged women [17].

In visceral hypersensitivity, lowered thresholds of nociceptor activation leads to increased intensity of sensation, particularly evoked pain. Visceral pain is generally diffuse and poorly localized in comparison with somatic pain, e.g., in patients with IBS, the pain on colon distension pain was felt in the hypogastrium in 40%, in one or both iliac fossae in 31%, and in the rectum in 21%; the other 8% felt pain in the back or elsewhere [3]. Chronic pain in the viscera is observed in functional gastrointestinal disorders (e.g., non-cardiac chest pain, chronic idiopathic dyspepsia, functional abdominal pain, IBS) and chronic pelvic pain in PBS/IC [18]. The discomfort, pain, and altered sensations (e.g., to intraluminal contents) that characterize the hypersensitivity may arise in the absence of evident tissue insult or inflammation [19, 20]. Peripheral and central afferent pathways are both involved in chronic visceral hypersensitivity and increased afferent processing in pathways, as showed in IBS patients [21]. The cause of visceral hypersensitivity is still uncertain in many subjects but includes genetic, environmental, and immunological factors [22]. Visceral nociceptors may become sensitized to non-noxious stimuli during an inflammatory process, and during nerve regeneration. Enhanced responses evoked in sensitized nociceptors can be sustained by the peripheral drive and eventually increase of the excitability of central nociceptive systems and sensitization [23]. Despite extensive research to understand mechanisms of visceral hypersensitivity, the specific roles of psychological and organic processes remain unclear. These include sensitization of gastrointestinal afferent nerves, sensitization of spinal cord dorsal horn neurons, altered descending excitatory or inhibitory influences to the spinal cord nociceptive neurons (which may be influenced by psychological processes), and misinterpretation of non-noxious sensation as noxious due to cognitive and emotional biasing (hypervigilance), the result of psychiatric/psychological disorders [24].

12.2 Visceral Pain and Hypersensitivity Mechanisms

The visceral nervous system and the brain communicate through neural, neuroendocrine (hypothalamo-pituitary-adrenal axis), and neuroimmune pathways, which interact to generate facilitatory or inhibitory responses to painful visceral stimuli [25–27]. Neuropeptides [28, 29] are differentially expressed in somatic and visceral systems [30] and show a distinctive and preferential pattern of distribution in the central autonomic pathways [31, 32]. Tissue damage leads to sensitization of primary afferent nerves due to the release of mediators including neurotrophic factors and cytokines [33, 34] and induces increased expression of sodium channels such as Na_v 1.8, the vanilloid receptor transient receptor potential vanilloid type 1 (TRPV1), the purinergic receptor P_2X_3, and acid-sensing ion channels (ASICs) [35–38]. These mediators also reduce the transduction threshold of primary afferents and recruit previously silent nociceptors. Cytokines secreted by macrophages and mast cells also contribute indirectly to nerve sensitization by upregulating the expression of nerve growth factor (NGF) [39] and cause the release of cyclooxygenase metabolites [40, 41] and sympathomimetic amines, with a consequent increase in pain sensitivity at the site of inflammation.

The TRPV1 receptor has been shown to be upregulated along with other peptides, cytokines, and neurotrophic factors in the urinary bladder and rectal biopsies [42] of patients with chronic hypersensitivity states without overt evidence of current inflammation. An upregulation of TRPV1, P_2X_3, and ASICs receptors has also been documented in the inflamed human gastrointestinal tract [35, 36, 38], which may become persistent in chronic disorders. Recent studies showed patients with esophagitis to have an upregulation of both TRPV1 [43] and cytokine interleukin-8 [44]. Patients with idiopathic vulvodynia showed an upregulation of TRPV1 in vulvodynia biopsy tissues [45], suggesting TRPV1 receptor is an important marker and play a role in pathophysiology. In the intradermal injection model of TRPV1 activation by capsaicin, there is an intense but transient pain, followed by heightened sensitivity to pinprick outside the region of the capsaicin injection (secondary mechanical hyperalgesia) and to low-threshold mechanosensory (brush) inputs (secondary mechanical allodynia). These indicate the presence of central sensitization [46–48].

Enhanced nociceptor input activates intracellular signaling cascades within spinal dorsal horn neurons with the consequent amplification of responses to both noxious and innocuous inputs, due to facilitated excitatory synaptic responses and depressed inhibition [49–51]. The amplification of nociceptive signals leads to the presynaptic release of neurotransmitters and neuromodulators, responsible for the increase of calcium influx and release of calcium from intracellular stores. Calcium-dependent activation of protein kinase A (PKA), protein kinase C (PKC), and tyrosine kinases in turn leads to changes of *N*-methyl-D-aspartic acid (NMDA) receptor kinetics [34], with an increase of its responsiveness to glutamate and synaptic strength, which enable previously subthreshold inputs to activate the cell [52, 53].

The gain of function in both peripheral and central nervous pathways causes a spread of tissue hypersensitivity beyond the site of injury and also generates an activity-dependent change of transcription in dorsal root ganglion and dorsal horn neurons [54]. After inflammation, non-nociceptive afferents begin to express substance P and brain-derived neurotrophic factor (BDNF) [55], and other neurotrophic factors such as NGF and neurotrophin-3 (NT-3) are upregulated in target tissues [56]. In patients with BPS, NGF plays a role in several clinical and pathological features, including sensitization of nociceptor fibers and increased numbers of mast cells [57]. These changes take hours to manifest but, when established, lead to long-lasting changes in normal stimulus-response characteristics. The increased efficacy includes both nociceptor terminal synapses activated by the conditioning stimulus and also synapses of low-threshold mechanosensitive Aβ fibers on dorsal horn neurons [58, 59]. Low-threshold sensory fibers, activated by innocuous stimuli such as light touch, can now activate normally high-threshold nociceptive neurons at the dorsal horn, contributing to a reduction in pain threshold such that non-painful stimuli are now perceived as pain (allodynia), which is a consequence of increased excitability of central nervous system (CNS) neurons. Although this pain may be referred to the periphery, it arises from within the CNS. This central facilitation manifests within seconds of an appropriate nociceptive conditioning stimulus and can outlast the stimulus for several hours [60]. Inputs from urinary bladder and colon may converge and contribute to the cross talk that occurs between visceral organs [61, 62]. Indeed, irritable bowel syndrome may occur concurrently with other visceral disorders such as the overactive bladder syndrome [63] or interstitial cystitis/painful bladder syndrome [64, 65]. The augmentation of the level of central sensitization has also been considered in pain states such as the fibromyalgia syndrome. Visceral pain has been shown to enhance fibromyalgia [66, 67] without peripheral sensitization [68]. Some patients with fibromyalgia show an impairment of small fiber function on quantitative sensory testing (QST), and of reduced intraepidermal nerve fiber density (IEND) [69], the latter attributed to factors modulating nerve degeneration and regeneration [70]. Pain-related evoked potentials to electrical stimuli applied to the skin showed reduced amplitudes and increased latencies after stimulation of face, hands, and feet, indicating abnormalities of small fibers or their central afferents [71].

12.3 Gastrointestinal Disorders

Functional gastrointestinal disorders (FGIDs) are a group of conditions characterized by a variety of abdominal symptoms, including changes in bowel habit, indigestion, nausea, early satiety, and, most commonly, pain [72]. The FGIDs are classified into six major groups: esophageal, gastroduodenal, bowel, functional abdominal pain syndrome (FAPS), biliary, and anorectal type [73]. Patients with FGIDs may have visceral pain hypersensitivity [74–76]. Inflammatory bowel disease (IBD) includes ulcerative colitis (UC) and Crohn's disease (CD) and shows some symptomatic and pathological features which overlap with FGIDs.

In inflamed human bowel, neurotrophin-3 and inwardly rectifying potassium channel IK1 were found both decreased in myenteric plexus in the inflamed colon from patients with UC and CD, which could explain the diarrhea and colicky abdominal pain produced by both IBDs and the IK1-blocking pyridine drugs prescribed for neuromuscular disorders as an undesirable side effect [77]. Acid-sensing ion channels (ASICs) expressed by sensory neurons which are modulated by changes in pH may mediate pain associated with tissue acidosis after inflammation [38]. The ATP-gated cation receptor P_2X_3 was found to be increased in the inflamed intestine suggesting a potential role in dysmotility and pain in human IBDs [35, 78].

TRPV1 sensory nerve fibers were found in human esophageal mucosa both in health and were increased in patients with gastroesophageal reflux disease (GORD) in the inflamed esophagus [43]. TRPV1 nerve fibers were also found to be greatly increased in the colon of patients with active IBD [36]. Further, increased TRPV1 (+) nerve fibers have been shown in patients with IBS [79] (Fig. 12.1) and in patients who had similar persistent abdominal pain despite quiescent IBDs [80]. This increase in TRPV1-expressing nerve fibers, along with NGF receptor TrkA-expressing nerve fibers, was reported in patients suffering from severe rectal hypersensitivity and fecal urgency; the increase in the TRPV1-expressing nerve fibers

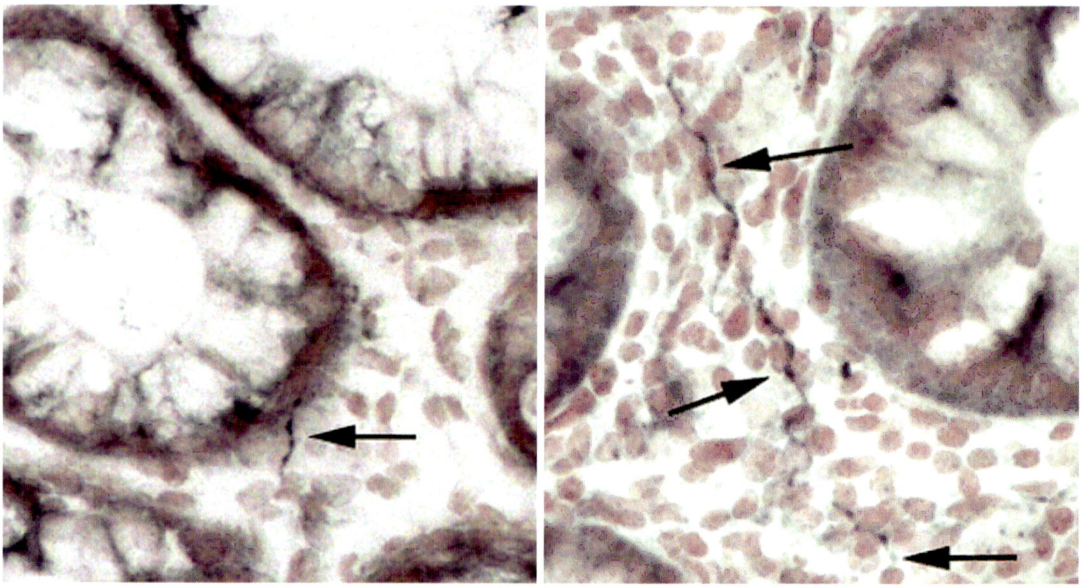

Fig. 12.1 TRPV1(+) nerve fibers in the sigmoid colon of irritable bowel syndrome. TRPV1 nerve fibers (*arrows*) in sigmoid colon of a control subject (*left*) and from a patient with irritable bowel syndrome (IBS, *right*). Magnification × 80

was correlated significantly with a decrease in the threshold to rectal heat and distension [42].

Voltage-gated sodium channels involved, along with potassium channels, influence the excitability of sensory neurons and were increased in mucosal, submucosal, and muscle layers in rectal biopsies of patients with rectal hypersensitivity and paroxysmal extreme pain disorder [81]. Conversely, patients with functional constipation who are diagnosed by symptoms not referred to pain or discomfort [82] including those with idiopathic slow-transit chronic constipation and Hirschsprung's disease show findings on quantitative sensory testing suggestive of small fiber neuropathy, along with abnormalities of visceral nerve fibers [83–86].

12.4 Genitourinary Disorders

In patients with vulvodynia, symptoms of burning sensation, allodynia, and hyperalgesia in the region of the vulval vestibule have been shown to be associated with an overexpression of TRPV1 [45]. Vulvodynia may coexist with burning mouth syndrome (BMS) [87], which is a chronic pain state dominated by oral pain, and also associated with an increase in TRPV1 and $Na_v1.8$ nerve fibers, along with increased NGF fibers and basal epithelial cell NGF staining [88] (Figs. 12.2 and 12.3).

In patients with BPS/IC and idiopathic detrusor overactivity (IDO), changes have been reported in regional nerve density and a number

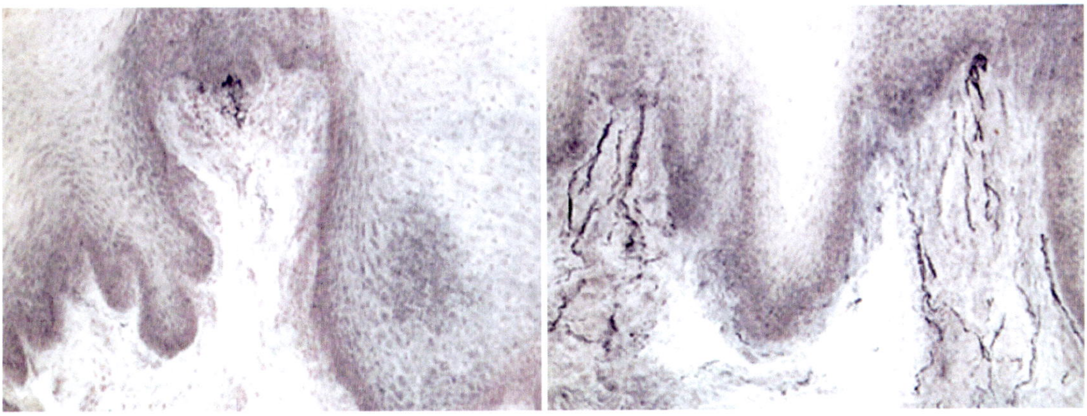

Fig. 12.2 TRPV1 expression in the tongue of burning mouth syndrome. Subepithelial TRPV1 fiber staining of a control tongue biopsy (*left*) and tongue biopsy from a patient with burning mouth syndrome (BMS, *right*). Magnification × 10

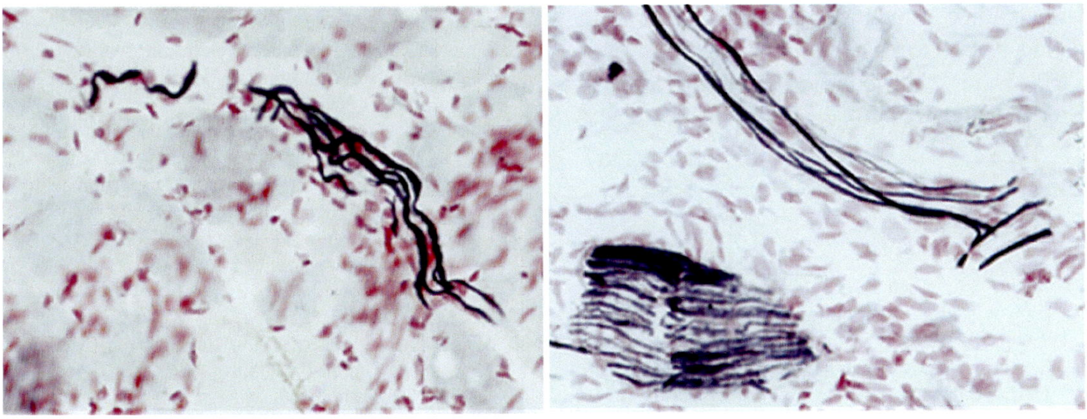

Fig. 12.3 $Na_v1.8$ expression in the tongue of burning mouth syndrome. $Na_v1.8$ fibers in control tongue biopsy (*left*) and tongue biopsy from a patient with burning mouth syndrome (BMS, *right*). Magnification × 40

of ion channels and receptors, which represent novel drug targets. The muscarinic receptors were found to be increased, and the changes correlated with clinical symptoms [89]. Cannabinoid receptor 1 (CB1) is known to play a role in models of inflammatory pain [90], and nerve fibers expressing this receptor were also found to be increased in these patients [91]. Further, the magnitude of the increase of the cool and menthol receptor TRPM8-expressing nerve fibers correlated with the frequency and pain [92]. TRPM8 has also been considered to play a role in evoking uninhibited detrusor contractions produced by intravesical instillation of cold saline [93]. In patients with BPS, the increased TRPV1 fibers were found to correlate with the pain score [94]. A decrease of TRPV1 nerve fibers was reported in patients who were responders to resiniferatoxin (RTX), after treatment; a similar decrease in protein gene product 9.5 (PGP9.5) fibers (pan-neuronal marker) suggested that the nerve fiber changes were mainly of sensory origin [95]. Patients with neurogenic detrusor overactivity (NDO) also showed increased TRPV1 immunoreactivity, which was reduced after resiniferatoxin (RTX) [96]. P_2X_3-immunoreactive nerve fibers were found increased in the suburothelium of patients with bladder over-reactivity and significantly decreased in the patients who responded to RTX [97]. Decreased levels of sensory receptors P_2X_3 and/or TRPV1 have been observed after intradetrusor injections of botulinum toxin (BoNT/A) in patients detrusor overactivity (DO), along with an improvement of both clinical and urodynamic parameters, suggesting that P_2X_3 and/or TRPV1 may contribute to the clinical effect of BoNT/A in detrusor overactivity [98].

References

1. Mayer EA, Gebhart GF. Basic and clinical aspects of visceral hyperalgesia. Gastroenterology. 1994;107:271–93.
2. Sarkar S, Aziz Q, Woolf CJ, Hobson AR, Thompson DG. Contribution of central sensitisation to the development of non-cardiac chest pain. Lancet. 2000;356:1154–9.
3. Ritchie J. Pain from distension of the pelvic colon by inflating a balloon in the irritable colon syndrome. Gut. 1973;14:125–32.
4. Salet GA, Samsom M, Roelofs JM, van Berge Henegouwen GP, Smout AJ, Akkermans LM. Responses to gastric distension in functional dyspepsia. Gut. 1998;42:823–9.
5. Shembalkar P, Anand P, Junaid I, Fowler C, Williams NS. Neuropathic pain with vesical and rectal hyperreflexia and cocontraction after pelvic surgery. J Neurol Neurosurg Psychiatry. 2001;70:410–1.
6. Talley NJ, Gabriel SE, Harmsen WS, Zinsmeister AR, Evans RW. Medical costs in community subjects with irritable bowel syndrome. Gastroenterology. 1995;109:1736–41.
7. Fullerton S. Functional digestive disorders (FDD) in the year 2000-economic impact. Eur J Surg Suppl. 1998;582:62–4.
8. Klinkman MS, Stevens D, Gorenflo DW. Episodes of care for chest pain: a preliminary report from MIRNET. Michigan Research Network. J Fam Pract. 1994;38:345–52.
9. Drossman DA, Whitehead WE, Camilleri M. Irritable bowel syndrome: a technical review for practice guideline development. Gastroenterology. 1997;112:2120–37.
10. Canavan C, West J, Card T. The epidemiology of irritable bowel syndrome. Clin Epidemiol. 2014;6:71–80.
11. Ahangari A. Prevalence of chronic pelvic pain among women: an updated review. Pain Physician. 2014;17:E141–7.
12. Berry SH, Elliott MN, Suttorp M, Bogart LM, Stoto MA, Eggers P, Nyberg L, et al. Prevalence of symptoms of bladder pain syndrome/interstitial cystitis among adult females in the United States. J Urol. 2011;186:540–4.
13. Berry SH, Bogart LM, Pham C, Liu K, Nyberg L, Stoto M, et al. Development, validation and testing of an epidemiological case definition of interstitial cystitis/painful bladder syndrome. J Urol. 2010;183:1848–52.
14. Hall SA, Link CL, Pulliam SJ, Hanno PM, Eggers PW, Kusek JW, et al. The relationship of common medical conditions and medication use with symptoms of painful bladder syndrome: results from the Boston area community health survey. J Urol. 2008;180:593–8.
15. Smith CP. Male chronic pelvic pain: an update. Indian J Urol. 2016;32:34–9.
16. Rottenstreich M, Glick Y, Gofrit ON. Chronic scrotal pain in young adults. BMC Res Notes. 2017;10:241.
17. Pukall CF, Goldstein AT, Bergeron S, Foster D, Stein A, Kellogg-Spadt S, et al. Vulvodynia: definition, prevalence, impact, and pathophysiological factors. J Sex Med. 2016;13:291–304.
18. Sengupta JN. Visceral pain: the neurophysiological mechanism. Handb Exp Pharmacol. 2009;194:31–74.
19. Gebhart GF. Pathobiology of visceral pain: molecular mechanisms and therapeutic implications IV. Visceral afferent contributions to the pathobiology of visceral

pain. Am J Physiol Gastrointest Liver Physiol. 2000;278:G834–8.
20. Al-Chaer ED, Kawasaki M, Pasricha PJ. A new model of chronic visceral hypersensitivity in adult rats induced by colon irritation during postnatal development. Gastroenterology. 2000;119:1276–85.
21. Verne GN, Himes NC, Robinson ME, Gopinath KS, Briggs RW, Crosson B, et al. Central representation of visceral and cutaneous hypersensitivity in the irritable bowel syndrome. Pain. 2003;103:99–110.
22. Bouma G, Strober W. The immunological and genetic basis of inflammatory bowel disease. Nat Rev Immunol. 2003;3:521–33.
23. Cervero F. Visceral pain: mechanisms of peripheral and central sensitization. Ann Med. 1995;27:235–9.
24. Anand P, Aziz Q, Willert R, van Oudenhove L. Peripheral and central mechanisms of visceral sensitization in man. Neurogastroenterol Motil. 2007;19:29–46.
25. Watkins LR, Maier SF. The pain of being sick: implications of immune-to-brain communication for understanding pain. Annu Rev Psychol. 2000;51:29–57.
26. Jones MP, Dilley JB, Drossman D, Crowell MD. Brain-gut connections in functional GI disorders: anatomic and physiologic relationships. Neurogastroenterol Motil. 2006;18:91–103.
27. Van Oudenhove L, Demyttenaere K, Tack J, Aziz Q. Central nervous system involvement in functional gastrointestinal disorders. Best Pract Res Clin Gastroenterol. 2004;18:663–80.
28. Gibson SJ, Polak JM, Anand P, Blank MA, Yiangou Y, Su HC, et al. A VIP/PHI-containing pathway links urinary bladder and sacral spinal cord. Peptides. 1986;7:205–19.
29. Anand P, Gibson SJ, Yiangou Y, Christofides ND, Polak JM, Bloom SR. PHI-like immunoreactivity co-locates with the VIP-containing system in human lumbosacral spinal cord. Neurosci Lett. 1984;46:191–6.
30. Anand P, Ghatei MA, Christofides ND, Blank MA, McGregor GP, Morrison JF, et al. Differential neuropeptide expression after visceral and somatic nerve injury in the cat and rat. Neurosci Lett. 1991;128:57–60.
31. Anand P, Gibson SJ, McGregor GP, Blank MA, Ghatei MA, Bacarese-Hamilton AJ, et al. A VIP-containing system concentrated in the lumbosacral region of human spinal cord. Nature. 1983;305:143–5.
32. Gooneratne ML, Facer P, Knowles CH, Chan CL, Lunniss PJ, Scott SM, et al. Normalization of substance P levels in rectal mucosa of patients with faecal incontinence treated successfully by sacral nerve stimulation. Br J Surg. 2008;95:477–83.
33. Costigan M, Woolf CJ. Pain: molecular mechanisms. J Pain. 2000;1:35–44.
34. Woolf CJ, Salter MW. Neuronal plasticity: increasing the gain in pain. Science. 2000;288:1765–9.
35. Yiangou Y, Facer P, Baecker PA, Ford AP, Knowles CH, Chan CL, et al. ATP-gated ion channel P2X(3) is increased in human inflammatory bowel disease. Neurogastroenterol Motil. 2001;13:365–9.
36. Yiangou Y, Facer P, Dyer NH, Chan CL, Knowles C, Williams NS, et al. Vanilloid receptor 1 immunoreactivity in inflamed human bowel. Lancet. 2001;357:1338–9.
37. Yiangou Y, Facer P, Ford A, Brady C, Wiseman O, Fowler CJ, et al. Capsaicin receptor VR1 and ATP-gated ion channel P2X3 in human urinary bladder. BJU Int. 2001;87:774–9.
38. Yiangou Y, Facer P, Smith JA, Sangameswaran L, Eglen R, Birch R, et al. Increased acid-sensing ion channel ASIC-3 in inflamed human intestine. Eur J Gastroenterol Hepatol. 2001;13:891–6.
39. Woolf CJ. Phenotypic modification of primary sensory neurons: the role of nerve growth factor in the production of persistent pain. Philos Trans R Soc Lond Ser B Biol Sci. 1996;351:441–8.
40. Sciberras DG, Goldenberg MM, Bolognese JA, James I, et al. Inflammatory responses to intradermal injection of platelet activating factor, histamine and prostaglandin E2 in healthy volunteers: a double blind investigation. Br J Clin Pharmacol. 1987;24:753–61.
41. Seibert K, Zhang Y, Leahy K, Hauser S, Masferrer J, Perkins W, et al. Pharmacological and biochemical demonstration of the role of cyclooxygenase 2 in inflammation and pain. Proc Natl Acad Sci U S A. 1994;91:12013–7.
42. Chan CLH, Facer P, Davis JB, Smith GD, Egerton J, Bountra C, et al. Sensory fibres expressing capsaicin receptor TRPV1 in patients with rectal hypersensitivity and faecal urgency. Lancet. 2003;361:385–91.
43. Matthews PJ, Aziz Q, Facer P, Davis JB, Thompson DG, Anand P. Increased capsaicin receptor TRPV1 nerve fibres in the inflamed human oesophagus. Eur J Gastroenterol Hepatol. 2004;16:897–902.
44. Yoshida N, Uchiyama K, Kuroda M, Sakuma K, KoKura S, Ichikawa H, Naito Y, et al. Interleukin-8 expression in the esophageal mucosa of patients with gastroesophageal reflux disease. Scand J Gastroenterol. 2004;39:816–22.
45. Tympanidis P, Casula MA, Yiangou Y, Terenghi G, Dowd P, Anand P. Increased vanilloid receptor VR1 innervation in vulvodynia. Eur J Pain. 2004;8:129–33.
46. LaMotte RH, Lundberg LE, Torebjörk HE. Pain, hyperalgesia and activity in nociceptive C units in humans after intradermal injection of capsaicin. J Physiol. 1992;448:749–64.
47. Simone DA, Baumann TK, LaMotte RH. Dose-dependent pain and mechanical hyperalgesia in humans after intradermal injection of capsaicin. Pain. 1989;38:99–107.
48. Koltzenburg M, Lundberg LE, Torebjörk HE. Dynamic and static components of mechanical hyperalgesia in human hairy skin. Pain. 1992;51:207–19.
49. Woolf CJ. Evidence for a central component of post-injury pain hypersensitivity. Nature. 1983;306:686–8.
50. Woolf CJ. Generation of acute pain: central mechanisms. Br Med Bull. 1991;47:523–33.
51. Woolf CJ. An overview of the mechanisms of hyperalgesia. Pulm Pharmacol. 1995;8:161–7.

52. Woolf CJ, Thompson SW. The induction and maintenance of central sensitization is dependent on N-methyl-D-aspartic acid receptor activation; implications for the treatment of post-injury pain hypersensitivity states. Pain. 1991;44:293–9.
53. Ji RR, Woolf CJ. Neuronal plasticity and signal transduction in nociceptive neurons: implications for the initiation and maintenance of pathological pain. Neurobiol Dis. 2001;8:1–10.
54. Woolf CJ, Costigan M. Transcriptional and posttranslational plasticity and the generation of inflammatory pain. Proc Natl Acad Sci U S A. 1999;96:7723–30.
55. Neumann S, Doubell TP, Leslie T, Woolf CJ. Inflammatory pain hypersensitivity mediated by phenotypic switch in myelinated primary sensory neurons. Nature. 1996;384:360–4.
56. Oddiah D, Anand P, McMahon SB, Rattray M. Rapid increase of NGF, BDNF and NT-3 mRNAs in inflamed bladder. Neuroreport. 1998;9:1455–8.
57. Lowe EM, Anand P, Terenghi G, Williams-Chestnut RE, Sinicropi DV, Osborne JL. Increased nerve growth factor levels in the urinary bladder of women with idiopathic sensory urgency and interstitial cystitis. Br J Urol. 1997;79:572–7.
58. Woolf CJ, King AE. Dynamic alterations in the cutaneous mechanoreceptive fields of dorsal horn neurons in the rat spinal cord. J Neurosci. 1990;10:2717–26.
59. Simone DA, Baumann TK, Collins JG, LaMotte RH. Sensitization of cat dorsal horn neurons to innocuous mechanical stimulation after intradermal injection of capsaicin. Brain Res. 1989;486:185–9.
60. Woolf CJ, Wall PD. Relative effectiveness of C primary afferent fibers of different origins in evoking a prolonged facilitation of the flexor reflex in the rat. J Neurosci. 1986;6:1433–42.
61. Qin C, Foreman RD. Viscerovisceral convergence of urinary bladder and colorectal inputs to lumbosacral spinal neurons in rats. Neuroreport. 2004;15: 467–71.
62. Christianson JA, Liang R, Ustinova EE, Davis BM, Fraser MO, Pezzone MA. Convergence of bladder and colon sensory innervation occurs at the primary afferent level. Pain. 2007;128:235–43.
63. Abrams P, Cardozo L, Fall M, Griddiths D, Rosier P, Ulmsten U, et al. The standardisation of terminology in lower urinary tract function: report from the standardisation sub-committee of the International Continence Society. Urology. 2003;61:37–49.
64. Novi JM, Jeronis S, Srinivas S, Srinivasan R, Morgan MA, Arya LA. Risk of irritable bowel syndrome and depression in women with interstitial cystitis: a case-control study. J Urol. 2005;174:937–40.
65. Nickel JC, Tripp DA, Pontari M, Moldwin R, Mayer R, Carr LK, et al. Interstitial cystitis/painful bladder syndrome and associated medical conditions with an emphasis on irritable bowel syndrome, fibromyalgia and chronic fatigue syndrome. J Urol. 2010;184:1358–63.
66. Costantini R, Affaitati G, Wesselmann U, Cazkanski P, Giamberardino MA. Visceral pain as a triggering factor for fibromyalgia symptoms in comorbid patients. Pain. 2017;158:1925–37.
67. Gracely RH, Petzke F, Wolf JM, Clauw DJ. Functional magnetic resonance imaging evidence of augmented pain processing in fibromyalgia. Arthritis Rheum. 2002;46:1333–43.
68. Üçeyler N, Kewenig S, Kafke W, Kittel-Schneider S, Sommer C. Skin cytokine expression in patients with fibromyalgia syndrome is not different from controls. BMC Neurol. 2014;14:185.
69. Doppler K, Rittner HL, Deckart M, Sommer C. Reduced dermal nerve fiber diameter in skin biopsies of patients with fibromyalgia. Pain. 2015;156:2319–25.
70. Leinders M, Doppler K, Klein T, Deckart M, Rittner H, Sommer C, et al. Increased cutaneous miR-let-7d expression correlates with small nerve fiber pathology in patients with fibromyalgia syndrome. Pain. 2016;157:2493–503.
71. Üçeyler N, Zeller D, Kahn AK, Kewenig S, Kittel-Schneider S, Schmid A, et al. Small fibre pathology in patients with fibromyalgia syndrome. Brain. 2013;136:1857–67.
72. Hobson AR, Aziz Q. Central nervous system processing of human visceral pain in health and disease. News Physiol Sci. 2003;18:109–14.
73. Drossman DA. The functional gastrointestinal disorders and the Rome III process. Gastroenterology. 2006;130:1377–90.
74. Farmer AD, Aziz Q. Visceral pain hypersensitivity in functional gastrointestinal disorders. Br Med Bull. 2009;91:123–36.
75. Whitehead WE, Holtkotter B, Enck P, Hoelzl R, Holmes KD, Anthony J, et al. Tolerance for rectosigmoid distention in irritable bowel syndrome. Gastroenterology. 1990;98:1187–92.
76. Lémann M, Dederding JP, Flourié B, Franchisseur C, Rambaud JC, Jian R. Abnormal perception of visceral pain in response to gastric distension in chronic idiopathic dyspepsia. The irritable stomach syndrome. Dig Dis Sci. 1991;36:1249–54.
77. Arnold SJ, Facer P, Yiangou Y, Chen MX, Plumpton C, Tate SN, et al. Decreased potassium channel IK1 and its regulator neurotrophin-3 (NT-3) in inflamed human bowel. Neuroreport. 2003;14:191–5.
78. Yiangou Y, Facer P, Birch R, Sangameswaran L, Eglen R, Anand P. P2X3 receptor in injured human sensory neurons. Neuroreport. 2000;11:993–6.
79. Akbar A, Yiangou Y, Facer P, Walters JR, Anand P, Ghosh S. Increased capsaicin receptor TRPV1-expressing sensory fibres in irritable bowel syndrome and their correlation with abdominal pain. Gut. 2008;57:923–9.
80. Akbar A, Yiangou Y, Facer P, Brydon WG, Walters JR, Anand P, et al. Expression of the TRPV1 receptor differs in quiescent inflammatory bowel disease with or without abdominal pain. Gut. 2010;59:767–74.
81. Yiangou Y, Facer P, Chessell IP, Bountra C, Chan C, Fertleman C, et al. Voltage-gated ion channel Nav1.7 innervation in patients with idiopathic

rectal hypersensitivity and paroxysmal extreme pain disorder (familial rectal pain). Neurosci Lett. 2007;427:77–82.
82. Siah KTH, Wong RK, Whitehead WE. Chronic constipation and constipation-predominant IBS: separate and distinct disorders or a spectrum of disease? Gastroenterol Hepatol (N Y). 2016;12:171–8.
83. Knowles CH, Scott SM, Wellmer A, Misra VP, Pilot MA, Williams NS, et al. Sensory and autonomic neuropathy in patients with idiopathic slow-transit constipation. Br J Surg. 1999;86:54–60.
84. Cheng W, Au DK, Knowles CH, Anand P, Tam PK. Hirschsprung's disease: a more generalised neuropathy? J Pediatr Surg. 2001;36:296–300.
85. Facer P, Knowles CH, Tam PK, Ford AP, Dyer N, Baecker PA, et al. Novel capsaicin (VR1) and purinergic (P2X3) receptors in Hirschsprung's intestine. J Pediatr Surg. 2001;36:1679–84.
86. Facer P, Knowles CH, Thomas PK, Tam PK, Williams NS, Anand P. Decreased tyrosine kinase C expression may reflect developmental abnormalities in Hirschsprung's disease and idiopathic slow-transit constipation. Br J Surg. 2001;88:545–52.
87. Gaitonde P, Rostron J, Longman L, Field EA. Burning mouth syndrome and vulvodynia coexisting in the same patient: a case report. Dent Update. 2002;29:75–6.
88. Yilmaz Z, Renton T, Yiangou Y, Zakrzewska J, Chessell IP, Bountra C, et al. Burning mouth syndrome as a trigeminal small fibre neuropathy: increased heat and capsaicin receptor TRPV1 in nerve fibres correlates with pain score. J Clin Neurosci. 2007;14:864–71.
89. Mukerji G, Yiangou Y, Grogono J, Underwood J, Agarwal SK, Khullar V, et al. Localization of M2 and M3 muscarinic receptors in human bladder disorders and their clinical correlations. J Urol. 2006;176:367–73.
90. Anand P, Whiteside G, Fowler CJ, Hohmann AG. Targeting CB2 receptors and the endocannabinoid system for the treatment of pain. Brain Res Rev. 2009;60:255–66.
91. Mukerji G, Yiangou Y, Agarwal SK, Anand P. Increased cannabinoid receptor 1-immunoreactive nerve fibers in overactive and painful bladder disorders and their correlation with symptoms. Urology. 2010;75:1514.e15–20.
92. Mukerji G, Yiangou Y, Corcoran SL, Selmer IS, Smith GD, Benham CD, et al. Cool and menthol receptor TRPM8 in human urinary bladder disorders and clinical correlations. BMC Urol. 2006;6:6.
93. Mukerji G, Waters J, Chessell IP, Bountra C, Agarwal SK, Anand P. Pain during ice water test distinguishes clinical bladder hypersensitivity from overactivity disorders. BMC Urol. 2006;6:31.
94. Mukerji G, Yiangou Y, Agarwal SK, Anand P. Transient receptor potential vanilloid receptor subtype 1 in painful bladder syndrome and its correlation with pain. J Urol. 2006;176:797–801.
95. Brady CM, Apostolidis AN, Harper M, Yiangou Y, Beckett A, Jacques TS, et al. Parallel changes in bladder suburothelial vanilloid receptor TRPV1 and pan-neuronal marker PGP9.5 immunoreactivity in patients with neurogenic detrusor overactivity after intravesical resiniferatoxin treatment. BJU Int. 2004;93:770–6.
96. Apostolidis A, Brady CM, Yiangou Y, Davis J, Fowler CJ, Anand P. Capsaicin receptor TRPV1 in urothelium of neurogenic human bladders and effect of intravesical resiniferatoxin. Urology. 2005;65:400–5.
97. Brady CM, Apostolidis A, Yiangou Y, Baecker PA, Ford AP, Freeman A, et al. P2X3-immunoreactive nerve fibres in neurogenic detrusor overactivity and the effect of intravesical resiniferatoxin. Eur Urol. 2004;46:247–53.
98. Apostolidis A, Popat R, Yiangou Y, Cockayne D, Ford AP, Davis JB, et al. Decreased sensory receptors P2X3 and TRPV1 in suburothelial nerve fibers following intradetrusor injections of botulinum toxin for human detrusor overactivity. J Urol. 2005;174: 977–82; discussion 982.

in these neurodegenerative disorders than what was previously thought and that patients also suffer from symptoms from various systems [3–5]. In amyotrophic lateral sclerosis (ALS), a motoneuron disease, sensory and autonomic involvement has been discovered [6]. Together with the necessity to assess and treat pain in these patients [7], this has spurred interest in the role of small fiber involvement in these disorders.

13.2 Small Fiber Involvement in Parkinson's Disease

13.2.1 The Autonomic System

Of the non-motor symptoms of Parkinson's disease, autonomic nervous system abnormalities were among the earliest to attract attention, since they may be an important source of discomfort and disability in the patients. Autonomic nervous system dysfunction may lead to orthostatic hypotension, delayed gastric emptying, constipation, urinary dysfunction and impaired sweating, lacrimation, and sebum production, all of which have been described in patients with Parkinson's disease [8]. This observation inspired others to investigate the peripheral autonomic system. The peripheral autonomic system comprises the innervation of dermal blood vessels, erector pili muscles, sweat glands, and hair follicles in the skin. Dabby et al. were the first to investigate these structures in patients with Parkinson's disease [9]. Using 4-μm-thin paraffin sections and comparing 22 patients with Parkinson's disease to 19 control subjects, they found reduced innervation of blood vessels, erector pili muscles, and sweat glands in the patients. Disease duration was the most important denominator of the denervation. The presence of orthostatic hypotension did not correlate with denervation of blood vessels [9]. The authors concluded from these findings that abnormal sympathetic innervation is an independent disease phenomenon in Parkinson's disease, and they speculated that autonomic denervation might precede the clinical appearance of autonomic symptoms. In the meantime, numerous studies focused on the involvement of the peripheral autonomic system in Parkinson's disease. Autonomic denervation has been described in skin biopsies [10–12], and autonomic dysfunction was shown by sudomotor, vasomotor, and pilomotor tests [12–14].

13.2.2 The Somatosensory System

Dermal sheet preparations were used to evaluate dermal nerve density in six patients with Parkinson's disease compared to six patients with multiple system atrophy (MSA) [15]. This method allows visualization and quantification of dermal nerve fibers. The authors hypothesized that in MSA, there would be only CNS neuronal degeneration, thus normal dermal innervation, whereas in Parkinson's disease, there would also be a reduction of peripheral innervation. Using immunohistochemistry with an antibody to the NGF-receptor p75 as a neuronal marker, the authors found reduced innervation in Parkinson's disease compared to MSA. However, normal controls were not included, such that it did not become clear if the MSA samples had normal results or not [15].

Nolano and coauthors combined quantitative sensory testing (QST) and skin biopsy analysis to study the somatosensory skin innervation in 18 patients with Parkinson's disease and 30 healthy controls [14]. QST showed functional impairment of large and small fibers, indicated by an increase in both tactile and thermal detection thresholds and a reduction in mechanical pain perception at the hands and feet of the patients. Skin biopsies were taken from the thigh, distal leg, and fingertip, and 50-μm sections were processed for immunofluorescence and confocal microscopy. The authors give a very differentiated picture of skin innervation in Parkinson's disease, with not only fiber loss but also pathological fiber abnormalities and abnormal innervation patterns in some regions. As in small fiber neuropathy, they show a reduction of intraepidermal nerve fiber density, which was less than half of that of controls at the fingertip, about half of that of controls at the lower leg, and about 2/3 of controls at the thigh. There was evidence of fiber pathology indicated by fiber swellings and of fiber remodeling, indicated by clusters and an increase in nerve branches. The authors also found evidence of unsuccessful attempts at regeneration, indicated by nerve fiber sprouts that were directed toward the epidermis but did not penetrate the basement membrane [14]. The reduction of intraepidermal nerve fiber density (IENFD) did not correlate with age, duration, or severity of disease. Interestingly, in ten patients with bilateral biopsies, IENFD was lower on the side more affected by Parkinson's disease.

The numbers of Meissner's corpuscles were reduced in the patients, and this loss correlated with disease severity. While myelinated fibers innervating the remaining Meissner's corpuscles were not significantly reduced in patients, these fibers showed axonal swellings and myelin abnormalities. Some fibers were enlarged. Myelin abnormalities included paranodal, distal, and intermodal demyelination [14]. The authors concluded from their findings that peripheral deafferentation might account for the impairment in sensory function in patients with Parkinson's disease. Further studies confirmed the finding of small fiber pathology in Parkinson's disease, including skin biopsy and corneal confocal microscopy (CCM) studies [16–18] as well as QST, contact heat evoked potential (CHEP), and quantitative sudomotor axon reflex test (QSART) [19, 20]. Besides small fiber pathology, large fiber neuropathy has also been described in Parkinson's disease, and it has been hypothesized in several studies that peripheral neuropathy in Parkinson's disease may be caused by treatment with levodopa that induces vitamin B12 and B6 deficiency [21–23]. However, a recent study by Nolano et al. as well as some former smaller studies gives evidence that there is no correlation between small fiber loss and treatment with levodopa, whereas large fiber neuropathy may be caused by vitamin deficiency [16, 24, 25]. The idea of small fiber involvement as an intrinsic factor of the disease is supported by evidence of a very early involvement of small fiber pathology during the course of disease: small fiber loss was detected by CCM in newly diagnosed patients, and a reduction of IENFD was even found in patients with REM sleep behavior disorder (RBD), a group with a very high risk of developing manifest Parkinson's disease [26, 27].

13.2.3 Detection of Alpha-Synuclein Deposits in Skin Nerve Fibers

Parkinson's disease, like Lewy body dementia and MSA, belongs to the synucleinopathies. This means that these diseases are characterized by the abnormal accumulation of aggregates of the protein alpha-synuclein in neurons, nerve fibers, or glial cells. While most researchers agree that alpha-synuclein is essential in the pathophysiology of Parkinson's disease, the debate on mechanisms is still open [28–33]. However, the crucial role of alpha-synuclein in Parkinson's disease has led to the call to use it as a biomarker. Parkinson's disease is a clinical diagnosis, and the definite diagnosis is made postmortem by pathology. Thus, it seemed attractive and necessary to find alpha-synuclein deposits, the pathological hallmark of Parkinson's disease by a simple method that would allow a premortem definite diagnosis. The first group to investigate alpha-synuclein in patients' skin used immunoblotting and found alpha-synuclein in most patients but also in most controls, concluding that this was not useful as a biomarker [34].

While alpha-synuclein is a protein with physiological functions, its phosphorylation at Ser129 is a modification that may enhance its toxicity and may be relevant for the pathogenesis of Parkinson's disease [35]. Phosphorylation may enhance the protein's tendency to build aggregates [36]. Thus, phospho-alpha-synuclein might be a better biomarker for Parkinson's disease than total alpha-synuclein (Fig. 13.1).

Using the skin from a large series of autopsies and immunohistochemistry on 6-μm paraffin sections with antibodies to phosphorylated alpha-synuclein, Ikemura et al. later showed deposits of phosphorylated alpha-synuclein in dermal nerve fibers [37]. The nerve fibers were identified by double labeling with antibodies to neurofilaments or to tyrosine hydroxylase, the latter for adrenergic fibers. The detection rate increased with the disease stage, and disease controls with progressive supra-

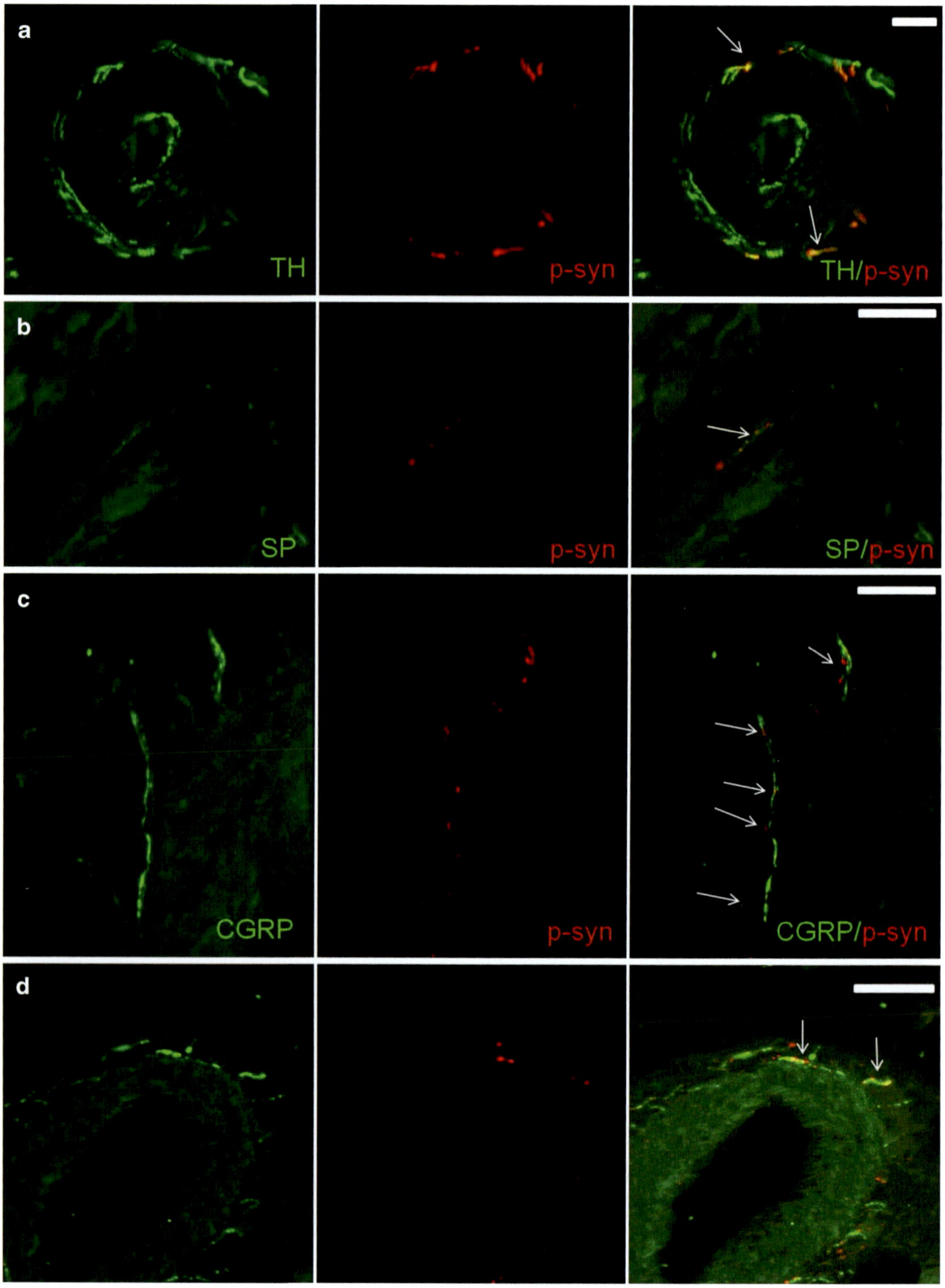

Fig. 13.1 Phosphorylated alpha-synuclein in the skin from Parkinson's disease. Photomicrographs of skin biopsies of patients with Parkinson's disease showing double labeling with anti-phospho-alpha-synuclein and anti-TH (**a**), anti-SP (**b**), anti-CGRP (**c**), and anti-VIP (**d**) P-alpha-synuclein deposits can be seen within TH-positive (**a**), SP-positive (**b**), CGRP-positive (**c**), and VIP-positive (**d**) fibers of the dermis. Arrows mark colocalization. Bars 20 μm [16]

nuclear palsy (PSP) and corticobasal degeneration (CBD, both tauopathies) were negative, indicating 100% specificity. Overall, there was positive immunoreactivity for phosphorylated alpha-synuclein in the skin in 23.5% of patients who also had such deposits in the brain and in none of those who did not. In an accompanying retrospective study, the sensitivity of the positive skin immunoreactivity was 70% [37]. Miki et al. aimed to transfer these promising autopsy results to premortem skin biopsies, but only achieved a sensitivity of 10% using paraffin sections of the skin of the chest and leg [38]. In contrast to these studies that all used antibodies against phospho-alpha-synuclein, Wang et al. applied an antibody that is directed against native alpha-synuclein and discovered immunoreactivity in dermal nerve fibers of patients as well as controls but at a higher intensity in patients with Parkinson's disease, correlating with the stage of disease and autonomic dysfunctions [12]. Recently, evidence has accumulated that phospho-alpha-synuclein deposition in autonomic nerve fibers may be a good biomarker of Parkinson's disease and dementia with Lewy bodies, as several studies, mostly using cryosections and immunohistochemistry with autoantibodies against phospho-alpha-synuclein, reported a high specificity of 100% and sensitivities of 50–100% [16, 17, 39, 40]. Two most recent studies demonstrated the detection of phospho-alpha-synuclein in patients with RBD, which is a high-risk factor and presumably a premotor stage of Parkinson's disease, giving evidence of an early involvement of autonomic nerve fibers in this disease [41, 42].

13.3 Small Nerve Fiber Involvement in Other Synucleinopathies

Research on the involvement of small nerve fibers and particularly on phospho-alpha-synuclein deposition in autonomic fibers has not been restricted to Parkinson's disease but was also studied in other synucleinopathies, namely, MSA and pure autonomic failure. Loss of unmyelinated fibers was described in early studies using autopsy material of sural nerves of both pure autonomic failure and MSA [43, 44]. In contrast, later studies supported the idea of a predominantly central autonomic dysfunction in MSA in contrast to pure autonomic failure and Parkinson's disease, disorders that had marked peripheral autonomic involvement [15, 45–47]. Accordingly, deposition of phospho-alpha-synuclein was described in dermal nerve fibers of patients with pure autonomic failure, first in single-case reports [48–50] and then in larger cohorts [51, 52]. More recently, there is new evidence of an involvement of the peripheral nervous system in MSA as well: sudomotor denervation was described in skin biopsies of patients [53], and deposition of phospho-alpha-synuclein was found in autonomic and somatosensory nerve fibers of skin biopsies of patients with MSA [39, 54] and within Schwann cells of cranial and spinal nerves and dorsal root ganglia in other studies [55, 56]. However, two other studies did not detect phospho-alpha-synuclein in dermal nerve fibers of patients with MSA, possibly due to methodological differences [57, 58], so the question is still open.

13.4 Small Fiber Involvement in Amyotrophic Lateral Sclerosis (ALS)

Amyotrophic lateral sclerosis (ALS) is considered a motor neuron disease affecting the upper motor neuron in the cerebral cortex as well as the lower motor neuron in the spinal cord. Characteristic symptoms include progressive paresis, bulbar symptoms, fasciculations, and muscle atrophy, resulting in immobility and respiratory insufficiency. However, sensory symptoms have been described in several studies, and involvement of the sensory nervous system was detected in electrophysiological as well as histology studies [59–61]. Whereas earlier histological studies gave evidence of a predominance of involvement of sensory large fibers [62, 63], more recent studies reported involvement of small fibers as well [64, 65]. The establishment of standardized evaluation of IENFD as a marker of a reduction of small somatosensory fibers paved the way for a systematic comparison between patients with ALS and control subjects. Several studies with different study designs and number of patients were published in the recent years. A reduction of the IENFD was reported by some of these studies, in some studies associated with mild sensory symptoms [59, 65]; in another study no correlation between IENFD and clinical data was observed [64]. In addition to histological evidence of small fiber loss, dysfunction of small somatosensory fibers was investigated by QST and revealed small fiber dysfunction in one study [59] but not in others [66, 67]. CHEP was studied in another study and did not show any alterations in patients with ALS [60]. In addition to skin biopsy, small fiber loss was recently also reported by CCM, another method of noninvasive small fiber quantification [68]. In summary, there is evidence of mild small fiber involvement in patients with ALS, though not detectable in all studies. A reason for divergent results may be differences of the patient cohorts: a recent study showed that small fiber involvement is mostly seen in patients with a spinal-onset type of ALS [59]. Indeed, ALS is a heterogeneous disease with different clinical presentations, like predominantly lower motor neuron involvement, bulbar forms, or forms associated with cognitive impairment, especially frontal dementia. Previous studies on the molecular pathology of ALS suggested different subtypes based on accumulation of certain protein aggregates within neurons as well as underlying mutations and sporadic forms [69, 70]. Hereditary ALS, a small subgroup of patients, also involves different genes with presumably different pathogenic mechanisms leading to similar but not identical clinical phenotypes [71, 72]. The heterogeneity of the disease may account for differences of small fiber involvement between patient groups. TDP-43, one of the most common protein aggregates in ALS, and neuronal loss were detected in sensory neurons in one study, but not in sympathetic ganglia [73]. This may explain the involvement of small somatosensory nerve fibers in ALS. The so-called Bunina bodies, a histopathological hallmark of ALS, were so far also restricted to the central nervous system [74]. However, several studies gave evidence of an increase of TDP-43 protein levels in skin biopsies or skin fibroblasts of patients with ALS, thus showing that protein aggregation is probably not restricted to the motor nervous system in this disease [61, 75, 76]. Besides protein aggregation, another pathophysiological mechanism leading to the involvement of the sensory nervous system may be of inflammatory origin. Vasculitic inflammatory infiltrates have been described in sural nerve biopsies of patients with ALS without any other abnormality of nerve fibers [77]. So although somatosensory and particularly small fiber involvement has been detected in many studies, its prevalence and clinical significance in ALS patients is still unclear, and the underlying pathophysiology is largely unknown.

13.5 Small Fiber Involvement in Other Neurodegenerative Disorders

Small fiber function and pathology have also been studied in a number of neurodegenerative diseases that do not primarily affect small fibers. This may reflect the vulnerability of the most distal and unmyelinated nerve endings to neurodegeneration in general. However, in most of these diseases, only single mono-center studies with a small number of cases are available, so that larger studies would be needed to strengthen the evidence of small fiber involvement. In 2001, Nolano et al. provided the first and so far only description of an involvement of cutaneous small somatosensory and autonomic nerve fiber loss and small fiber dysfunction in patients with Friedreich's ataxia, a trinucleotide repeat disorder with the major pathology in the dorsal root ganglion (large fiber) and spinal cord (spinocerebellar tract, lateral corticospinal tract, and posterior columns) [78]. Fourteen patients were included, and the study design included skin biopsy and QST, both demonstrating small fiber involvement in Friedreich's ataxia [78]. Another rare hereditary neurological disorder where small fiber involvement was studied is Kennedy's disease, a hereditary X-linked spinobulbar muscular atrophy of trinucleotide repeat disorder: in 2007, Manganelli et al. described two subjects with Kennedy's disease who showed a non-length-dependent loss of myelinated, unmyelinated, and autonomic dermal nerve fibers and abnormal findings in the QST and sweat tests [79]. Although motor symptoms are predominant in spinobulbar muscular atrophy, sensory involvement also including dysfunction of small nerve fibers has been described before using laser-evoked potentials and sural nerve biopsy [80] or nerve conduction studies that measure large sensory fibers [81]. Cerebral autosomal dominant arteriopathy with subcortical infarcts and leukoencephalopathy (CADASIL) is another genetic disorder that has been discussed to be associated with peripheral neuropathy. In a small study including 11 patients, peripheral neuropathy as detected by nerve conduction studies was found in 7 patients [82]. However, in another study, peripheral neuropathy was only detected in a minority of 7/43 patients with CADASIL by nerve conduction studies and was explained by other causes of neuropathy in 6 of these patients, thus challenging the concept of peripheral neuropathy as part of the clinical phenotype of CADASIL [83]. A previous study by Nolano et al. from 2016 gave evidence of a small fiber involvement in CADASIL: by performing skin biopsies in 14 patients with CADASIL, they detected a non-length-dependent reduction of the IENFD as well as autonomic innervation in the patients with CADASIL, supporting the idea of an involvement of the peripheral nervous system in this disease [84].

Another rare disorder that is accompanied by loss of autonomic innervation is Ross syndrome, a disease of unknown cause that is characterized by anhidrosis, tonic pupils, and hyporeflexia. There is evidence that anhidrosis goes along with autonomic denervation. Loss of sudomotor innervation and selective loss of cholinergic autonomic fibers was described in small cohorts of patients with Ross syndrome using skin punch biopsies [62, 85]. In addition to autonomic denervation, loss of unmyelinated and myelinated nerve fibers was described in subsequent studies [63, 86], accompanied by abnormal autonomic testing and QST [86].

13.6 Conclusion

Small nerve fibers are involved in a number of neurodegenerative diseases, and in some instances, peripheral pathology may closely mirror that in the CNS. Thus, the skin as "window to the brain" may help to better understand the pathophysiology of neurodegenerative diseases, to make an early definite diagnosis, and may provide a measure for follow-up in the natural course of disease and under treatment.

References

1. Galvin JE. Interaction of alpha-synuclein and dopamine metabolites in the pathogenesis of Parkinson's disease: a case for the selective vulnerability of the substantia nigra. Acta Neuropathol. 2006;112:115–26.
2. Song P, Li S, Wu H, Gao R, Rao G, Wang D, et al. Parkin promotes proteasomal degradation of p62: implication of selective vulnerability of neuronal cells in the pathogenesis of Parkinson's disease. Protein Cell. 2016;7:114–29.
3. Munhoz RP, Moro A, Silveira-Moriyama L, Teive HA. Non-motor signs in Parkinson's disease: a review. Arq Neuropsiquiatr. 2015;73:454–62.
4. Jellinger KA. Neuropathobiology of non-motor symptoms in Parkinson disease. J Neural Transm (Vienna). 2015;122:1429–40.
5. Borghammer P, Knudsen K, Brooks DJ. Imaging systemic dysfunction in Parkinson's disease. Curr Neurol Neurosci Rep. 2016;16:51.
6. Nolano M, Provitera V, Manganelli F, Iodice R, Caporaso G, Stancanelli A, et al. Non-motor involvement in amyotrophic lateral sclerosis: new insight from nerve and vessel analysis in skin biopsy. Neuropathol Appl Neurobiol. 2017;43:119–32.
7. Bartolo M, Chio A, Ferrari S, Tassorelli C, Tamburin S, Avenali M, et al. Assessing and treating pain in movement disorders, amyotrophic lateral sclerosis, severe acquired brain injury, disorders of consciousness, dementia, oncology and neuroinfectivology. Evidence and recommendations from the Italian consensus conference on pain in neurorehabilitation. Eur J Phys Rehabil Med. 2016;52:841–54.

8. van Dijk JG, Haan J, Zwinderman K, Kremer B, van Hilten BJ, Roos RA. Autonomic nervous system dysfunction in Parkinson's disease: relationships with age, medication, duration, and severity. J Neurol Neurosurg Psychiatry. 1993;56:1090–5.
9. Dabby R, Djaldetti R, Shahmurov M, Treves TA, Gabai B, Melamed E, et al. Skin biopsy for assessment of autonomic denervation in Parkinson's disease. J Neural Transm (Vienna). 2006;113:1169–76.
10. Navarro-Otano J, Casanova-Molla J, Morales M, Valls-Sole J, Tolosa E. Cutaneous autonomic denervation in Parkinson's disease. J Neural Transm (Vienna). 2015;122:1149–55.
11. Giannoccaro MP, Donadio V, Incensi A, Pizza F, Cason E, Di Stasi V, et al. Skin biopsy and I-123 MIBG scintigraphy findings in idiopathic Parkinson's disease and parkinsonism: a comparative study. Mov Disord. 2015;30:986–9.
12. Wang N, Gibbons CH, Lafo J, Freeman R. Alpha-Synuclein in cutaneous autonomic nerves. Neurology. 2013;81:1604–10.
13. Siepmann T, Frenz E, Penzlin AI, Goelz S, Zago W, Friehs I, et al. Pilomotor function is impaired in patients with Parkinson's disease: a study of the adrenergic axon-reflex response and autonomic functions. Parkinsonism Relat Disord. 2016;31:129–34.
14. Nolano M, Provitera V, Estraneo A, Selim MM, Caporaso G, Stancanelli A, et al. Sensory deficit in Parkinson's disease: evidence of a cutaneous denervation. Brain. 2008;131:1903–11.
15. Novak P, Marya NB, Whren K, Bhawan J. Dermal sheet preparations in the evaluation of dermal innervation in Parkinson's disease and multiple system atrophy. J Cutan Pathol. 2009;36:296–301.
16. Doppler K, Ebert S, Üceyler N, Trenkwalder C, Ebentheuer J, Volkmann J, et al. Cutaneous neuropathy in Parkinson's disease: a window into brain pathology. Acta Neuropathol. 2014;128:99–109.
17. Donadio V, Incensi A, Leta V, Giannoccaro MP, Scaglione C, Martinelli P, et al. Skin nerve alpha-synuclein deposits: a biomarker for idiopathic Parkinson disease. Neurology. 2014;82:1362–9.
18. Kass-Iliyya L, Javed S, Gosal D, Kobylecki C, Marshall A, Petropoulos IN, et al. Small fiber neuropathy in Parkinson's disease: a clinical, pathological and corneal confocal microscopy study. Parkinsonism Relat Disord. 2015;21:1454–60.
19. Lin CH, Chao CC, Wu SW, Hsieh PC, Feng FP, Lin YH, et al. Pathophysiology of small-fiber sensory system in Parkinson's disease: skin innervation and contact heat evoked potential. Medicine (Baltimore). 2016;95:e3058.
20. Akaogi Y, Asahina M, Yamanaka Y, Koyama Y, Hattori T. Sudomotor, skin vasomotor, and cardiovascular reflexes in 3 clinical forms of Lewy body disease. Neurology. 2009;73:59–65.
21. Toth C, Brown MS, Furtado S, Suchowersky O, Zochodne D. Neuropathy as a potential complication of levodopa use in Parkinson's disease. Mov Disord. 2008;23:1850–9.
22. Toth C, Breithaupt K, Ge S, Duan Y, Terris JM, Thiessen A, et al. Levodopa, methylmalonic acid, and neuropathy in idiopathic Parkinson disease. Ann Neurol. 2010;68:28–36.
23. Rajabally YA, Martey J. Neuropathy in Parkinson disease: prevalence and determinants. Neurology. 2011;77:1947–50.
24. Nolano M, Provitera V, Manganelli F, Iodice R, Stancanelli A, Caporaso G, et al. Loss of cutaneous large and small fibers in naive and l-dopa-treated PD patients. Neurology. 2017;89:776–84.
25. Nolano M, Provitera V, Lanzillo B, Santoro L. Neuropathy in idiopathic Parkinson disease: an iatrogenic problem? Ann Neurol. 2011;69:427–8.
26. Podgorny PJ, Suchowersky O, Romanchuk KG, Feasby TE. Evidence for small fiber neuropathy in early Parkinson's disease. Parkinsonism Relat Disord. 2016;28:94–9.
27. Schrempf W, Katona I, Dogan I, Felbert VV, Wienecke M, Heller J, et al. Reduced intraepidermal nerve fiber density in patients with REM sleep behavior disorder. Parkinsonism Relat Disord. 2016;29:10–6.
28. Kalia LV, Kalia SK. Alpha-Synuclein and Lewy pathology in Parkinson's disease. Curr Opin Neurol. 2015;28:375–81.
29. Xu Y, Deng Y, Qing H. The phosphorylation of alpha-synuclein: development and implication for the mechanism and therapy of the Parkinson's disease. J Neurochem. 2015;135:4–18.
30. Benskey MJ, Perez RG, Manfredsson FP. The contribution of alpha synuclein to neuronal survival and function—implications for Parkinson's disease. J Neurochem. 2016;137:331–59.
31. Del Tredici K, Braak H. Review: sporadic Parkinson's disease: development and distribution of alpha-synuclein pathology. Neuropathol Appl Neurobiol. 2016;42:33–50.
32. Longhena F, Faustini G, Missale C, Pizzi M, Spano P, Bellucci A. The contribution of alpha-Synuclein spreading to Parkinson's disease Synaptopathy. Neural Plast. 2017;2017:5012129.
33. Rocha EM, De Miranda B, Sanders LH. Alpha-synuclein: pathology, mitochondrial dysfunction and neuroinflammation in Parkinson's disease. Neurobiol Dis. 2018;109:249–57.
34. Michell AW, Luheshi LM, Barker RA. Skin and platelet alpha-synuclein as peripheral biomarkers of Parkinson's disease. Neurosci Lett. 2005;381:294–8.
35. Chen L, Periquet M, Wang X, Negro A, McLean PJ, Hyman BT, et al. Tyrosine and serine phosphorylation of alpha-synuclein have opposing effects on neurotoxicity and soluble oligomer formation. J Clin Invest. 2009;119:3257–65.
36. Fujiwara H, Hasegawa M, Dohmae N, Kawashima A, Masliah E, Goldberg MS, et al. Alpha-Synuclein is phosphorylated in synucleinopathy lesions. Nat Cell Biol. 2002;4:160–4.
37. Ikemura M, Saito Y, Sengoku R, Sakiyama Y, Hatsuta H, Kanemaru K, et al. Lewy body pathology involves

cutaneous nerves. J Neuropathol Exp Neurol. 2008;67:945–53.
38. Miki Y, Tomiyama M, Ueno T, Haga R, Nishijima H, Suzuki C, et al. Clinical availability of skin biopsy in the diagnosis of Parkinson's disease. Neurosci Lett. 2010;469:357–9.
39. Doppler K, Weis J, Karl K, Ebert S, Ebentheuer J, Trenkwalder C, et al. Distinctive distribution of phospho-alpha-synuclein in dermal nerves in multiple system atrophy. Mov Disord. 2015;30:1688–92.
40. Donadio V, Incensi A, Rizzo G, Capellari S, Pantieri R, Stanzani Maserati M, et al. A new potential biomarker for dementia with Lewy bodies: skin nerve alpha-synuclein deposits. Neurology. 2017;89:318–26.
41. Doppler K, Jentschke HM, Schulmeyer L, Vadasz D, Janzen A, Luster M, et al. Dermal phospho-alpha-synuclein deposits confirm REM sleep behaviour disorder as prodromal Parkinson's disease. Acta Neuropathol. 2017;133:535–45.
42. Antelmi E, Donadio V, Incensi A, Plazzi G, Liguori R. Skin nerve phosphorylated alpha-synuclein deposits in idiopathic REM sleep behavior disorder. Neurology. 2017;88:2128–31.
43. Kanda T, Tomimitsu H, Yokota T, Ohkoshi N, Hayashi M, Mizusawa H. Unmyelinated nerve fibers in sural nerve in pure autonomic failure. Ann Neurol. 1998;43:267–71.
44. Kanda T, Tsukagoshi H, Oda M, Miyamoto K, Tanabe H. Changes of unmyelinated nerve fibers in sural nerve in amyotrophic lateral sclerosis, Parkinson's disease and multiple system atrophy. Acta Neuropathol. 1996;91:145–54.
45. Orimo S, Oka T, Miura H, Tsuchiya K, Mori F, Wakabayashi K, et al. Sympathetic cardiac denervation in Parkinson's disease and pure autonomic failure but not in multiple system atrophy. J Neurol Neurosurg Psychiatry. 2002;73:776–7.
46. Orimo S, Suzuki M, Inaba A, Mizusawa H. 123I-MIBG myocardial scintigraphy for differentiating Parkinson's disease from other neurodegenerative parkinsonism: a systematic review and meta-analysis. Parkinsonism Relat Disord. 2012;18:494–500.
47. Donadio V, Cortelli P, Elam M, Di Stasi V, Montagna P, Holmberg B, et al. Autonomic innervation in multiple system atrophy and pure autonomic failure. J Neurol Neurosurg Psychiatry. 2010;81:1327–35.
48. Kaufmann H, Hague K, Perl D. Accumulation of alpha-synuclein in autonomic nerves in pure autonomic failure. Neurology. 2001;56:980–1.
49. Igari R, Wada M, Wakabayashi K, Kato T. Alpha-Synuclein in the skin nerve of pure autonomic failure. Intern Med. 2011;50:3049–50.
50. Shishido T, Ikemura M, Obi T, Yamazaki K, Terada T, Sugiura A, et al. Alpha-synuclein accumulation in skin nerve fibers revealed by skin biopsy in pure autonomic failure. Neurology. 2010;74:608–10.
51. Donadio V, Incensi A, Cortelli P, Giannoccaro MP, Jaber MA, Baruzzi A, et al. Skin sympathetic fiber alpha-synuclein deposits: a potential biomarker for pure autonomic failure. Neurology. 2013;80:725–32.
52. Donadio V, Incensi A, Piccinini C, Cortelli P, Giannoccaro MP, Baruzzi A, et al. Skin nerve misfolded alpha-synuclein in pure autonomic failure and Parkinson disease. Ann Neurol. 2016;79:306–16.
53. Provitera V, Nolano M, Caporaso G, Stancanelli A, Manganelli F, Iodice R, et al. Postganglionic sudomotor denervation in patients with multiple system atrophy. Neurology. 2014;82:2223–9.
54. Doppler K, Volkmann J, Sommer C. Skin biopsies in the differential diagnosis of parkinsonism: are we ready for simplified protocols? Brain. 2016;139:e5.
55. Nakamura K, Mori F, Tanji K, Miki Y, Toyoshima Y, Kakita A, et al. Alpha-Synuclein pathology in the cranial and spinal nerves in Lewy body disease. Neuropathology. 2016;36:262–9.
56. Nakamura K, Mori F, Kon T, Tanji K, Miki Y, Tomiyama M, et al. Filamentous aggregations of phosphorylated alpha-synuclein in Schwann cells (Schwann cell cytoplasmic inclusions) in multiple system atrophy. Acta Neuropathol Commun. 2015;3:29.
57. Haga R, Sugimoto K, Nishijima H, Miki Y, Suzuki C, Wakabayashi K, et al. Clinical utility of skin biopsy in differentiating between Parkinson's disease and multiple system atrophy. Parkinsons Dis. 2015;2015:167038.
58. Zange L, Noack C, Hahn K, Stenzel W, Lipp A. Phosphorylated alpha-synuclein in skin nerve fibres differentiates Parkinson's disease from multiple system atrophy. Brain. 2015;138:2310–21.
59. Truini A, Biasiotta A, Onesti E, Di Stefano G, Ceccanti M, La Cesa S, et al. Small-fibre neuropathy related to bulbar and spinal-onset in patients with ALS. J Neurol. 2015;262:1014–8.
60. Xu YS, Zhang J, Zheng JY, Zhang S, Kang DX, Fan DS. Fully intact contact heat evoked potentials in patients with amyotrophic lateral sclerosis. Muscle Nerve. 2009;39:735–8.
61. Yang S, Zhang KY, Kariawasam R, Bax M, Fifita JA, Ooi L, et al. Evaluation of skin fibroblasts from amyotrophic lateral sclerosis patients for the rapid study of pathological features. Neurotox Res. 2015;28:138–46.
62. Sommer C, Lindenlaub T, Zillikens D, Toyka KV, Naumann M. Selective loss of cholinergic sudomotor fibers causes anhidrosis in Ross syndrome. Ann Neurol. 2002;52:247–50.
63. Perretti A, Nolano M, De Joanna G, Tugnoli V, Iannetti G, Provitera V, et al. Is Ross syndrome a dysautonomic disorder only? An electrophysiologic and histologic study. Clin Neurophysiol. 2003;114:7–16.
64. Dalla Bella E, Lombardi R, Porretta-Serapiglia C, Ciano C, Gellera C, Pensato V, et al. Amyotrophic lateral sclerosis causes small fiber pathology. Eur J Neurol. 2016;23:416–20.
65. Weis J, Katona I, Muller-Newen G, Sommer C, Necula G, Hendrich C, et al. Small-fiber neuropathy in patients with ALS. Neurology. 2011;76:2024–9.
66. Deepika J, Manvir B, Sumit S, Vinay G, Trilochan S, Garima S, et al. Quantitative thermal sensory testing in patients with amyotrophic lateral sclerosis using

reaction time exclusive method of levels (MLE). Electromyogr Clin Neurophysiol. 2006;46:145–8.
67. Isak B, Pugdahl K, Karlsson P, Tankisi H, Finnerup NB, Furtula J, et al. Quantitative sensory testing and structural assessment of sensory nerve fibres in amyotrophic lateral sclerosis. J Neurol Sci. 2017;373:329–34.
68. Ferrari G, Grisan E, Scarpa F, Fazio R, Comola M, Quattrini A, et al. Corneal confocal microscopy reveals trigeminal small sensory fiber neuropathy in amyotrophic lateral sclerosis. Front Aging Neurosci. 2014;6:278.
69. Blokhuis AM, Groen EJ, Koppers M, van den Berg LH, Pasterkamp RJ. Protein aggregation in amyotrophic lateral sclerosis. Acta Neuropathol. 2013;125:777–94.
70. Ince PG, Highley JR, Kirby J, Wharton SB, Takahashi H, Strong MJ, et al. Molecular pathology and genetic advances in amyotrophic lateral sclerosis: an emerging molecular pathway and the significance of glial pathology. Acta Neuropathol. 2011;122:657–71.
71. Renton AE, Chio A, Traynor BJ. State of play in amyotrophic lateral sclerosis genetics. Nat Neurosci. 2014;17:17–23.
72. Sabatelli M, Conte A, Zollino M. Clinical and genetic heterogeneity of amyotrophic lateral sclerosis. Clin Genet. 2013;83:408–16.
73. Nishihira Y, Tan CF, Onodera O, Toyoshima Y, Yamada M, Morita T, et al. Sporadic amyotrophic lateral sclerosis: two pathological patterns shown by analysis of distribution of TDP-43-immunoreactive neuronal and glial cytoplasmic inclusions. Acta Neuropathol. 2008;116:169–82.
74. Wakabayashi K, Mori F, Tanji K, Orimo S, Takahashi H. Involvement of the peripheral nervous system in synucleinopathies, tauopathies and other neurodegenerative proteinopathies of the brain. Acta Neuropathol. 2010;120:1–12.
75. Abe K, Ohkubo T, Yokota T. TDP-43 in the skin of amyotrophic lateral sclerosis patients. J Med Dent Sci. 2017;64:9–17.
76. Suzuki M, Mikami H, Watanabe T, Yamano T, Yamazaki T, Nomura M, et al. Increased expression of TDP-43 in the skin of amyotrophic lateral sclerosis. Acta Neurol Scand. 2010;122:367–72.
77. Devigili G, Üçeyler N, Beck M, Reiners K, Stoll G, Toyka KV, et al. Vasculitis-like neuropathy in amyotrophic lateral sclerosis unresponsive to treatment. Acta Neuropathol. 2011;122:343–52.
78. Nolano M, Provitera V, Crisci C, Saltalamacchia AM, Wendelschafer-Crabb G, Kennedy WR, et al. Small fibers involvement in Friedreich's ataxia. Ann Neurol. 2001;50:17–25.
79. Manganelli F, Iodice V, Provitera V, Pisciotta C, Nolano M, Perretti A, et al. Small-fiber involvement in spinobulbar muscular atrophy (Kennedy's disease). Muscle Nerve. 2007;36:816–20.
80. Antonini G, Gragnani F, Romaniello A, Pennisi EM, Morino S, Ceschin V, et al. Sensory involvement in spinal-bulbar muscular atrophy (Kennedy's disease). Muscle Nerve. 2000;23:252–8.
81. Suzuki K, Katsuno M, Banno H, Takeuchi Y, Atsuta N, Ito M, et al. CAG repeat size correlates to electrophysiological motor and sensory phenotypes in SBMA. Brain. 2008;131:229–39.
82. Sicurelli F, Dotti MT, De Stefano N, Malandrini A, Mondelli M, Bianchi S, et al. Peripheral neuropathy in CADASIL. J Neurol. 2005;252:1206–9.
83. Kang SY, Oh JH, Kang JH, Choi JC, Lee JS. Nerve conduction studies in cerebral autosomal dominant arteriopathy with subcortical infarcts and leukoencephalopathy. J Neurol. 2009;256:1724–7.
84. Nolano M, Provitera V, Donadio V, Caporaso G, Stancanelli A, Califano F, et al. Cutaneous sensory and autonomic denervation in CADASIL. Neurology. 2016;86:1039–44.
85. Bergmann I, Dauphin M, Naumann M, Flachenecker P, Mullges W, Koltzenburg M, et al. Selective degeneration of sudomotor fibers in Ross syndrome and successful treatment of compensatory hyperhidrosis with botulinum toxin. Muscle Nerve. 1998;21:1790–3.
86. Nolano M, Provitera V, Perretti A, Stancanelli A, Saltalamacchia AM, Donadio V, et al. Ross syndrome: a rare or a misknown disorder of thermoregulation? A skin innervation study on 12 subjects. Brain. 2006;129:2119–31.

Part IV

Neuropathic Pain in Small Fiber Neuropathy: Mechanisms and Therapy

transcranial direct current stimulation (tDCS) have shown promising advances in pain treatment. These techniques may reset nociceptive processes in the brain by modulating synaptic plasticity in the pain-processing areas. Nevertheless, large randomized clinical trials are still warranted to evaluate the efficacy of these alternative treatment methods.

Keywords
Central sensitization · Maladaptive plasticity · Sensory deafferentation · Axonal degeneration · Neuroimaging · Functional magnetic resonance imaging (fMRI) · Transcranial magnetic stimulation (TMS) · Transcranial direct current stimulation (tDCS) · Anterior cingulate cortex · Amygdala · Limbic system

14.1 Introduction

Small fiber neuropathy (SFN) is a constellation of peripheral nerve disorders that affect small myelinated Aδ and unmyelinated C fibers [1]. The etiologies of SFN are diverse; they may be of metabolic, inflammatory, or genetic origin or related to neoplasm or chemotherapy. In about 30% of patients, SFN is idiopathic [2]. The manifestations of SFN can be divided into negative and positive symptoms. Negative symptoms, for example, hypesthesia (a decrease or even loss of sensation) or autonomic dysfunction (manifesting as orthostatic hypotension and gastroparesis), are due to functional loss in small nerve fibers. Positive symptoms include spontaneous and stimulus-evoked pain, collectively known as neuropathic pain. Spontaneous neuropathic pain may be continuous or paroxysmal. Stimulus-evoked pain is exemplified by hyperalgesia, which means increased sensitivity to nociceptive stimuli, and allodynia, pain evoked by normally innocuous mechanical or thermal stimuli [3, 4]. Neuropathic pain is caused by enhanced nerve excitability due to denervation or dysfunction in small nociceptive fibers, presenting mostly as a burning or tingling sensation in the lower limbs. It is the most common and disabling symptom in SFN [2].

SFN should be suspected when patients present with the above sensory and/or autonomic symptoms and have no signs or symptoms of large fiber involvement. Quantitative sensory testing (QST) provides psychophysical recordings of subjective perceptions of thermal stimuli. Most SFN patients exhibit elevated warm and cold thresholds under QST [5, 6]. More objective tests such as laser-evoked potential (LEP) or contact heat evoked potential (CHEP) studies provide neurophysiological evidence of SFN, such as prolongation in the latency or reduction in the amplitude of evoked potentials recorded over the sensory areas of the brain [7]. Finally, the diagnosis of SFN can be confirmed by skin punch biopsy, a minimally invasive procedure that circularly excises a small piece of the skin (usually 3 mm in diameter) for quantitation of epidermal innervation. Individual intraepidermal nerve fibers (IENFs) in the biopsied skin become visible through immunohistochemical staining using antibodies against protein gene product 9.5 (PGP 9.5), a protein expressed in neuronal cell bodies and axons [8]. SFN patients exhibit a reduction of density or even total loss of IENFs in the distal legs [8]. The severity of skin denervation may to some extent reflect the degree of functional impairments. For example, reduced IENF density was found to be associated with abnormal warm and mechanical detection thresholds measured by QST in SFN patients [2]. Nevertheless, the intensity of neuropathic pain does not parallel the severity of epidermal denervation even though SFN is pathologically characterized by depletion of IENF [2, 9], indicating that IENF damage is not the only factor responsible for pain symptoms. Morphologically normal appearing IENF may produce spontaneous activity, which is a functional correlate of pain. Furthermore, several lines of evidence suggest that maladaptive plasticity in the central nervous system, which develops as a consequence of peripheral denervation, also contributes to the generation of neuropathic pain [6, 7, 10, 11]. The involvement of the central nervous system complicates the mechanisms underlying neuropathic pain in SFN and also imposes challenges on the optimization of treatment strategies.

In the following sections, we will discuss postulated peripheral and central neural mechanisms underlying neuropathic pain in SFN and review current major and alternative treatment strategies.

14.2 Pathogenesis of Neuropathic Pain

14.2.1 Peripheral Nerve Damage

Neuropathic pain in SFN has been thought to be a direct consequence of damage to IENFs, which is the pathological hallmark of SFN [1, 12–14]. In an animal model of painful neuropathy, rats with chronic constriction injury (CCI) at the sciatic nerve showed a reduction of IENF density in the footpad, and the maxima of pain-related behaviors occurred at the time of maximal reduction of the IENF density [15]. Nevertheless, the severity of IENF depletion does not correlate with the intensity of pain [2, 9]. In fact, complete skin denervation and insensitivity to pain can sometimes coexist in the same patients [16]. As such, instead of being a predictive biomarker for neuropathic pain in SFN, depletion of IENF per se may relate more to the loss of integrity in the peripheral nociceptive system [17]. Functional impairments in the remaining IENFs due to abnormal local neurochemical environments are more likely to be responsible for neuropathic pain [15]. Several lines of evidence support this hypothesis. For example, nerve growth factor released after Wallerian degeneration triggers upregulation of voltage-gated sodium channels in uninjured nerve fibers, causing uninjured fibers to exhibit pathological spontaneous activities and consequent hypersensitivity [18]. Furthermore, damage to small nociceptive nerves also leads to upregulation of the transient receptor potential cation channel subfamily vanilloid type 1 (TRPV1), the main receptor for capsaicin [19, 20]. Overexpression of TRPV1 in uninjured nerve fibers facilitates the development of thermal hyperalgesia [21]. Finally, a recent study showed that overactivity of glutamate receptors triggered by nerve injury was associated with neuropathic pain. The expression of metabotropic glutamate receptor subtype 5 (mGluR5) in the subepidermal nerve fibers of footpad skin was intensely increased in rats showing neuropathic pain behaviors after CCI, and those CCI-induced neuropathic pain behaviors could be dose-dependently attenuated after intraplantar injection of 2-methyl-6-(phenylethynyl)-pyridine (MPEP), a selective mGluR5 antagonist [22].

Inflammation plays an important role in the generation and maintenance of neuropathic pain. It may be a primary factor causing nerve damage in inflammatory neuropathies such as vasculitic or autoimmune neuropathy or a secondary reaction to Wallerian degeneration. Proinflammatory cytokines released from activated macrophages, such as tumor necrosis factor alpha (TNF-α) or interleukin-2 (IL-2), induce pathological hypersensitivity of afferent nociceptors in injured or uninjured nerves [23]. The plasma level of TNF-α correlated with pain intensity in painful diabetic neuropathy [24]. Uçeyler et al. demonstrated that patients with SFN had elevated gene expression of IL-2, transforming growth factor (TGF)-β1, IL-6, and IL-8 in affected skin. Moreover, in the subgroup of patients who expressed a length-dependent gradient of epidermal denervation, there was an elevated expression of TNF, IL-1β, IL-6, and IL-8 in the distal skin as compared to the proximal skin [25].

14.2.2 Maladaptive Plasticity in the Central Nervous System

Maladaptive plasticity in the central nervous system induced by abnormal peripheral sensory input contributes to the development of neuropathic pain in SFN and also to the refractoriness of neuropathic pain to medical treatment [26–28]. One example of maladaptive plasticity in the central nervous system is *central sensitization*, where hypersensitivity in the central nociceptive system is induced by long-term peripheral overstimulation. The abnormal peripheral sensory input leads to dysfunction in the inhibitory neurons such that low-intensity innocuous stimuli (e.g., light touch) activate the high-threshold pain circuit [26, 29].

Central sensitization was first demonstrated in the dorsal horn neurons of the spinal cord. Mendell and Wall documented that repeated stimulation of peripheral C fibers led to the *windup* response, that is, a progressive increase in action potential firing in the dorsal horn neurons during the course of stimulation [30]. Cook et al. further showed that brief electrical stimulation of peripheral unmyelinated afferent fibers induced a prolonged increase in the synaptic efficacy of the spinal dorsal horn neurons, lasting for tens of minutes after the end of stimulation [31]. Torebjork et al. provided additional lines of direct evidence linking peripheral overstimulation and central sensitization, in which an intradermal injection of capsaicin induced an area of tactile allodynia far outside the injection site, which was attributed to secondary central hypersensitivity subsequent to peripheral noxious stimulation. Moreover, electrical intraneural microstimulation of Aβ mechanoreceptive fibers, which normally elicited non-painful tactile sensation, began to produce pain when the receptive field of these fibers was within the allodynia zone, and the pain could not be reversed by lidocaine anesthesia of the cutaneous innervation territory of the stimulated fibers [32]. Central sensitization of pain is mediated by glutamate signaling in the spinal cord, where injury-induced activity of peripheral Aδ or C fibers in turn activates the N-methyl-D-aspartate (NMDA) receptors in the dorsal horn neurons [33–35]. Substance P and brain-derived neurotrophic factor (BDNF) also mediate the excitatory synaptic responses in the nociceptive circuits. Substance P released from the small-diameter dorsal root ganglion (DRG) neurons upon noxious stimulation binds to the neurokinin-1 (NK1) receptor expressed in spinal dorsal horn neurons. This causes a long-lasting membrane depolarization and enhances the glutamatergic excitatory synaptic transmission [27, 36]. BDNF is released from spinal dorsal horn neurons upon strong noxious stimulation of C fibers, which facilitates the release of glutamate and substance P at spinal synapses [36].

Supraspinal structures play a key role in the development of central sensitization. Animal models showed that sustained nociceptive input induces descending facilitation from the rostroventromedial medulla (RVM) to spinal dorsal horn neurons, which enhances spinal nociceptive input to the brain. This descending facilitation amplifies pain from the primary injury site and also sensory input from adjacent uninjured regions, causing secondary hyperalgesia [37, 38]. The anterior cingulate cortex (ACC) is a critical area for central sensitization of pain. Digit amputation in rats caused long-lasting facilitation of the excitatory responses of the ACC, measured as enhanced excitatory postsynaptic potential (EPSP) elicited by nociceptive electrical stimulation delivered to the normal hindpaw. The potentiation of EPSP in the ACC was also induced by focal cortical stimulation within the ACC [39]. The amygdala is another unique area for the development of central sensitization. In rats with joint pain due to kaolin/carrageenan-induced arthritis (a common arthritis pain model), the amplitude of excitatory postsynaptic currents (EPSCs) evoked in neurons in the laterocapsular division of the central nucleus of the amygdala (CeLC) was significantly higher than that in the control group, indicating that pain facilitated synaptic plasticity in the amygdala. Moreover, changes in synaptic plasticity in the amygdala were associated with an increase in pain-related behaviors and were mediated by NMDA-glutamate signaling [40]. The pain-induced sensitization in the amygdala may be reversed by potentiation of γ-aminobutyric acid (GABA) transmission. In rats with CCI at the sciatic nerve, bilateral injection of muscimol, a GABA agonist, into the central amygdala attenuated pain-related behaviors in the sensory-discriminative domain, such as hindpaw mechanical hypersensitivity, as well as the affective-motivational domain, such as escape or avoidance behaviors [41]. Taken together, the above experimental findings suggest that the development of central sensitization in the brain may be attributed to deficits in GABAergic signaling pathways or an imbalance between glutamate and GABA neurotransmission.

Taking advantage of noninvasive neuroimaging techniques, central sensitization could be demonstrated in vivo in the human brain, where augmented reaction patterns in the pain circuitry contributed to chronification of neuropathic pain.

Schweinhardt et al. studied eight patients with traumatic nerve injury or plexus avulsion, finding that innocuous brush stimuli applied to the affected body site evoked pain, a sign of mechanical allodynia. Compared to stimulation to the contralateral unaffected body side, innocuous brush stimulation to the affected side also induced a stronger blood-oxygen-level dependent (BOLD) response in functional magnetic resonance imaging (fMRI), the primary (S1) and secondary (S2) somatosensory cortices, the insular cortex, the ACC, the thalamus, and the prefrontal cortex. These brain structures are involved in the sensory-discriminative and affective-emotional aspects of pain and are sometimes known as the "pain matrix" (or more accurately, pain networks) [42–44]. Moreover, brush-evoked allodynia also activated brain structures not normally involved in these pain networks, such as the supplementary motor and premotor cortices, the lentiform nucleus, and the cerebellum [45]. Becerra et al. studied six patients with chronic unilateral facial pain due to traumatic facial neuropathy or postherpetic neuralgia, documenting that brush-evoked allodynia in the affected side of the face activated brain structures in the pain matrix, such as the S1 and S2 cortices, the insular cortex, the ACC, the thalamus, and the prefrontal cortex, and also brain areas outside the pain matrix, such as the putamen and globus pallidus [46]. Spillover of brain activations was also shown in allodynia evoked by cold stimuli, which induced a stronger BOLD response in the inferior parietal lobule, the globus pallidus, and the cerebellum [46]. Enhanced neuronal excitability in neuropathic pain may be associated with altered structural neuroplasticity. DaSilva et al. found that in patients with trigeminal neuropathic pain involving the maxillary division (V2), there was bilateral thinning in the craniofacial representation of the somatosensory cortex. Moreover, these brain areas with cortical thinning overlapped with areas showing BOLD activation in fMRI caused by brush-induced allodynia over the trigeminal V2 area [47]. All these studies were conducted in patients with unilateral nerve injury. A critical issue is whether there are alterations in functional and structural brain imaging in systemic peripheral nerve degeneration.

Our studies documented sensitization of the brain's nociceptive systems in the brain of patients with chronic neuropathic pain caused by pathologically confirmed small fiber damage. In the study by Chao et al. of diabetic patients with neuropathic pain and skin denervation, the psychophysical sensitivity to thermal stimuli was reduced, with elevated warm and cold thresholds measured by QST. The amplitude of the CHEP recorded over the sensory areas of the brain also became lower, with a prolonged latency to the onset of thermal stimulation to the distal leg. However, the excitability index, derived by calculating the ratio of the CHEP amplitude over the IENF density, was higher than that in control participants, indicating an increase in nociceptive activity in the brain even when peripheral sensory inputs were reduced [7, 10]. fMRI data acquired from patients with diabetic neuropathy with small fiber involvement also exhibited a robust pattern. Compared to patients without neuropathic pain, patients with neuropathic pain had augmented brain responses to noxious thermal stimuli measured as an elevated BOLD contrast in the limbic system, such as the ACC, the prefrontal cortex, the medial thalamus, and the anterior insular cortex. The BOLD contrast was also increased in the lentiform nucleus. Moreover, the magnitude of BOLD contrasts in the ACC and the lentiform nucleus were correlated with pain ratings to thermal stimulations and also the duration of neuropathic pain in the painful patients [11].

Although the concept of central sensitization well explains the causal linkage between peripheral overstimulation and exaggerated central responses in the generation of pain, neuropathic pain may still develop even when IENF completely disappear, in which the degree of peripheral nociceptive stimulation is greatly, if not totally, diminished [48]. This suggests that neuropathic pain may be a consequence of peripheral sensory deafferentation. Reduction or absence of sensory input due to peripheral denervation results in cortical reorganization or remodeling in the pain-processing brain regions, facilitating the development of chronic central pain [49, 50]. Our group further combined structural and functional neuroimaging to investigate whether there were alterations in gray matter volume of the brain and functional connectivity in patients with SFN, using tensor-based morphometry [51] and psychophysiological interaction (PPI) analysis [52]. We demonstrated that a decrease in IENF density was linearly correlated with the reduction of brain volume in the ACC bilaterally. The reduction of IENF density was correlated with the reduction of functional connectivity between the left ACC and the right amygdala and also between the right ACC and the right precuneus (Fig. 14.1). Moreover, the above associations of the IENF density with brain volume or with functional connectivity were independent of neuropathic pain symptoms [6]. These observations suggest that the structural and functional alterations in the brain in SFN patients are more likely associated with sensory deafferentation secondary to IENF depletion, rather than central sensitization induced by peripheral overstimulation. Taken together, the contribution of central maladaptive

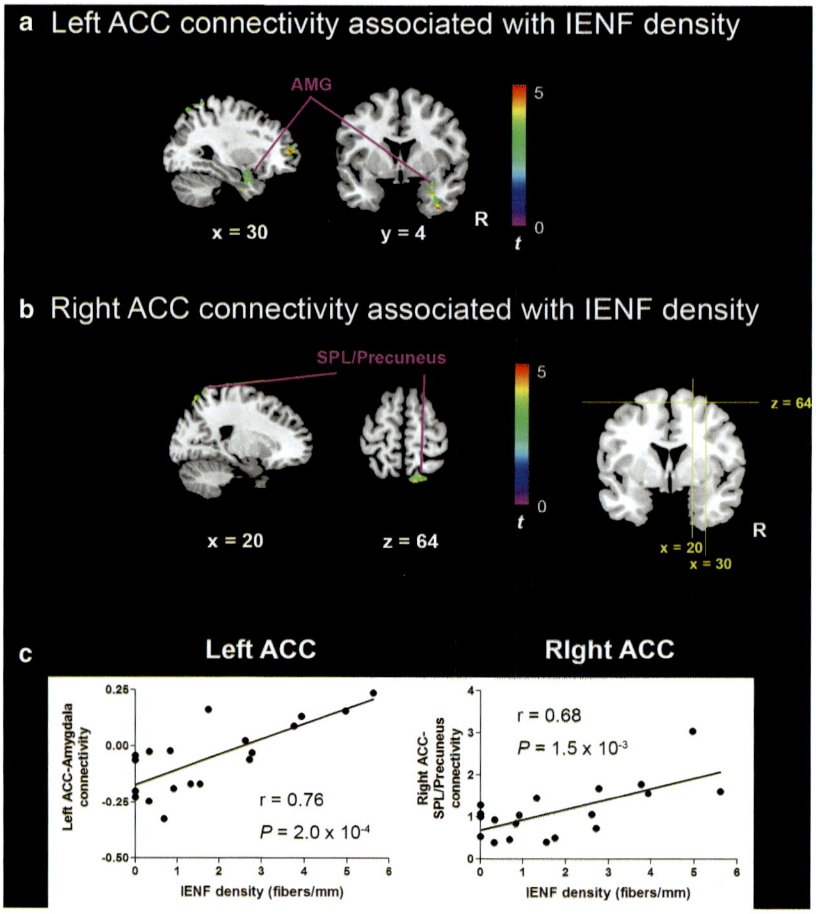

Fig. 14.1 Associations between IENF density and functional connectivity across pain-related brain regions in small fiber neuropathy. The brain maps show that functional connectivity between the left ACC and the right amygdala (**a**) and between the right ACC and the right superior parietal lobule (SPL) and precuneus (**b**) during noxious thermal stimulation of the distal leg was significantly associated with the IENF density in the distal leg. (**c**) The associations in (a) and (b) are also demonstrated in the scatter plots where the mean functional connectivity strength of the significant region (amygdala and SPL/precuneus, respectively) is regressed against the IENF density in the distal leg, showing that greater decrease in IENF density is correlated with greater reduction in functional connectivity. "R" indicates the right hemisphere. The MNI coordinate (mm) is shown at the bottom of each slice. Abbreviations: *ACC* anterior cingulate cortex, *AMG* amygdala, *SPL* superior parietal lobule. Reproduced with permission from [6]

neuroplasticity to neuropathic pain in SFN indicates that peripheral neuropathy is not purely peripheral; neuroimaging and electrophysiological studies of the brain to identify structural and functional alterations should be included in the repertoire of research strategies when assessing neuropathic pain in SFN.

14.3 Treatment of Neuropathic Pain

14.3.1 Pharmacological Treatment

Pharmacological treatment is the major strategy to alleviate pain in SFN, which will be detailed in Chap. 15 of this monograph. Evidence-based meta-analyses and consensus statements are available for pharmacotherapy of neuropathic pain in diabetic neuropathy [53]. Given that neuropathic pain in SFN is assumed to share similar mechanisms (even though the underlying causes of nerve injury may be different), pharmacological guidelines that are effective to relieve pain in diabetic neuropathy, the most common cause of SFN, are also applicable to neuropathic pain in SFN of other etiologies [54]. Medications for neuropathic pain can be classified into three major categories: antiepileptic drugs, antidepressants, and opioids. For example, pregabalin, an anticonvulsant that blocks calcium channels, is recommended as the first choice to alleviate neuropathic pain in diabetic neuropathy with level-A evidence [53]. It is also effective for neuropathic pain caused by peripheral nerve disorders other than diabetic neuropathy [55]. The effectiveness of antiepileptic drugs indicates that ion-channel blockade is a feasible therapeutic strategy for neuropathic pain and also corroborates the association between neuropathic pain and enhanced neuronal excitability in the peripheral and central nervous systems [11, 32].

14.3.2 Alternative Treatment Strategies

The development of chronic neuropathic pain in SFN involves complex interactions between enhanced peripheral nerve excitability and central maladaptive neuroplasticity as reviewed in the preceding sections, possibly explaining why up to 40% of patients with neuropathic pain respond suboptimally to pharmacological treatment [56]. Several nonpharmacological approaches have been proposed, including immunotherapy and modulatory stimulation of the peripheral nerves, spinal cord, or brain. These treatment approaches may be considered, when appropriate, as alternatives for patients who are nonresponsive or intolerable to medical treatment.

14.3.3 Immunotherapy

Intravenous immunoglobulin (IVIg) therapy has been reported to be effective for relieving neuropathic pain, particularly when the clinical picture suggests that immune processes are involved in the pathogenesis of pain. In a case series of five patients affected by painful sensory neuropathy associated with Sjögren's syndrome, IVIg therapy (0.4 g/kg/day for 5 days) ameliorated neuropathic pain from days 2–14 following treatment, with a 73.4% reduction in visual analog scale (VAS) on pain rating [57]. The same authors reported a patient with proximal diabetic neuropathy manifested by severe neuropathic pain and muscle weakness, whose neuropathic pain reduced remarkably (with VAS pain ratings reduced from 10 to 2) following two sessions of IVIg therapy (0.4 g/kg/day for 5 days on each session), accompanied by improvement in muscle strength (from 1 to 3 on manual muscle test) [58].

14.3.4 Percutaneous Electrical Nerve Stimulation

In percutaneous electrical nerve stimulation (PENS), percutaneously inserted needle electrodes are used to stimulate peripheral sensory nerves. The electric current interferes with pain signals, preventing the pain signals from reaching the brain [59]. PENS was recommended by the American Academy of Neurology for treating painful diabetic neuropathy, with level-B evidence [53]. A randomized clinical trial showed that in patients with diabetic neuropathic pain, those who received PENS for 3 weeks exhibited a 42% reduction in VAS, significantly greater than 14% in the sham-controlled patients [60]. However, other groups reported that electrical stimulation was of no analgesic effect in diabetic neuropathic pain [61].

14.3.5 Electrical Spinal Cord Stimulation

Electrical spinal cord stimulation (ESCS) stimulates the dorsal column of the spinal cord through electrodes implanted in the epidural space. Activation of A-beta fibers in the dorsal column interrupts pain input through Aδ or C fibers. The analgesic effect of ESCS may be also attributed to the improvement of nerve blood flow [62]. ESCS has been tested in patients with diabetic neuropathic pain unresponsive to conventional treatment and was found to improve both background and peak pain as well as exercise tolerance [63]. ESCS has also been reported to markedly reduce neuropathic pain in a patient with refractory neuropathic pain due to familial amyloid polyneuropathy [64].

14.3.6 Brain Stimulation Therapies

Given that maladaptive neuroplasticity of the brain contributes to the development and maintenance of neuropathic pain in SFN, electromagnetic stimulation of the brain is an attractive alternative to pharmacological therapy because it is likely to reset central nociceptive processes by inducing neuronal depolarization or modulating synaptic connectivity in the pain matrix. In recent years, transcranial magnetic stimulation (TMS) and transcranial direct current stimulation (tDCS) have become particularly favored because they provide a noninvasive, painless, and safe way to modulate neural activity in specific brain networks [65].

14.4 Noninvasive Brain Stimulations

14.4.1 Transcranial Magnetic Stimulation (TMS)

TMS depolarizes cortical neurons by generating a strong time-varying magnetic field that in turn induces an electric current in the brain [66]. TMS delivered in repetitive pulses, a technique known as repetitive TMS (rTMS), may induce plastic changes in the brain. Low-frequency ($\leq$ 1 Hz) rTMS generally inhibits the excitability of the motor cortex, an effect similar to long-term depression (LTD). By contrast, high-frequency (5–20 Hz) rTMS usually enhances motor cortex excitability, which is similar to long-term potentiation (LTP) [67]. Application of rTMS to the primary motor cortex (M1) or the dorsolateral prefrontal cortex (DLPFC) has been shown to alleviate experimental pain in normal volunteers [68, 69] or pathological pain in neuropathic patients, which may be attributed to a descending modulation mechanism through the corticothalamic pathway [70]. For patients with chronic neuropathic pain, 5–20 Hz rTMS at the M1 or DLPFC is shown to be more effective than low-frequency stimulation [71]. Onesti et al. showed that targeting 20-Hz rTMS to the lower limb representation in the motor cortex through the Hesed (H)-coil effectively alleviated drug-resistant neuropathic pain in the lower limbs in patients with diabetic neuropathy. The VAS rating of pain severity decreased significantly after 1500 pulses of rTMS, and the analgesic effect persisted until 3 weeks after stimulation [72]. The same high-frequency setting (20-Hz rTMS to M1 at the dosage of 2000 pulses per day) also successfully alleviated deafferentation pain in a patient with brachial plexus avulsion injury that was pharmacologically intractable for 25 years, leading to a 34.6% reduction in VAS pain rating after 20 days of the daily rTMS treatment [73]. The analgesic effect of rTMS may be attributable to activation of endogenous opioid systems because the injection of the opioid receptor antagonist significantly decreased the analgesic effect of rTMS [74, 75]. GABA neurotransmission is also involved in rTMS-induced analgesia. It was found that neuropathic pain was associated with motor cortex disinhibition due to impaired GABAergic transmission, and therefore the analgesic effect of rTMS was likely through modulation of intracortical neural circuits so that inhibitory GABAergic interneurons were reinforced [76].

14.4.2 Transcranial Direct Current Stimulation (tDCS)

tDCS is a noninvasive technique where the cerebral cortex is stimulated by a weak direct current delivered through scalp electrodes, leading to modulation of synaptic plasticity [77]. The analgesic effect of tDCS in painful diabetic neuropathy was demonstrated in a randomized, sham-controlled study where anodal tDCS, i.e., tDCS with the positive current passing through the target electrode, applied to the primary motor cortex for five daily sessions significantly reduced VAS pain score in patients, compared to the same stimulation applied to the DLPFC or sham-tDCS. The analgesic effect persisted for 2–4 weeks [78]. Nevertheless, the mechanism of tDCS-induced analgesia is not totally clear, although some studies have suggested that it might be related to enhanced intracortical inhibition [79] or an increase in endogenous opioid release [80].

14.5 Conclusion

Neuropathic pain is the most disabling symptom in SFN. It arises as a direct consequence of damage in nociceptive nerves of the skin, where degenerative or inflammatory processes induce local pain-triggering reactions. Maladaptive plastic changes in the central nociceptive system induced by peripheral overstimulation or deafferentation also contribute to the development of pain. Pharmacological treatment, including anticonvulsants, antidepressants, and opioids, is the standard approach to alleviate pain symptoms. Unfortunately, the treatment effect is often unsatisfactory, and most patients remain symptomatic even under multiple drug therapies. Alternative approaches, such as electromagnetic stimulation of the peripheral sensory nerves, spinal cord, or brain, have been proposed, although additional randomized controlled clinical trials and subsequent meta-analyses are required to validate their effectiveness. SFN is traditionally considered a peripheral nervous system disorder. Nevertheless, the involvement of the central nervous system in the generation of neuropathic pain indicates that the pathological process of SFN is not limited to the peripheral nerves. Future research is warranted to understand the neural mechanisms through which peripheral denervation induces central maladaptive plasticity and to identify biomarkers that help predict the analgesic response with respect to specific treatment modalities.

References

1. Hsieh ST. Pathology and functional diagnosis of small-fiber painful neuropathy. Acta Neurol Taiwanica. 2010;19:82–9.
2. Devigili G, Tugnoli V, Penza P, Camozzi F, Lombardi R, Melli G, et al. The diagnostic criteria for small fibre neuropathy: from symptoms to neuropathology. Brain. 2008;131:1912–25.
3. Baron R, Forster M, Binder A. Subgrouping of patients with neuropathic pain according to pain-related sensory abnormalities: a first step to a stratified treatment approach. Lancet Neurol. 2012;11:999–1005.
4. Truini A, Garcia-Larrea L, Cruccu G. Reappraising neuropathic pain in humans—how symptoms help disclose mechanisms. Nat Rev Neurol. 2013;9:572–82.
5. Lin YH, Hsieh SC, Chao CC, Chang YC, Hsieh ST. Influence of aging on thermal and vibratory thresholds of quantitative sensory testing. J Peripher Nerv Syst. 2005;10:269–81.
6. Hsieh PC, Tseng MT, Chao CC, Lin YH, Tseng WY, Liu KH, et al. Imaging signatures of altered brain responses in small-fiber neuropathy: reduced functional connectivity of the limbic system after peripheral nerve degeneration. Pain. 2015;156:904–16.
7. Chao CC, Hsieh SC, Tseng MT, Chang YC, Hsieh ST. Patterns of contact heat evoked potentials (CHEP) in neuropathy with skin denervation: correlation of CHEP amplitude with intraepidermal nerve fiber density. Clin Neurophysiol. 2008;119:653–61.
8. Chien HF, Tseng TJ, Lin WM, Yang CC, Chang YC, Chen RC, et al. Quantitative pathology of cutaneous nerve terminal degeneration in the human skin. Acta Neuropathol. 2001;102:455–61.

9. Chiang MC, Lin YH, Pan CL, Tseng TJ, Lin WM, Hsieh ST. Cutaneous innervation in chronic inflammatory demyelinating polyneuropathy. Neurology. 2002;59:1094–8.
10. Chao CC, Tseng MT, Lin YJ, Yang WS, Hsieh SC, Lin YH, et al. Pathophysiology of neuropathic pain in type 2 diabetes: skin denervation and contact heat-evoked potentials. Diabetes Care. 2010;33:2654–9.
11. Tseng MT, Chiang MC, Chao CC, Tseng WY, Hsieh ST. fMRI evidence of degeneration-induced neuropathic pain in diabetes: enhanced limbic and striatal activations. Hum Brain Mapp. 2013;34:2733–46.
12. Holland NR, Stocks A, Hauer P, Cornblath DR, Griffin JW, McArthur JC. Intraepidermal nerve fiber density in patients with painful sensory neuropathy. Neurology. 1997;48:708–11.
13. Kennedy WR, Said G. Sensory nerves in skin: answers about painful feet? Neurology. 1999;53:1614–5.
14. Periquet MI, Novak V, Collins MP, Nagaraja HN, Erdem S, Nash SM, et al. Painful sensory neuropathy: prospective evaluation using skin biopsy. Neurology. 1999;53:1641–7.
15. Lindenlaub T, Sommer C. Epidermal innervation density after partial sciatic nerve lesion and pain-related behavior in the rat. Acta Neuropathol. 2002;104:137–43.
16. Nolano M, Crisci C, Santoro L, Barbieri F, Casale R, Kennedy WR, et al. Absent innervation of skin and sweat glands in congenital insensitivity to pain with anhidrosis. Clin Neurophysiol. 2000;111:1596–601.
17. Lauria G, Lombardi R. Skin biopsy in painful and immune-mediated neuropathies. J Peripher Nerv Syst. 2012;17(Suppl 3):38–45.
18. Lai J, Hunter JC, Porreca F. The role of voltage-gated sodium channels in neuropathic pain. Curr Opin Neurobiol. 2003;13:291–7.
19. Hong S, Wiley JW. Early painful diabetic neuropathy is associated with differential changes in the expression and function of vanilloid receptor 1. J Biol Chem. 2005;280:618–27.
20. Hudson LJ, Bevan S, Wotherspoon G, Gentry C, Fox A, Winter J. VR1 protein expression increases in undamaged DRG neurons after partial nerve injury. Eur J Neurosci. 2001;13:2105–14.
21. Davis JB, Gray J, Gunthorpe MJ, Hatcher JP, Davey PT, Overend P, et al. Vanilloid receptor-1 is essential for inflammatory thermal hyperalgesia. Nature. 2000;405:183–7.
22. Ko MH, Hsieh YL, Hsieh ST, Tseng TJ. Nerve demyelination increases metabotropic glutamate receptor subtype 5 expression in peripheral painful mononeuropathy. Int J Mol Sci. 2015;16:4642–65.
23. Marchand F, Perretti M, McMahon SB. Role of the immune system in chronic pain. Nat Rev Neurosci. 2005;6:521–32.
24. Purwata TE. High TNF-alpha plasma levels and macrophages iNOS and TNF-alpha expression as risk factors for painful diabetic neuropathy. J Pain Res. 2011;4:169–75.
25. Uçeyler N, Kafke W, Riediger N, He L, Necula G, Toyka KV, et al. Elevated proinflammatory cytokine expression in affected skin in small fiber neuropathy. Neurology. 2010;74:1806–13.
26. Woolf CJ. Central sensitization: implications for the diagnosis and treatment of pain. Pain. 2011;152:S2–15.
27. Latremoliere A, Woolf CJ. Central sensitization: a generator of pain hypersensitivity by central neural plasticity. J Pain. 2009;10:895–926.
28. Baron R. Mechanisms of disease: neuropathic pain—a clinical perspective. Nat Clin Pract Neurol. 2006;2:95–106.
29. von Hehn CA, Baron R, Woolf CJ. Deconstructing the neuropathic pain phenotype to reveal neural mechanisms. Neuron. 2012;73:638–52.
30. Mendell LM, Wall PD. Responses of single dorsal cord cells to peripheral cutaneous unmyelinated fibres. Nature. 1965;206:97–9.
31. Cook AJ, Woolf CJ, Wall PD, McMahon SB. Dynamic receptive field plasticity in rat spinal cord dorsal horn following C-primary afferent input. Nature. 1987;325:151–3.
32. Torebjork HE, Lundberg LE, LaMotte RH. Central changes in processing of mechanoreceptive input in capsaicin-induced secondary hyperalgesia in humans. J Physiol. 1992;448:765–80.
33. Devor M. Ectopic discharge in Abeta afferents as a source of neuropathic pain. Exp Brain Res. 2009;196:115–28.
34. Koltzenburg M, Lundberg LE, Torebjork HE. Dynamic and static components of mechanical hyperalgesia in human hairy skin. Pain. 1992;51:207–19.
35. Woolf CJ, Thompson SW. The induction and maintenance of central sensitization is dependent on N-methyl-D-aspartic acid receptor activation; implications for the treatment of post-injury pain hypersensitivity states. Pain. 1991;44:293–9.
36. Luo C, Kuner T, Kuner R. Synaptic plasticity in pathological pain. Trends Neurosci. 2014;37:343–55.
37. Porreca F, Ossipov MH, Gebhart GF. Chronic pain and medullary descending facilitation. Trends Neurosci. 2002;25:319–25.
38. Urban MO, Gebhart GF. Supraspinal contributions to hyperalgesia. Proc Natl Acad Sci U S A. 1999;96:7687–92.
39. Wei F, Zhuo M. Potentiation of sensory responses in the anterior cingulate cortex following digit amputation in the anaesthetised rat. J Physiol. 2001;532:823–33.
40. Fu Y, Han J, Ishola T, Scerbo M, Adwanikar H, Ramsey C, et al. PKA and ERK, but not PKC, in the amygdala contribute to pain-related synaptic plasticity and behavior. Mol Pain. 2008;4:26.
41. Pedersen LH, Scheel-Kruger J, Blackburn-Munro G. Amygdala GABA-A receptor involvement in mediating sensory-discriminative and affective-motivational pain responses in a rat model of peripheral nerve injury. Pain. 2007;127:17–26.

42. Vogt BA. Pain and emotion interactions in subregions of the cingulate gyrus. Nat Rev Neurosci. 2005;6:533–44.
43. Derbyshire SW. Exploring the pain "neuromatrix". Curr Rev Pain. 2000;4:467–77.
44. Peyron R, Laurent B, Garcia-Larrea L. Functional imaging of brain responses to pain. A review and meta-analysis (2000). Neurophysiol Clin. 2000;30:263–88.
45. Schweinhardt P, Glynn C, Brooks J, McQuay H, Jack T, Chessell I, et al. An fMRI study of cerebral processing of brush-evoked allodynia in neuropathic pain patients. Neuroimage. 2006;32:256–65.
46. Becerra L, Morris S, Bazes S, Gostic R, Sherman S, Gostic J, et al. Trigeminal neuropathic pain alters responses in CNS circuits to mechanical (brush) and thermal (cold and heat) stimuli. J Neurosci. 2006;26:10646–57.
47. DaSilva AF, Becerra L, Pendse G, Chizh B, Tully S, Borsook D. Colocalized structural and functional changes in the cortex of patients with trigeminal neuropathic pain. PLoS One. 2008;3:e3396.
48. Lauria G, Morbin M, Lombardi R, Capobianco R, Camozzi F, Pareyson D, et al. Expression of capsaicin receptor immunoreactivity in human peripheral nervous system and in painful neuropathies. J Peripher Nerv Syst. 2006;11:262–71.
49. MacIver K, Lloyd DM, Kelly S, Roberts N, Nurmikko T. Phantom limb pain, cortical reorganization and the therapeutic effect of mental imagery. Brain. 2008;131:2181–91.
50. Flor H. Maladaptive plasticity, memory for pain and phantom limb pain: review and suggestions for new therapies. Expert Rev Neurother. 2008;8:809–18.
51. Chiang MC, Dutton RA, Hayashi KM, Lopez OL, Aizenstein HJ, Toga AW, et al. 3D pattern of brain atrophy in HIV/AIDS visualized using tensor-based morphometry. Neuroimage. 2007;34:44–60.
52. Friston KJ, Buechel C, Fink GR, Morris J, Rolls E, Dolan RJ. Psychophysiological and modulatory interactions in neuroimaging. Neuroimage. 1997;6:218–29.
53. Bril V, England J, Franklin GM, Backonja M, Cohen J, Del Toro D, et al. Evidence-based guideline: treatment of painful diabetic neuropathy: report of the American Academy of Neurology, the American Association of Neuromuscular and Electrodiagnostic Medicine, and the American Academy of Physical Medicine and Rehabilitation. Neurology. 2011;76:1758–65.
54. Finnerup NB, Otto M, McQuay HJ, Jensen TS, Sindrup SH. Algorithm for neuropathic pain treatment: an evidence based proposal. Pain. 2005;118:289–305.
55. Moore A, Wiffen P, Kalso E. Antiepileptic drugs for neuropathic pain and fibromyalgia. JAMA. 2014;312:182–3.
56. Hansson PT, Attal N, Baron R, Cruccu G. Toward a definition of pharmacoresistant neuropathic pain. Eur J Pain. 2009;13:439–40.
57. Morozumi S, Kawagashira Y, Iijima M, Koike H, Hattori N, Katsuno M, et al. Intravenous immunoglobulin treatment for painful sensory neuropathy associated with Sjogren's syndrome. J Neurol Sci. 2009;279:57–61.
58. Kawagashira Y, Watanabe H, Oki Y, Iijima M, Koike H, Hattori N, et al. Intravenous immunoglobulin therapy markedly ameliorates muscle weakness and severe pain in proximal diabetic neuropathy. J Neurol Neurosurg Psychiatry. 2007;78:899–901.
59. White PF, Li S, Chiu JW. Electroanalgesia: its role in acute and chronic pain management. Anesth Analg. 2001;92:505–13.
60. Hamza MA, White PF, Craig WF, Ghoname ES, Ahmed HE, Proctor TJ, et al. Percutaneous electrical nerve stimulation: a novel analgesic therapy for diabetic neuropathic pain. Diabetes Care. 2000;23:365–70.
61. Oyibo SO, Breislin K, Boulton AJ. Electrical stimulation therapy through stocking electrodes for painful diabetic neuropathy: a double blind, controlled crossover study. Diabet Med. 2004;21:940–4.
62. Jacobs MJ, Jorning PJ, Joshi SR, Kitslaar PJ, Slaaf DW, Reneman RS. Epidural spinal cord electrical stimulation improves microvascular blood flow in severe limb ischemia. Ann Surg. 1988;207:179–83.
63. Tesfaye S, Watt J, Benbow SJ, Pang KA, Miles J, MacFarlane IA. Electrical spinal-cord stimulation for painful diabetic peripheral neuropathy. Lancet. 1996;348:1698–701.
64. Miyazaki Y, Koike H, Akane A, Shibata Y, Nishiwaki K, Sobue G. Spinal cord stimulation markedly ameliorated refractory neuropathic pain in transthyretin Val30Met familial amyloid polyneuropathy. Amyloid. 2011;18:87–90.
65. Fregni F, Freedman S, Pascual-Leone A. Recent advances in the treatment of chronic pain with non-invasive brain stimulation techniques. Lancet Neurol. 2007;6:188–91.
66. Hallett M. Transcranial magnetic stimulation and the human brain. Nature. 2000;406:147–50.
67. Ziemann U. TMS induced plasticity in human cortex. Rev Neurosci. 2004;15:253–66.
68. Tamura Y, Okabe S, Ohnishi T, NS D, Arai N, Mochio S, et al. Effects of 1-Hz repetitive transcranial magnetic stimulation on acute pain induced by capsaicin. Pain. 2004;107:107–15.
69. Graff-Guerrero A, Gonzalez-Olvera J, Fresan A, Gomez-Martin D, Mendez-Nunez JC, Pellicer F. Repetitive transcranial magnetic stimulation of dorsolateral prefrontal cortex increases tolerance to human experimental pain. Brain Res Cogn Brain Res. 2005;25:153–60.
70. Nguyen JP, Nizard J, Keravel Y, Lefaucheur JP. Invasive brain stimulation for the treatment of neuropathic pain. Nat Rev Neurol. 2011;7:699–709.
71. Galhardoni R, Correia GS, Araujo H, Yeng LT, Fernandes DT, Kaziyama HH, et al. Repetitive transcranial magnetic stimulation in chronic pain: a review of the literature. Arch Phys Med Rehabil. 2015;96:S156–72.

72. Onesti E, Gabriele M, Cambieri C, Ceccanti M, Raccah R, Di Stefano G, et al. H-coil repetitive transcranial magnetic stimulation for pain relief in patients with diabetic neuropathy. Eur J Pain. 2013;17:1347–56.
73. Qiu YQ, Hua XY, Zuo CT, Li T, Zheng MX, Shen YD, et al. Deactivation of distant pain-related regions induced by 20-day rTMS: a case study of one-week pain relief for long-term intractable deafferentation pain. Pain Physician. 2014;17:E99–105.
74. Maarrawi J, Peyron R, Mertens P, Costes N, Magnin M, Sindou M, et al. Motor cortex stimulation for pain control induces changes in the endogenous opioid system. Neurology. 2007;69:827–34.
75. de Andrade DC, Mhalla A, Adam F, Texeira MJ, Bouhassira D. Neuropharmacological basis of rTMS-induced analgesia: the role of endogenous opioids. Pain. 2011;152:320–6.
76. Lefaucheur JP, Drouot X, Menard-Lefaucheur I, Keravel Y, Nguyen JP. Motor cortex rTMS restores defective intracortical inhibition in chronic neuropathic pain. Neurology. 2006;67:1568–74.
77. Filmer HL, Dux PE, Mattingley JB. Applications of transcranial direct current stimulation for understanding brain function. Trends Neurosci. 2014;37:742–53.
78. Kim YJ, Ku J, Kim HJ, Im DJ, Lee HS, Han KA, et al. Randomized, sham controlled trial of transcranial direct current stimulation for painful diabetic polyneuropathy. Ann Rehabil Med. 2013;37:766–76.
79. Portilla AS, Bravo GL, Miraval FK, Villamar MF, Schneider JC, Ryan CM, et al. A feasibility study assessing cortical plasticity in chronic neuropathic pain following burn injury. J Burn Care Res. 2013;34:e48–52.
80. DosSantos MF, Love TM, Martikainen IK, Nascimento TD, Fregni F, Cummiford C, et al. Immediate effects of tDCS on the mu-opioid system of a chronic pain patient. Front Psychiatry. 2012;3:93.

Therapy for Small Fiber Neuropathy

15

Ahmad R. Abuzinadah
and Christopher H. Gibbons

Abstract

Therapy of small fiber neuropathy (SFN) includes two strategies: (1) treat the underlying etiology and (2) symptomatic treatment, in particular for neuropathic pain. If a cause for a SFN is identified, the management will focus on treatment of the underlying disease. However, in the setting of neuropathic pain, symptomatic therapy may be required independent of the underlying disease. There are many medications and topical agents that can be used for relief of pain from neuropathy based on various levels of evidence including many randomized clinical trials. However, the evidence for neuropathic pain treatment is usually derived from trials designed for neuropathy (both large and small fiber), but is not specific for SFN. Therefore, the recommendations can only provide some general guidance on treatment of pain in SFN. Major classes of medications used to control neuropathic pain are antidepressants, antiepileptics, opioid (and opioid-like) medications, and topical agents. The selection of agents to treat neuropathic pain in SFN is based on several factors such as mechanism of action of the drug, the presumed mechanism of the pain, the patient's comorbidities, the medication's side effect profile, and the possible drug-drug interactions. Therefore, the choice of a particular therapy to treat neuropathic pain in SFN is a decision made by both the treating physician and the patient to achieve a realistic goal of pain relief.

Keywords

Neuropathic pain · Peripheral sensitization
Central sensitization · Clinical trials · Number needed to treat (NNT) · Antidepressants
Antiepileptic drugs · Sodium channel blockers · Cannabinoids · Tricyclic antidepressants · Serotonin norepinephrine reuptake inhibitors · Topical therapy · Capsaicin
Lidocaine

15.1 Introduction

There are many potential medical conditions that can lead to the development of a small fiber neuropathy (SFN), but detection and amelioration of the underlying disease are the primary focus. Despite significant advances in our diagnostic capability and our array of therapeutic interventions, there are still many individuals with a

A. R. Abuzinadah
Internal Medicine Department, Neurology Division, King Abdulaziz University, Jeddah, Saudi Arabia

C. H. Gibbons (✉)
Neurology Department, Harvard Medical School, Beth Israel Deaconess Medical Center, Boston, MA, USA
e-mail: cgibbons@bidmc.harvard.edu

S.-T. Hsieh et al. (eds.), *Small Fiber Neuropathy and Related Syndromes: Pain and Neurodegeneration*, https://doi.org/10.1007/978-981-13-3546-4_15

diagnosis of idiopathic or cryptogenic SFN. In these cryptogenic cases, we continue to investigate potential causes, but do not have potential disease modifying therapies.

In contrast, the majority of individuals with SFN do have an identifiable underlying condition, such as diabetes, that will require ongoing treatment and management. Despite the improvements in diagnostics, there are still significant limitations to our ability to alter the natural history of the disease. In most cases of SFN, patients will continue to have symptoms despite optimal medical management. Therefore symptomatic relief is often a focus of treatment in SFN.

A significant proportion of individuals with SFN report neuropathic pain of sufficient severity to require treatment. Although treatment of the underlying medical condition is a major step in disease management, treatment of neuropathic pain should be started concurrently. There are several potential mechanisms involved in the generation of neuropathic pain, and knowledge of these mechanisms may guide treatment selection. Several guidelines have been published on the approaches to neuropathic pain therapy. Unfortunately, the vast majority of treatments for neuropathic pain have been studied in mixed large and small fiber neuropathies. There are limited data available on isolated SFN. This chapter will focus on the medications that can be used in treatment of neuropathic pain in general and will discuss both efficacy and side effects. In addition, we will review specific aspects of treatment of neuropathic pain in SFN.

15.2 Therapies Targeting the Mechanism of Neuropathic Pain

Peripheral and central alterations in the nervous system occur in patients with SFN. In general, the mechanisms of neuropathic pain are classified according to (1) location in the peripheral or central nervous systems and (2) modifications occurring within the specified parts of the nervous system. Table 15.1 summarizes potential mechanisms through which various medications work.

Table 15.1 Assumed pathophysiological mechanism of neuropathic pain, against which the medications are directed

Mechanism of pain	Potentials drugs related to the mechanism of pain
Disturbance of analgesic system	Tricyclic antidepressant (TCAs) Selective serotonin reuptake inhibitors (SSRIs) Serotonin norepinephrine reuptake inhibitors (SNRI) Tramadol Opiates
Central sensitization	Gabapentin Lamotrigine Oxcarbazepine Levetiracetam Topiramate Ketamine Dextromethorphan Methadone
Ectopic firing due overactive/sensitized sodium channels	Carbamazepine Oxcarbazepine Topiramate Lamotrigine Mexiletine Lidocaine
Sympathetic sprouting/coupling	Clonidine Phentolamine

15.2.1 Peripheral Mechanisms

1. **Ectopic firing due to altered activity of sodium channels**: After experimental nerve injury, there is an accumulation of sodium channels as well as increased rate of spontaneous firing [1]. Up to 10% of SFN cases considered to be idiopathic may have a mutation in a voltage-gated sodium channel, leading to increased activity. Several medications work by blocking voltage-gated sodium channels and may help to alleviate neuropathic pain [2]. However, evidence to support their use through this exact mechanism is lacking. These medications include carbamazepine, oxcarbazepine, topiramate, lamotrigine, mexiletine, and lidocaine.
2. Neuroinflammation may sensitize and activate nociceptors [3], and pain is reduced with causal treatment of some of the inflammatory neuropathies.
3. Sympathetic sprouting/coupling: The sympathetic postganglionic cells encircle the cell

body of the dorsal root ganglia neurons in experimental settings [1, 4]. However, the clinical importance of this finding has not been shown, and there is no evidence that medications that work on this mechanism (e.g., clonidine, phentolamine) may help neuropathic pain.

15.2.2 Central Mechanisms

1. **Disturbance in the analgesic system**: There are two major descending modulatory pathways, also known as the "analgesic system." The first pathway originates in the periaqueductal gray (PAG) at the level of the midbrain and the nucleus raphae magnus (NRM) [1, 4–6]. The second pathway originates at the level of the locus ceruleus in the medulla [1, 4–6]. Certain neuropathic pain medications enhance the inhibitory signals by blocking reuptake of serotonin and noradrenaline at the level of the dorsal horn neurons. These medications include tricyclic antidepressants (TCAs), selective serotonin reuptake inhibitors (SSRIs), serotonin noradrenalin reuptake inhibitors (SNRI), tramadol, and opioids [1, 4–6].
2. **Central sensitization**: Central sensitization is an exaggerated response to a stimulus at a different level of the central nervous system. This exaggerated response occurs due to use-dependent enhancement of transmission across the synapse [5]. In the central nervous system, the major neurotransmitter responsible for the pain transmission is glutamate [4]. Medications that block N-type calcium channels or block NMDA receptors may partially alleviate the pain associated with central sensitization. These medications include N-type calcium channel antagonists such as gabapentin or pregabalin and medications that have secondary calcium channel effects such as lamotrigine, oxcarbazepine, and levetiracetam, while agents that can block NMDA receptors include topiramate, ketamine, dextromethorphan, and methadone [4]. Clinically, central sensitization may manifest as allodynia and hyperpathia.

15.3 Comparison of Medication Efficacy

There are few clinical trials that actually compare the efficacy of one drug to another in the treatment of neuropathic pain. Therefore, one of the means to measure the general efficacy of one drug to another is a ranking scale defined at the number needed to treat (NNT). The NNT is the number of patients one would need to treat to get one more responder with the active treatment than one would have gotten with placebo [7]. In pain research, often an NNT for 30% or 50% reduction in pain are given [8, 9]. Therefore, the lower the NNT, the more efficacious the drug is at reducing neuropathic pain, although some caveats apply [7]. Most medications are in the range of 2–8 for NNT, which means that somewhere between 2 and 8 individuals require treatment before 1 more person has 50% pain relief than with placebo treatment.

15.4 Medication Classes

The American Academy of Neurology (AAN) and the European Federation of Neurological Societies (EFNS) have reviewed the evidence supporting the use of different medications in the treatment of neuropathic pain [10, 11]. Also the Neuropathic Pain Special Interest Group (NeuPSIG) of International Association for the Study of Pain (IASP) have recently reviewed the evidence of different therapies for neuropathic pain and gave a graded recommendation [12]. A summary of these recommendations is provided in Table 15.2.

15.5 Antidepressants

15.5.1 Tricyclic Antidepressants

Tricyclic antidepressants are norepinephrine and serotonin reuptake inhibitors, although the selectivity for norepinephrine and serotonin does vary between drugs within this class [4]. The reuptake inhibition leads to an increase in synaptic norepinephrine and serotonin and

Table 15.2 Comparison of guidelines on pharmacological treatment of neuropathic pain

	AAN (for diabetic neuropathy) [10]	EFNS [11]	NeuPSIG [12]
Pregabalin	Level A	Level A (first line)	Strong (first line)
Gabapentin	Level B	Level A (first line)	Strong (first line)
TCA	–	Level A (first line)	Strong (first line)
Amitriptyline	Level B	–	Strong (first line)
Duloxetine	Level B	Level A (first line)	Strong (first line)
Venlafaxine	Level B	Level A (first line)	Strong (first line)
Valproic acid	Level B	No efficacy/conflicting data	Weak against
Tramadol	Level B	Level A (second line)	Weak (second line)
Oxycodone	Level B	Level A (second line)	Weak (third line)
Morphine sulfate	Level B	Level A (second line)	Weak (third line)
Capsaicin	Level B	No efficacy/conflicting data	Weak (second line)
Isosorbide dinitrate spray	Level B	Level A	
Carbamazepine	–	Level C	Inconclusive
Phenytoin	–	Level C	
Oxcarbazepine	Not recommended	No efficacy/conflicting data	
Lacosamide	Not recommended	No efficacy/conflicting data	Inconclusive
Clonidine	Not recommended	–	Inconclusive
Mexiletine	Not recommended	No efficacy/conflicting data	Strong against
Lamotrigine	Not recommended	No efficacy/conflicting data	Inconclusive

AAN American Academy of Neurology [10], *EFNS* European Federation of Neurological Societies [11], *NeuPSIG* Neuropathic Pain Special Interest Group of the International Association for the Study of Pain [12]
Level A: Recommendation directly based on Level I evidence
Level B: Recommendation directly based on Level II evidence or extrapolated recommendations from Level I evidence

enhances the descending inhibitory pain pathways [4, 13]. The NNTs for tricyclic antidepressants are 3.6 (96% CI 3–4.5) but did vary depending on the particular tricyclic antidepressant. They are among the lowest NNTs of any drug but rely on older, smaller studies with low placebo responses, which hamper comparability with the modern drugs [14, 15].

Desipramine was found to be effective in treating diabetic neuropathy in several studies where 61% of patients had moderate or greater reduction of pain in depressed and non-depressed patients with NNT of 3.4 [4, 16–18]. Desipramine was also effective in post-herpetic neuralgia [17]. However, a recent systematic review has questioned the quality of these studies as they did not account for subject withdrawal and the primary outcome was not robust [16].

Amitriptyline is a commonly used tricyclic antidepressant to treat neuropathic pain or pain disorder. A recent review of studies with a minimal 4-week duration showed a NNTs of 4.6 (96% CI 3.6–6.6) with 38% experience benefit from amitriptyline and 16% among patients received placebo [12, 19, 20]. The percentage of individuals who reported benefits of moderate to complete reduction of pain was 74% in one randomized blinded study. The most common side effects reported include dry mouth 63%, fatigue and somnolence 34%, constipation 8%, and orthostatic symptoms 5% [12].

Nortriptyline is a noradrenergic metabolite of amitriptyline. While amitriptyline is more balance norepinephrine and serotonin reuptake inhibitor, nortriptyline is a more of norepinephrine reuptake inhibitor [9, 21]. A recent review found no high-quality evidence for the use of nortriptyline in neuropathic pain syndromes [22]. However, the estimated NNT from the available studies (acknowledging the high risk of bias) ranges from 3 to 4 [22]. A combination of nortriptyline and gabapentin has been demonstrated to be more efficacious than either one alone in one study [22]. The side effects of nortriptyline include dry mouth (56%), somnolence (15%), constipation, and urinary retention.

15.6 Serotonin Norepinephrine Reuptake Inhibitor

Duloxetine is a serotonin and norepinephrine reuptake inhibitor. It has been proven to be effective in treatment of diabetic neuropathic pain in two randomized placebo-controlled clinical trials [23, 24]. The NNT to achieve a 50% pain reduction in diabetic neuropathy is 5.0 and for fibromyalgia is 8.0 (although some reviews have NNT numbers closer to 4) [23, 24]. Duloxetine was effective in both pain reduction and improving activities of daily living. The frequency with which duloxetine is discontinued due to side effects is 20% [23, 24]. The common side effects include nausea (20–30% dose-dependent), somnolence (8–15%), constipation (10%), and dizziness (10%) [23].

Venlafaxine, another drug in the SNRI class, has the advantage of a low affinity for the cholinergic receptors which make its side effect profile more favorable than tricyclic antidepressants [25]. The NNT to achieve a 50% pain reduction in diabetic neuropathic pain was 4.5 using 150 mg to 225 mg, while the 75 mg dose was not effective for pain relief. A Cochrane review estimated the NNT of venlafaxine to be closer to 3.1 [15]. Side effects include somnolence 15%, nausea 10%, and insomnia 10% [25].

15.7 Antiepileptic Drugs

15.7.1 α2δ Ligands of Voltage-Gated Calcium Channels

Pregabalin modulates GABAergic transmission and calcium influx through binding to the α2δ subunit of voltage-gated calcium channels. It has been proven to be effective in diabetic painful neuropathy with NNTs to achieve 50% pain reduction of 7.5–11 for 300 mg/day and 5–6.3 for 600 mg, although other reviews have estimated the NNT to be closer to 4 [26]. A dose of pregabalin at 150 mg/day was equivalent to placebo. Dizziness was the most common side effect, which occurred in 38% of patients taking 600 mg per day [27]. In another study, 67% of patients with painful diabetic neuropathy improved with 300 mg pregabalin versus 39% in the placebo group, and side effects of dizziness occurred in 35% [28].

Gabapentin is also a ligand of the α2δ subunit of voltage-activated calcium channels [29]. Gabapentin was specifically studied in SFN in an enrichment crossover design and reduced the pain intensity score by 1.12 compared to active placebo (diphenhydramine) [30]. In patients with painful diabetic neuropathy, the chance to achieve moderate or greater pain reduction was 59% with gabapentin (900 mg to 3600 mg) versus 33% with placebo in a randomized clinical trial [29]. In a meta-analysis, the chance of 50% pain reduction was 38% with gabapentin versus 21% with placebo with a NNT = 5.9 [31]. Other analyses have suggested the NNT is lower than 5 [11]. In post-herpetic neuralgia, the chance of 50% pain reduction was 34% with gabapentin versus 21% with placebo with an NNT = 8 [31]. In fibromyalgia, a randomized clinical trial showed that the chance to achieve 30% or more pain reduction was 51% with gabapentin (1200 mg to 2400 mg) and 31% with placebo, which was statistically significant [32]. Side effects of gabapentin include dizziness (19%), somnolence (14%), gait disturbance (9%), and confusion (8%) [31].

15.7.2 Sodium Channel Blockers

Carbamazepine is an antiepileptic drug that blocks voltage-gated sodium channels. Despite the presence of several positive randomized placebo-controlled trials of carbamazepine in painful diabetic neuropathy, the short duration (<4 weeks) and small size (<200 participants) of the trials limit the weight of evidence for its use [33, 34]. Thus, the current quality of evidence is relatively low for treatment of pain in diabetic neuropathy because the studies are subject to a bias that overestimates the efficacy of the drug. The majority of evidence of the efficacy of carbamazepine in pain comes from trials of trigeminal neuralgia [35, 36].

Oxcarbazepine is a related nonspecific sodium channel blocker. There are several studies with conflicting results about the efficacy of oxcarbazepine in painful diabetic neuropathy. One randomized placebo-controlled study including 146 patients showed that the chance to achieve >50% pain reduction was 35.2% with oxcarbazepine versus 18.4% in placebo. The dose used was 300 mg per day that was titrated up to 1800 mg per day and study duration was 16 weeks [37]. In another 16-week placebo-controlled study of 1200 mg per day oxcarbazepine failed to show efficacy in the treatment of painful diabetic neuropathy [38]. Side effects of oxcarbazepine include vertigo (8%), somnolence (5.9%), and nausea (5.1%) [39].

Topiramate is an antiepileptic drug which blocks voltage-gated sodium and calcium channels and has an effect on glutamate transmission. A randomized clinical trial for topiramate in painful diabetic polyneuropathy showed that 50% of the topiramate group had 30% pain reduction, while it was only 34% in the placebo group. 50% pain reduction occurred in 36% of topiramate group and in 21% of the placebo group. The dose was titrated up to 400 mg or maximum tolerated dose [40]. Despite this positive result, the trial was criticized because of an attrition bias with >10% withdrawal in the active treatment group [41]. In another randomized clinical trial of painful diabetic neuropathy (100 mg, 200 mg, and 400 mg arms), topiramate did not demonstrate efficacy over placebo. The withdrawal rate due to side effect was again high at 24%. With such a high withdrawal rate, there is a high risk of attrition bias despite the large sample size (>200 patients in each group) [42]. Side effects of topiramate included weight loss (>10% of body weight) in 52%, somnolence in 10%, difficulty concentrating in 5%, loss of appetite in 10.9%, and nausea in 9.5% [40]. There are no trials investigating the use of topiramate in SFN.

Lamotrigine is a voltage-gated sodium channel antagonist that also inhibits glutamate release. A randomized placebo-controlled trial showed that 50% pain reduction was achieved in 44% of patients with lamotrigine (200 mg, 300 mg, and 400 mg) versus 19% with placebo in painful diabetic neuropathy [43]. However, this study included <50 patients in each group, and there was significant amounts of missing data (83% with lamotrigine versus 73% with placebo completed the study) that were not carried forward, which may have placed this trial at a high risk of bias. Similarly, in HIV neuropathy, lamotrigine was effective in a randomized placebo-controlled trial, but the study was criticized for missing data and small sample size [44–46]. Overall, there is no good-quality evidence to support the use of lamotrigine in neuropathic pain [47].

Lacosamide is voltage-gated sodium channel blocker which enhances the slow inactivation of voltage-gated sodium channels without any impact on the fast activated sodium channels. In a randomized placebo-controlled trial, lacosamide-treated patients had a 50% drop in their pain score versus a 36% drop in the placebo group with an NNT of 8–10 [48]. Side effects include headache (18%), dizziness (15%), and nausea (12%) [48].

15.7.3 Cannabinoid Derivatives

Nabilone is a synthetic cannabinoid that works as an agonist of cannabinoid receptors. It was approved for use by the Federal Drug Administration (FDA) as an antiemetic and has been investigated in a randomized placebo-controlled clinical trial in painful diabetic neuropathy and found to be effective in pain relief, with the chance of achieving 50% pain reduction of 31% with nabilone versus 8% with placebo, while 30% pain reduction occurred in 85% with nabilone and 31% with placebo [49]. However, this study was criticized for the enriched enrollment design, as they only randomized responders and excluded non-responders. Furthermore, the very low placebo response and the small sample size (<50 participants in each group) give rise to a significant risk of bias. Potential unmasking of the treatment could be as high as 62% in both groups, another major confounder of the results [49]. Another trial compared nabilone with oxycodone and found a better efficacy with oxycodone [50].

15.8 Opioids

The use of opioids for control of neuropathic pain has become an increasingly controversial topic. The risk of abuse and diversion and the huge impact on society have created a public health crisis that requires physicians to be very cognizant of the risk-benefit ratio to both the individual and to society [51, 52]. Therefore, as a general rule, opioids should never be considered as first-line therapy for neuropathic pain and in general should be reserved for specific situations where the risk-benefit ratio has been fully explored [52]. One of the major criticisms for the use of opioids in neuropathic pain is the short duration of most studies, where many of the harmful side effects, and tolerance to the drug, develop beyond the endpoint of the study, therefore inflating the potential benefit while reducing the apparent risks associated with their use.

Tramadol is a μ-opioid receptor agonist. In placebo-controlled trials, tramadol is effective in treating neuropathic pain with an NNT to achieve 50% pain reduction of 3.8 and a number needed to cause harm (NNH) of 8.3 [53]. In a randomized placebo-controlled trial, the pain intensity score was lower with tramadol compared to placebo at 42 days of treatment [53]. The average dose was 210 mg per day. The most common side effects were nausea, constipation, headache, and somnolence [53].

Oxycodone is a μ-opioid receptor agonist. In a recent review and meta-analysis, a 30% pain reduction was seen in 44% of patients treated with oxycodone and 27% with placebo providing an NNT of 5.7 in painful diabetic neuropathy. The quality of reported outcome measures was not optimal, and the overall quality of evidence is low for use of oxycodone [54]. In another placebo-controlled randomized clinical trial for oxycodone, the NNT to achieve 50% pain reduction was 2.7, and the chance of side effects was 27% for constipation, 40% for somnolence, 36% for nausea, and 32% for dizziness [55].

Methadone is an NMDA receptor antagonist and amine reuptake inhibitor in addition to being a synthetic opioid. According to the Cochrane review, there was a single randomized controlled trial that compared placebo versus methadone (20 mg) and found that methadone reduced the average pain intensity score in neuropathic pain [56]. Methadone has a very long drug half-life that puts patients at risk for central nervous system depression several days after a dose adjustment. However, the use of methadone for pain requires multiple daily dosing of the drug. This combination of a short duration of pain relief with a long duration of central nervous system depression places patients at extremely high risk of death from overdose and requires prescribing physicians to be extremely careful with the use of methadone.

Fentanyl is another μ-opioid receptor agonist, often prescribed as a patch for outpatient use. Despite its use in clinical practice, there is insufficient evidence to support its use in neuropathic pain [57].

15.9 Topical Agents

Lidocaine can be applied in a 5% topical patch for focal peripheral neuropathic pain syndromes such as post-herpetic neuralgia and is sometimes used off-label in painful diabetic neuropathy. The lidocaine patch was found to be effective in a randomized placebo-controlled trial with a 50% pain reduction in 31% of patients with post-herpetic neuralgia treated with topical lidocaine versus 8% in the placebo-treated group with an NNT of 4.4 [58, 59]. The placebo group had a very low responder rate compared to most other studies of neuropathic pain. There was no difference in side effect profile versus placebo, and commonly reported side effects include skin irritation, itching, or burning sensation [58]. The lidocaine patch was also investigated in post-herpetic neuralgia in another randomized placebo-controlled trial and found to be effective [60]. These two studies of the lidocaine patch were not considered high quality as the withdrawal rates were >10% and there were small numbers of participants in each group (<50 patients per group) [61].

Capsaicin acts on the transient receptor potential of subfamily vanilloid receptor type 1 (TRPV1) receptor and is the active ingredient in hot chili peppers. Capsaicin can be used as an over-the-counter cream application or as a high-concentration 8% capsaicin patch [10, 62, 63]. The number needed to treat to achieve 50% pain reduction using the 8% capsaicin patch compared to placebo was 7 for post-herpetic neuralgia [64]. Trials for diabetic neuropathy and HIV neuropathy were also positive, but the quality of the studies was considered lower by a recent meta-analysis [64].

Topical phenytoin and other topical formulations of the antidepressants or antiepileptics (such as gabapentin and amitriptyline) can be developed by a compounding pharmacy with the advantage of fewer side effects than oral administration. However, there is no good-quality evidence to suggest that topical formulations of these drugs are efficacious in the treatment of neuropathic pain [65].

15.10 Combination of Therapy

In general, polypharmacy should be avoided unless necessary. However, in some cases the use of more than one agent can provide a synergistic effect on pain control, but this should include an analysis of the risk-benefit ratio with a focus on side effects and drug-drug interactions.

A combination of standard doses of duloxetine (60 mg/day) and pregabalin (300 mg/day) was compared against a higher-dose monotherapy of each, duloxetine (120 mg/day), or pregabalin (600 mg/day) in a study of painful diabetic neuropathy. Despite the absence of statistically significant primary outcome, the combination therapy was favored in primary and secondary outcomes and suggested that combination therapy may be appropriate in certain circumstances [66].

A combination of gabapentin and nortriptyline was superior to using one drug alone in patients with painful diabetic neuropathy and post-herpetic neuralgia. There was a 48% pain reduction with gabapentin, 45% with nortriptyline, and 63% with the combination [67, 68].

Despite the positive results suggested by studies of combination therapies for neuropathic pain, a Cochrane review of the topic determined that the level of evidence for any particular combination was quite low and that additional studies needed to include comparisons of the combination therapy with both drugs alone and with placebo before a level of evidence was sufficient to make clinical recommendations [35, 67].

15.11 Approaches to Treatment of Neuropathic Pain in Small Fiber Neuropathy

Most guidelines suggest starting with a first-line therapy, which includes tricyclic antidepressants (amitriptyline, nortriptyline, or desipramine), an $\alpha 2\delta$ voltage-gated calcium channel antagonist (gabapentin or pregabalin), or serotonin-norepinephrine reuptake inhibitors (duloxetine or venlafaxine) [69–71]. Duloxetine is given once daily, and tricyclic antidepressants can be given once daily if pain is predominant at night, which is better for compliance. However, tricyclic antidepressants do cause more anticholinergic side effects (e.g., sedation, dry mouth, and dizziness) and are not recommended for individuals over 65 years of age. Secondary amines (nortriptyline and desipramine) have fewer anticholinergic effects than primary amines (amitriptyline and imipramine). A history of arrhythmia is a contraindication for use of tricyclic antidepressants [71].

The gabapentinoids (pregabalin and gabapentin) require more than once daily dosing, although can be used only at night in patients with neuropathic pain that occurs predominantly in the evening or night. Somnolence and sedation are reported more with amitriptyline (30%), while it is less with nortriptyline, duloxetine, venlafaxine, pregabalin, and gabapentin (15%). The use of sedating agents at night in patients that have trouble sleeping due to neuropathic pain can be a valuable way to incorporate a drug side effect into a treatment success. Gastrointestinal symptoms are common during the first 3 days of duloxetine dosing, such that adding an antiemetic initially may be advisable.

The decision to add a new agent to an existing drug or simply to switch agents is often a challenge. One approach is to add a second agent if there is a partial response to the first agent but to switch to another agent if there was no response to the first agent or if there were intolerable side effects. However, many "failed" trials of drugs occur in the setting of inadequate dosing. Table 15.3 provides the suggested minimum dose necessary to determine that a drug is not efficacious in the treatment of neuropathic pain.

Second-line therapies include tramadol and opioid analgesics (see Table 15.3 [69, 71]). The indications for using opioid analgesics are (1) episodic exacerbation of neuropathic pain and (2) during titrating up to the maintenance dose of neuropathic pain medications. As suggested earlier in this chapter, there are several concerns with using opioids including physical and psychological dependence. With chronic use of opioids, some individuals may develop both hyperalgesia and tolerance. Opioid use should be monitored and minimized as possible. Tolerance and physical dependence are not synonymous for addiction, but rather addiction is the compulsive use of substance despite lack of benefits or presence of harm [72].

If first- and second-line medications have failed to provide significant pain relief, a third line agent can be tried. These include antiepileptic medications described in Table 15.3, such as topiramate and carbamazepine. Despite the low-quality randomized clinical trials that support the use of third-line agents, there is still clearly a minority of patients that responds to these drugs because of significant variability across individuals in responding to treatments.

Topical agents such as capsaicin and lidocaine are indicated for severe focal pain and are considered as third line. The efficacy was reviewed by Cochrane database for capsaicin [64] and lidocaine [61]. Lidocaine patches are applied only for 12 hours and then should be taken off for another 12 hours before they can be applied again. An adequate trial of lidocaine is 2 weeks. Another topical agent frequently successful is high-concentration 8% capsaicin patch, with a single dose applied for 30 min at the feet and for 60 min in other body regions, to be repeated every 3 months, if necessary [64].

Table 15.3 Starting and maintenance dose of drugs used in neuropathic pain

Medications	Starting dose	Increment dose	Maintenance dose	Maximum dose
First line				
Amitriptyline	10 mg at bed time	10 mg weekly	75–100 mg/d	150 mg/day
Nortriptyline	25 mg at bed time	25 mg weekly	75–100 mg/d	150 mg/day
Pregabalin	75 mg TID	75 mg TID	150 mg BID	600 mg/day
Gabapentin	300 QD to TID	300 mg TID	2400 mg total	3600 mg total
Duloxetine	30 mg a day	30 mg weekly	60 mg/d	120 mg
Venlafaxine	75 mg	75 mg weekly	150 mg/d	225 mg
Second line				
Tramadol	50 mg TID	50 mg	100 mg TID	400 mg per day
Oxycodone	5 mg QD to TID	10 mg	30 mg BID	60 mg BID
Morphine	15 mg BID	15 mg	30 mg BID	120 mg BID
Third line				
Topiramate	25 mg at bed time	50 mg	25–100 mg BID	200 mg BID
Carbamazepine	200 mg QD	200 mg	200–400 mg BID	1600 mg per day
Fourth line				
Nabilone	0.5 mg BID	0.5 mg	2 mg BID	6 mg per day

It should be noted that the maintenance dose is the effective dose for pain relief. Therefore if pain is achieved at a lower dose than is recommended in this table, there is no need to titrate upward to a "maintenance dose" range

15.11.1 Cause- or Molecular Target-Specific Treatment for Small Fiber Neuropathy

Specific treatments that might be helpful in sarcoidosis associated with SFN include intravenous immunoglobulin (IVIG) and antitumor necrosis factor directed agents [73]. Similarly immunomodulatory therapy including IVIG, methotrexate, and hydroxychloroquine may help SFN associated with antibodies to voltage-gated potassium channels (VGKC) [74]. These immune-mediated diseases causing SFN are discussed in more detail in Chap. 10.

Treatment of SFN associated with Fabry's disease may include enzyme replacement with agalsidase beta or agalsidase alpha to control the progression of underlying pathology [75]. A randomized trial showed an efficacy of enzyme replacement in treating Fabry's disease [76, 77]. However, whether the data show really benefits or not is still controversial [78]. Specific discussion of Fabry's disease is available in Chap. 7.

Hereditary erythromelalgia is linked to sodium channel mutations (channelopathy) and is associated with neuropathic pain (discussed in Chap. 7). Because flushing associated with this condition is caused by vasodilation, the use of topical midodrine (which is a vasoconstrictor though α_1 agonist mechanism) has been reported to relieve the symptoms in non-randomized study [79].

15.11.2 Neuroprotection

Vitamin E was found in a randomized placebo-controlled clinical trial to reduce cisplatin-induced neuropathy from 41.7% to 5.9% which provided class II evidence for neuroprotection [80]. Erythropoietin receptors are present in the peripheral nerve axons as well as in the dorsal root ganglia. Erythropoietin has been shown to reduce the risk of developing ulcers in a mouse model of resiniferatoxin-induced SFN [81]. Clinically, erythropoietin was tried in cancer patients, and the neuroprotection was found to be dose-dependent [82]. The proposed mechanism by which erythropoietin provides neuroprotection in chemotherapy-induced neuropathy is by making the target more resistant to DNA damage caused by platination [83]. N-palmitoylethanolamine (PEA) injection in animal models of neuropathy reduces macrophage infiltration and leads to an increase in myelin amount and axonal diameter [84]. Resveratrol was found in studies to provide neuroprotection in experimental diabetic neuropathy by upregulation of Nrf2 and SIRT1 and inhibition of NF-κB and AP-1 [85]. The blocking of muscarinic acetylcholine receptor type 1 protects in animal models of SFN is another potential area of neuroprotection that will be explored [86].

15.11.3 Specific Sodium Channel Subtype Blockers

Voltage-gated sodium channel 1.7 ($Na_v1.7$) is expressed mainly in the nociceptive neurons with minimal expression in the brain [87]. Several molecules developed from sulfonamide compounds and spider venom are undergoing clinical trials by different pharmaceutical companies [87]. The voltage-gated sodium channel 1.8 ($Na_v1.8$) has a similar expression in nociceptive neurons [87], and sodium channel blockers may be candidates for neuropathic pain therapy, for example, ambroxol, an $Na_v1.8$ blocker, could be effective in a mouse model of pain [88]. A previous clinical trial for $Na_v1.8$ blocker (PF-04531083) failed in dental pain [87], but other trials for $Na_v1.8$ blockers are ongoing.

15.12 Conclusion

Treatment of neuropathic pain in SFN is a complex and challenging task. Possible mechanisms underlying the neuropathic pain in an individual patient are one of the factors to consider in selection of a category of treatment. Additional items to consider include possible side effects, drug-drug interactions, topical or systemic treatment, duration of therapy, and cost. The American Academy of Neurology, the European Federation of Neurological Societies, and the NeuPSIG of IASP have provided guidance on selection of agents to use for treatment of neuropathic pain. Although this information is helpful as a reference, it does not supplant the time and effort required to develop a reasonable goal of care in conjunction with the patient. Additional studies specific to SFN are needed to define specific adjustments of treatment regimes in this population.

References

1. Bridges D, Thompson SW, Rice AS. Mechanisms of neuropathic pain. Br J Anaesth. 2001;87:12–26.
2. Hoeijmakers JG, Faber CG, Lauria G, Merkies IS, Waxman SG. Small-fibre neuropathies—advances in diagnosis, pathophysiology and management. Nat Rev Neurol. 2012;8:369–79.
3. Sommer C, Leinders M, Uceyler N. Inflammation in the pathophysiology of neuropathic pain. Pain. 2018;159:595–602.
4. Coluzzi F, Mattia C. Mechanism-based treatment in chronic neuropathic pain: the role of antidepressants. Curr Pharm Des. 2005;11:2945–60.
5. Woolf CJ. Central sensitization: implications for the diagnosis and treatment of pain. Pain. 2011;152:S2–15.
6. Cheng C, Webber CA, Wang J, Xu Y, Martinez JA, Liu WQ, et al. Activated RHOA and peripheral axon regeneration. Exp Neurol. 2008;212:358–69.
7. Katz N, Paillard FC, Van Inwegen R. A review of the use of the number needed to treat to evaluate the efficacy of analgesics. J Pain. 2015;16:116–23.
8. Sindrup SH, Jensen TS. Pharmacologic treatment of pain in polyneuropathy. Neurology. 2000;55:915–20.
9. Finnerup NB, Otto M, McQuay HJ, Jensen TS, Sindrup SH. Algorithm for neuropathic pain treatment: an evidence based proposal. Pain. 2005;118:289–305.
10. Bril V, England J, Franklin GM, Backonja M, Cohen J, Del Toro D, et al. Evidence-based guideline: treatment of painful diabetic neuropathy: report of the American Academy of Neurology, the American Association of Neuromuscular and Electrodiagnostic Medicine, and the American Academy of Physical Medicine and Rehabilitation. Neurology. 2011;76:1758–65.
11. Attal N, Cruccu G, Baron R, Haanpaa M, Hansson P, Jensen TS, et al. EFNS guidelines on the pharmacological treatment of neuropathic pain: 2010 revision. Eur J Neurol. 2010;17:1113–e1188.
12. Finnerup NB, Attal N, Haroutounian S, McNicol E, Baron R, Dworkin RH, et al. Pharmacotherapy for neuropathic pain in adults: a systematic review and meta-analysis. Lancet Neurol. 2015;14:162–73.
13. Stillman M. Clinical approach to patients with neuropathic pain. Cleve Clin J Med. 2006;73:726–39.
14. Saarto T, Wiffen PJ. Antidepressants for neuropathic pain: a Cochrane review. J Neurol Neurosurg Psychiatry. 2010;81:1372–3.
15. Saarto T, Wiffen PJ. Antidepressants for neuropathic pain. Cochrane Database Syst Rev. 2007:CD005454.
16. Hearn L, Moore RA, Derry S, Wiffen PJ, Phillips T. Desipramine for neuropathic pain in adults. Cochrane Database Syst Rev. 2014:CD011003.
17. Sindrup SH, Gram LF, Skjold T, Grodum E, Brosen K, Beck-Nielsen H. Clomipramine vs desipramine vs placebo in the treatment of diabetic neuropathy symptoms. A double-blind cross-over study. Br J Clin Pharmacol. 1990;30:683–91.
18. Max MB, Lynch SA, Muir J, Shoaf SE, Smoller B, Dubner R. Effects of desipramine, amitriptyline, and fluoxetine on pain in diabetic neuropathy. N Engl J Med. 1992;326:1250–6.
19. Moore RA, Derry S, Aldington D, Cole P, Wiffen PJ. Amitriptyline for neuropathic pain in adults. Cochrane Database Syst Rev. 2015:CD008242.
20. Rico-Villademoros F, Slim M, Calandre EP. Amitriptyline for the treatment of fibromyalgia: a comprehensive review. Expert Rev Neurother. 2015;15:1123–50.
21. Gilron I, Watson CP, Cahill CM, Moulin DE. Neuropathic pain: a practical guide for the clinician. CMAJ. 2006;175:265–75.
22. Derry S, Wiffen PJ, Aldington D, Moore RA. Nortriptyline for neuropathic pain in adults. Cochrane Database Syst Rev. 2015;1:CD011209.
23. Wernicke JF, Pritchett YL, D'Souza DN, Waninger A, Tran P, Iyengar S, et al. A randomized controlled trial of duloxetine in diabetic peripheral neuropathic pain. Neurology. 2006;67:1411–20.
24. Goldstein DJ, Lu Y, Detke MJ, Lee TC, Iyengar S. Duloxetine vs. placebo in patients with painful diabetic neuropathy. Pain. 2005;116:109–18.
25. Rowbotham MC, Goli V, Kunz NR, Lei D. Venlafaxine extended release in the treatment of painful diabetic neuropathy: a double-blind, placebo-controlled study. Pain. 2004;110:697–706.
26. Moore RA, Straube S, Wiffen PJ, Derry S, McQuay HJ. Pregabalin for acute and chronic pain in adults. Cochrane Database Syst Rev. 2009:CD007076.
27. Richter RW, Portenoy R, Sharma U, Lamoreaux L, Bockbrader H, Knapp LE. Relief of painful diabetic peripheral neuropathy with pregabalin: a randomized, placebo-controlled trial. J Pain. 2005;6:253–60.
28. Rosenstock J, Tuchman M, LaMoreaux L, Sharma U. Pregabalin for the treatment of painful diabetic peripheral neuropathy: a double-blind, placebo-controlled trial. Pain. 2004;110:628–38.
29. Backonja M, Beydoun A, Edwards KR, Schwartz SL, Fonseca V, Hes M, et al. Gabapentin for the symptomatic treatment of painful neuropathy in patients with diabetes mellitus: a randomized controlled trial. JAMA. 1998;280:1831–6.
30. Ho TW, Backonja M, Ma J, Leibensperger H, Froman S, Polydefkis M. Efficient assessment of neuropathic pain drugs in patients with small fiber sensory neuropathies. Pain. 2009;141:19–24.
31. Moore RA, Wiffen PJ, Derry S, Toelle T, Rice AS. Gabapentin for chronic neuropathic pain and

fibromyalgia in adults. Cochrane Database Syst Rev. 2014:CD007938.
32. Arnold LM, Goldenberg DL, Stanford SB, Lalonde JK, Sandhu HS, Keck PE Jr, et al. Gabapentin in the treatment of fibromyalgia: a randomized, double-blind, placebo-controlled, multicenter trial. Arthritis Rheum. 2007;56:1336–44.
33. Rull JA, Quibrera R, Gonzalez-Millan H, Lozano CO. Symptomatic treatment of peripheral diabetic neuropathy with carbamazepine (Tegretol): double blind crossover trial. Diabetologia. 1969;5:215–8.
34. Wilton TD. Tegretol in the treatment of diabetic neuropathy. S Afr Med J. 1974;48:869–72.
35. Wiffen PJ, Derry S, Moore RA, Aldington D, Cole P, Rice AS, et al. Antiepileptic drugs for neuropathic pain and fibromyalgia—an overview of Cochrane reviews. Cochrane Database Syst Rev. 2013:CD010567.
36. Wiffen PJ, Derry S, Moore RA, Kalso EA. Carbamazepine for chronic neuropathic pain and fibromyalgia in adults. Cochrane Database Syst Rev. 2014:CD005451.
37. Dogra S, Beydoun S, Mazzola J, Hopwood M, Wan Y. Oxcarbazepine in painful diabetic neuropathy: a randomized, placebo-controlled study. Eur J Pain. 2005;9:543–54.
38. Grosskopf J, Mazzola J, Wan Y, Hopwood M. A randomized, placebo-controlled study of oxcarbazepine in painful diabetic neuropathy. Acta Neurol Scand. 2006;114:177–80.
39. Magenta P, Arghetti S, Di Palma F, Jann S, Sterlicchio M, Bianconi C, et al. Oxcarbazepine is effective and safe in the treatment of neuropathic pain: pooled analysis of seven clinical studies. Neurol Sci. 2005;26:218–26.
40. Raskin P, Donofrio PD, Rosenthal NR, Hewitt DJ, Jordan DM, Xiang J, et al. Topiramate vs placebo in painful diabetic neuropathy: analgesic and metabolic effects. Neurology. 2004;63:865–73.
41. Wiffen PJ, Derry S, Lunn MP, Moore RA. Topiramate for neuropathic pain and fibromyalgia in adults. Cochrane Database Syst Rev. 2013:CD008314.
42. Thienel U, Neto W, Schwabe SK, Vijapurkar U, Topiramate Diabetic Neuropathic Pain Study Group. Topiramate in painful diabetic polyneuropathy: findings from three double-blind placebo-controlled trials. Acta Neurol Scand. 2004;110:221–31.
43. Eisenberg E, Lurie Y, Braker C, Daoud D, Ishay A. Lamotrigine reduces painful diabetic neuropathy: a randomized, controlled study. Neurology. 2001;57:505–9.
44. Simpson DM, Olney R, McArthur JC, Khan A, Godbold J, Ebel-Frommer K. A placebo-controlled trial of lamotrigine for painful HIV-associated neuropathy. Neurology. 2000;54:2115–9.
45. Vinik AI, Tuchman M, Safirstein B, Corder C, Kirby L, Wilks K, et al. Lamotrigine for treatment of pain associated with diabetic neuropathy: results of two randomized, double-blind, placebo-controlled studies. Pain. 2007;128:169–79.
46. Simpson DM, McArthur JC, Olney R, Clifford D, So Y, Ross D, et al. Lamotrigine for HIV-associated painful sensory neuropathies: a placebo-controlled trial. Neurology. 2003;60:1508–14.
47. Wiffen PJ, Derry S, Moore RA. Lamotrigine for chronic neuropathic pain and fibromyalgia in adults. Cochrane Database Syst Rev. 2013:CD006044.
48. Rauck RL, Shaibani A, Biton V, Simpson J, Koch B. Lacosamide in painful diabetic peripheral neuropathy: a phase 2 double-blind placebo-controlled study. Clin J Pain. 2007;23:150–8.
49. Toth C, Mawani S, Brady S, Chan C, Liu C, Mehina E, et al. An enriched-enrolment, randomized withdrawal, flexible-dose, double-blind, placebo-controlled, parallel assignment efficacy study of nabilone as adjuvant in the treatment of diabetic peripheral neuropathic pain. Pain. 2012;153:2073–82.
50. Frank B, Serpell MG, Hughes J, Matthews JN, Kapur D. Comparison of analgesic effects and patient tolerability of nabilone and dihydrocodeine for chronic neuropathic pain: randomised, crossover, double blind study. BMJ. 2008;336:199–201.
51. Esquibel AY, Borkan J. Doctors and patients in pain: conflict and collaboration in opioid prescription in primary care. Pain. 2014;155:2575–82.
52. Breivik H, Stubhaug A. Burden of disease is often aggravated by opioid treatment of chronic pain patients: etiology and prevention. Pain. 2014;155:2441–3.
53. Hollingshead J, Duhmke RM, Cornblath DR. Tramadol for neuropathic pain. Cochrane Database Syst Rev. 2006:CD003726.
54. Gaskell H, Derry S, Stannard C, Moore RA. Oxycodone for neuropathic pain in adults. Cochrane Database Syst Rev. 2016;7:CD010692.
55. Gimbel JS, Richards P, Portenoy RK. Controlled-release oxycodone for pain in diabetic neuropathy: a randomized controlled trial. Neurology. 2003;60:927–34.
56. Haroutiunian S, McNicol ED, Lipman AG. Methadone for chronic non-cancer pain in adults. Cochrane Database Syst Rev. 2012;11:CD008025.
57. Derry S, Stannard C, Cole P, Wiffen PJ, Knaggs R, Aldington D, et al. Fentanyl for neuropathic pain in adults. Cochrane Database Syst Rev. 2016;10:CD011605.
58. Meier T, Wasner G, Faust M, Kuntzer T, Ochsner F, Hueppe M, et al. Efficacy of lidocaine patch 5% in the treatment of focal peripheral neuropathic pain syndromes: a randomized, double-blind, placebo-controlled study. Pain. 2003;106:151–8.
59. Chai J, Herrmann DN, Stanton M, Barbano RL, Logigian EL. Painful small-fiber neuropathy in Sjogren syndrome. Neurology. 2005;65:925–7.
60. Rowbotham MC, Davies PS, Verkempinck C, Galer BS. Lidocaine patch: double-blind controlled study of a new treatment method for post-herpetic neuralgia. Pain. 1996;65:39–44.
61. Derry S, Wiffen PJ, Moore RA, Quinlan J. Topical lidocaine for neuropathic pain in adults. Cochrane Database Syst Rev. 2014:CD010958.
62. Watson CP, Evans RJ, Watt VR. Post-herpetic neuralgia and topical capsaicin. Pain. 1988;33:333–40.

63. Bernstein JE, Korman NJ, Bickers DR, Dahl MV, Millikan LE. Topical capsaicin treatment of chronic postherpetic neuralgia. J Am Acad Dermatol. 1989;21:265–70.
64. Derry S, Rice AS, Cole P, Tan T, Moore RA. Topical capsaicin (high concentration) for chronic neuropathic pain in adults. Cochrane Database Syst Rev. 2017;1:CD007393.
65. Kopsky DJ, Keppel Hesselink JM. Phenytoin in topical formulations augments pain reduction of other topically applied analgesics in the treatment of trigeminal neuralgia. J Clin Anesth. 2017;38:154–5.
66. Tesfaye S, Wilhelm S, Lledo A, Schacht A, Tolle T, Bouhassira D, et al. Duloxetine and pregabalin: high-dose monotherapy or their combination? The "COMBO-DN study"—a multinational, randomized, double-blind, parallel-group study in patients with diabetic peripheral neuropathic pain. Pain. 2013;154:2616–25.
67. Chaparro LE, Wiffen PJ, Moore RA, Gilron I. Combination pharmacotherapy for the treatment of neuropathic pain in adults. Cochrane Database Syst Rev. 2012:CD008943.
68. Gilron I, Bailey JM, Tu D, Holden RR, Jackson AC, Houlden RL. Nortriptyline and gabapentin, alone and in combination for neuropathic pain: a double-blind, randomised controlled crossover trial. Lancet. 2009;374:1252–61.
69. Dworkin RH, O'Connor AB, Backonja M, Farrar JT, Finnerup NB, Jensen TS, et al. Pharmacologic management of neuropathic pain: evidence-based recommendations. Pain. 2007;132:237–51.
70. Moulin DE, Clark AJ, Gilron I, Ware MA, Watson CP, Sessle BJ, et al. Pharmacological management of chronic neuropathic pain—consensus statement and guidelines from the Canadian Pain Society. Pain Res Manag. 2007;12:13–21.
71. Moulin D, Boulanger A, Clark AJ, Clarke H, Dao T, Finley GA, et al. Pharmacological management of chronic neuropathic pain: revised consensus statement from the Canadian Pain Society. Pain Res Manag. 2014;19:328–35.
72. Jovey RD, Ennis J, Gardner-Nix J, Goldman B, Hays H, Lynch M, et al. Use of opioid analgesics for the treatment of chronic noncancer pain—a consensus statement and guidelines from the Canadian Pain Society, 2002. Pain Res Manag. 2003;8(Suppl A):3A–28A.
73. Tavee JO, Karwa K, Ahmed Z, Thompson N, Parambil J, Culver DA. Sarcoidosis-associated small fiber neuropathy in a large cohort: clinical aspects and response to IVIG and anti-TNF alpha treatment. Respir Med. 2017;126:135–8.
74. Klein CJ, Lennon VA, Aston PA, McKeon A, Pittock SJ. Chronic pain as a manifestation of potassium channel-complex autoimmunity. Neurology. 2012;79:1136–44.
75. Politei JM, Bouhassira D, Germain DP, Goizet C, Guerrero-Sola A, Hilz MJ, et al. Pain in Fabry disease: practical recommendations for diagnosis and treatment. CNS Neurosci Ther. 2016;22:568–76.
76. Schiffmann R, Kopp JB, Austin HA 3rd, Sabnis S, Moore DF, Weibel T, et al. Enzyme replacement therapy in Fabry disease: a randomized controlled trial. JAMA. 2001;285:2743–9.
77. Wilcox WR, Banikazemi M, Guffon N, Waldek S, Lee P, Linthorst GE, et al. Long-term safety and efficacy of enzyme replacement therapy for Fabry disease. Am J Hum Genet. 2004;75:65–74.
78. Desnick RJ. Enzyme replacement therapy for Fabry disease: lessons from two alpha-galactosidase A orphan products and one FDA approval. Expert Opin Biol Ther. 2004;4:1167–76.
79. Davis MD, Morr CS, Warndahl RA, Sandroni P. Topically applied Midodrine, 0.2%, an alpha1-agonist, for the treatment of Erythromelalgia. JAMA Dermatol. 2015;151:1025–6.
80. Pace A, Giannarelli D, Galie E, Savarese A, Carpano S, Della Giulia M, et al. Vitamin E neuroprotection for cisplatin neuropathy: a randomized, placebo-controlled trial. Neurology. 2010;74:762–6.
81. Danigo A, Magy L, Richard L, Desmouliere A, Bourthoumieu S, Funalot B, et al. Neuroprotective effect of erythropoietin against pressure ulcer in a mouse model of small fiber neuropathy. PLoS One. 2014;9:e113454.
82. Bianchi R, Gilardini A, Rodriguez-Menendez V, Oggioni N, Canta A, Colombo T, et al. Cisplatin-induced peripheral neuropathy: neuroprotection by erythropoietin without affecting tumour growth. Eur J Cancer. 2007;43:710–7.
83. Yoon MS, Katsarava Z, Obermann M, Schafers M, Liedert B, Dzagnidze A, et al. Erythropoietin overrides the triggering effect of DNA platination products in a mouse model of cisplatin-induced neuropathy. BMC Neurosci. 2009;10:77.
84. Di Cesare ML, D'Agostino G, Pacini A, Russo R, Zanardelli M, Ghelardini C, et al. Palmitoylethanolamide is a disease-modifying agent in peripheral neuropathy: pain relief and neuroprotection share a PPAR-alpha-mediated mechanism. Mediat Inflamm. 2013;2013:328797.
85. Kumar A, Negi G, Sharma SS. Neuroprotection by resveratrol in diabetic neuropathy: concepts & mechanisms. Curr Med Chem. 2013;20:4640–5.
86. Calcutt NA, Smith DR, Frizzi K, Sabbir MG, Chowdhury SK, Mixcoatl-Zecuatl T, et al. Selective antagonism of muscarinic receptors is neuroprotective in peripheral neuropathy. J Clin Invest. 2017;127:608–22.
87. Yekkirala AS, Roberson DP, Bean BP, Woolf CJ. Breaking barriers to novel analgesic drug development. Nat Rev Drug Discov. 2017;16:545–64.
88. Gaida W, Klinder K, Arndt K, Weiser T. Ambroxol, a Nav1.8-preferring Na(+) channel blocker, effectively suppresses pain symptoms in animal models of chronic, neuropathic and inflammatory pain. Neuropharmacology. 2005;49:1220–7.

Index

S.-T. Hsieh et al. (eds.), *Small Fiber Neuropathy and Related Syndromes: Pain and Neurodegeneration*, https://doi.org/10.1007/978-981-13-3546-4

GPSR Compliance

The European Union's (EU) General Product Safety Regulation (GPSR) is a set of rules that requires consumer products to be safe and our obligations to ensure this.

If you have any concerns about our products, you can contact us on ProductSafety@springernature.com

In case Publisher is established outside the EU, the EU authorized representative is:

Springer Nature Customer Service Center GmbH
Europaplatz 3
69115 Heidelberg, Germany

Batch number: 10756384

Printed by Printforce, the Netherlands